Stung!

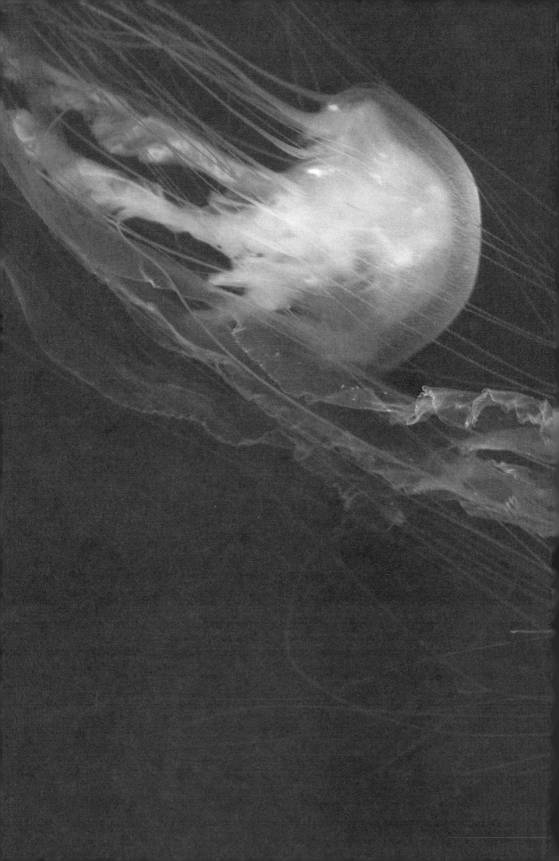

Stung!

On Jellyfish Blooms and the Future of the Ocean

Lisa-ann Gershwin

With a Foreword by Sylvia Earle

The University of Chicago Press | Chicago and London

Lisa-ann Gershwin is director of the Australian Marine Stinger Advisory Services. She was awarded a Fulbright in 1998 for her studies on jellyfish blooms and evolution, and she has discovered over 150 new species including at least sixteen types of jellyfish that are highly dangerous, as well as a new species of dolphin and has written for numerous scientific and popular publications.

The University of Chicago Press, Chicago 60637
The University of Chicago Press, Ltd., London
© 2013 by Lisa-ann Gershwin
All rights reserved. Published 2013.
Printed in the United States of America

22 21 20 19 18 17 16 15 14 13   1 2 3 4 5

ISBN-13: 978-0-226-02010-5 (cloth)
ISBN-13: 978-0-226-02024-2 (e-book)

Library of Congress Cataloging-in-Publication Data

Gershwin, Lisa-ann.
    Stung! : on jellyfish blooms and the future of the ocean / Lisa-ann Gershwin ; with a foreword by Sylvia Earle.
        pages ; cm
    Includes bibliographical references and index.
    ISBN 978-0-226-02010-5 (cloth : alk. paper)—ISBN 978-0-226-02024-2 (e-book)
1. Jellyfish blooms.   2. Jellyfishes—Ecology.   3. Marine ecology.   I. Earle, Sylvia A., 1935–   II. Title.
    QL377.S4G47 2013
    593.5′3-dc23

                                                                        2012043147

♾ This paper meets the requirements of ANSI/NISO Z39.48–1992 (Permanence of Paper).

To Tom and Tina McGlynn

Thanks can mean a lot . . . it does here.

For Patrick

"And till my ghastly tale is told,
This heart within me burns"
—Samuel Taylor Coleridge,
The Rime of the Ancient Mariner

[CONTENTS]

"You must see this! Jellyfish! Millions of them." Fifteen year-old Timmy Shriver and his cousin, John F. Kennedy Jr., called from the clear, blue water of Chuuk Lagoon, where they were assisting photographer Al Giddings and me with research for a *National Geographic* magazine article.

Still on the deck of our dive boat, Al looked over the side and said, "Kids! They always exaggerate." A few minutes later, camera finally ready, Al leaped in, took one look and yelled back to me, "Hurry and get in here! Jellyfish! There must be millions!"

A blizzard of moon jellies seemed to stretch to infinity in the exact place we had dived the day before where we had seen none. The sea appeared to be more jellyfish than seawater, a scene memorialized in a full-page photograph by Giddings in the May 1976 issue of *National Geographic*. It was my first experience as a witness to phenomena Lisa-ann Gershwin eloquently describes in this book, when certain species of jellyfish or members of another gelatinous group, the comb jellies, have population explosions that seem to magically arise from nowhere. Such events occur naturally, but there are increasing concerns about a distinctly unnatural increase in the number, size, and duration of jellyfish "blooms," their causes, and consequences.

During thousands of hours exploring the ocean with scuba and submersibles, and sometimes living under the sea for days or weeks at a time, I have witnessed hundreds of variations on the theme of the ancient, soft-bodied sea creatures collectively known as "jellies." I have admired their diaphanous beauty and wondered at how as a group they are at once exquisitely fragile and extraordinarily durable, with a history of survival through hundreds of millions of years of planetary changes. Entire categories of life have come and gone in the past half billion years, but jellies have persisted and prospered. Geologically speaking, humans are newcomers, our recognizable ancestors appearing just a few million years ago, with most of the technological advances underpinning modern society occurring in the past few centuries. The last century, in particular, has been a revolutionary time for both humans and jellies, a relationship that is explored with creative finesse in this volume.

With conversational prose, peppered with reader-friendly terms such as "gob-smacking", "yep", "hullabaloo" and "jellyfish goo and poo," Gershwin combines her own scholarly research and that of numerous other scientists with concerns that affect everyone, everywhere, in this thought-provoking overview of how the current and future affairs of humankind are linked to that of creatures many think of as a nuisance—if they think of them at all.

Reading this book should inspire heightened respect for these typically translucent creatures, some notable for their sophisticated stinging apparatus, some for their rainbow-colored bands of iridescent cilia, some for their ability to flash, sparkle or glow with their own living light—all, in a sense, "living fossils," considering their ancient lineage. They have found and have maintained a place for themselves in the liquid, three-dimensional realm that includes nearly all of the major divisions of animal life known. Carnivores by nature, they are a significant part of the great ocean food web with a dual role as both predator and prey.

I have watched a dozen triggerfish vying for bites of a tattered moon jelly in the Bahamas, and observed a *Mola mola* ocean sunfish nibbling on a lion's mane jelly near Monterey, California. Leatherback sea turtles dive more than a thousand feet to dine on deep-sea jellies, I'm told, and I have witnessed one of their cousins, a hawksbill, wrestling with a jellyfish fragment, tearing off and swallowing shimmering chunks near Cocos Island, Costa Rica. But, as Gershwin explains, those who eat jellyfish are no match for those eaten *by* jellyfish.

In *Stung!*, the connection between jellyfish blooms and the reduction in certain fish populations is explored, as well as the underlying causes of exceptional jellyfish population explosions that plague coastal resorts and fishing grounds. The apparent increase in size and frequency of such blooms is convincingly linked to human activity, from global warming, overfishing, and habitat destruction to the introduction of fertilizers, toxic chemicals, and plastic trash. Like a detective assembling clues that link one thing to another, Gershwin has gathered extensive data and reviewed research that demonstrate that we should be holding up the mirror to find the underlying causes of problems we attribute to jellies.

In the middle of the twentieth century, it seemed the ocean was too vast, too resilient, for humans to cause any harm. Now we know otherwise. Since the 1960s, 90 percent of many sought-after fish are gone, including sharks, tuna, swordfish, marlin, and many others. Half the coral reefs, mangrove forests, and seagrass meadows have been destroyed or are in a state of sharp decline. Hundreds of "dead zones" have formed in coastal waters, while phytoplankton has declined globally by as much as 40 percent. Excess carbon dioxide released into the atmosphere is causing the ocean to become more acidic. The increase in jellyfish blooms is one of many signs signaling a sea change.

There are many reasons for despair about the future of the ocean—and therefore, of humankind. The ocean is, in effect, our life-support system, driving climate and weather, governing the water cycle, stabilizing temperature, generating most of the oxygen in the atmosphere, taking up much of the carbon dioxide, shaping planetary chemistry. If the ocean is in trouble, so are we.

Armed with unprecedented knowledge, there are ways to resolve some of the thorniest issues and take action that can restore and protect the ocean systems vital to our survival and well-being. Ten percent of the sharks are still in the ocean. Half the coral reefs are in good shape. Most whale species have shown some recovery since commercial killing ended in 1986. Reducing the take of fish, oysters, lobsters, crabs, and other depleted species could help them recover, too. Some nations have begun to establish networks of safe havens—marine reserves—aimed at protecting and restoring regions under their jurisdiction and to consider policies for protection of the high seas, a global commons that embraces nearly half of the planet.

The naturalist writer John Muir famously observed, "When we try to pick out anything by itself, we find it hitched to everything else in the universe."

By picking out jellyfish and telling their stories, Lisa-ann Gershwin masterfully shows how they and we are hitched together—and to everything else in the universe.

Sylvia A. Earle

National Geographic Explorer in Residence and founder of Mission Blue Chief Scientist, National Oceanic and Atmospheric Agency, 1990–1992

It seems the ocean's chain of life is actually a fragile, silken web. If just one strand is removed, the whole thing unravels. And it may never be whole again.

—MARLA CONE, *Los Angeles Times* environmental writer

Standing in a sandstone quarry in central Wisconsin takes us back 500 million years to a time when life on earth was very different from what it is today (see plate 16). No bones, no claws, no teeth had yet evolved. Nothing with jaws. Creatures with shells were just beginning to form. Mostly jellyfish and their kin, and perhaps some worms. Soft-bodied drifting creatures, and some that stuck in the sand. Some perhaps with photosynthesizing symbionts, similar to today's corals. But no spectacular coral reefs. No vast filtering mussel beds. No sharks slicing through the water as schools of fish flee for their lives.

The tiny hamlet of Mosinee is the gateway to not one, but seven of history's most magnificent and unlikely events. Stacked one on top of the other like pages in a history book are seven successive bedding planes, each with hundreds of jellyfish fossils stranded together in seven separate accidents of timing. It is rare enough for jellyfish to fossilize, but to have an entire stranding event so perfectly preserved—seven times—is simply splendid.

The "footprints" they left behind through their "excavation" behavior tell the story of their fateful wrong turn, some larger, some smaller, some four-parted, some five-parted, all high and dry. But there is another thing this tells us, something more ominous: large swarms of jellyfish have been a frequent occurrence for a long, long time . . . and they probably aren't going to go away.

It is most likely, to some extent, just what jellyfish do.

But what if there was something that we humans are doing that was favorable to jellyfish? What if we are fishing out their predators and competitors and making the seas more toxic so that sensitive species couldn't survive? And what if warming waters speed up their metabolism and make them breed and grow faster while simultaneously stripping the ocean of oxygen so that heavy breathers like fish and crustaceans gasp for air? These are not actually "what-ifs"—these are real, and this is their story.

Imagine if fish vanished and shrimp ran out and oysters disappeared and what if, except for the occasional slug sliming its pathway along the seabed or the worms still thriving in the sediments, jellyfish dominated the oceans? If I offered evidence that jellyfish are displacing penguins in Antarctica—not someday, but now, today—what would you think? If I suggested that jellyfish could crash the world's fisheries, outcompete the tuna and swordfish, and starve the whales to extinction, would you believe me?

Yep, jellyfish. Most people have never spent more than a moment of their lives, if that, thinking about jellyfish. But things are changing. The climate is changing. Pollution is increasing. Fish stocks are vanishing. Oceans are becoming more acidic. Species composition is rearranging. And jellyfish populations are exploding into superabundances and exploiting these changes in ways that we could never have imagined—not only exploiting the changes, but in some cases driving them. As seas become stressed, the jellyfish are there, like an eagle to an injured lamb or golden staph to a postoperative patient—more than just a symptom of weakness, more like the angel of death.

From jellyfish that grow to the size of refrigerators to others as small as a few grains of sand, and from swarms that blanket hundreds of miles to species that kill healthy adults in two minutes flat, jellyfish are ubiquitous in marine environments. And with increasing frequency and force, they are making their presence known. Emergency shutdowns of nuclear power stations. Disabling of America's most powerful nuclear-powered supercarrier. Causing the disruption and relocation of filming of a recent major Hollywood movie—twice.

Nearly stopping an Olympic triathlon. These are just a few of the many recent inconveniences caused by jellyfish blooms.

Jellyfish blooms signal a much more serious problem, one with long-term consequences to our ecosystems and food security. Ecosystems are stressed, and jellyfish are taking advantage.

They have been around at least 565 million years, and probably far longer. And they haven't needed to change their body form or their lifestyle in all that time . . . because they work. Jellyfish are among the world's most successful organisms, having survived freezes, thaws, superheated conditions, shifting and rearranging of continents, mass extinctions, meteor strikes, predators, competitors, and even man. And all the while, as creatures around them evolved tails and feet and brains and learned to breathe and fly, jellyfish have persisted just as they are.

Yes, you can laugh about them being spineless and brainless with no visible means of support, but you've got to admit, these multimillennial survivors are doing something right . . . and lately, it seems that they've been doing a lot more of it than normal.

I began studying jellyfish on 22 December 1992, at a time when they were so completely unfashionable to work with that people would simply look at me and blink . . . and blink again, unsure of what to say. The "big news" in the very small jellyfish research community at the time was that Frank Zappa had just written a song about Nando Boero, an Italian scientist who had named a new species of jellyfish after Zappa a few years earlier.

In 1998, while working on my PhD at Berkeley, I was awarded a Fulbright Fellowship to examine the effects of jellyfish blooms on commercial fisheries in Australia. But it soon became clear that such a project was not yet possible, because most of the species were unidentified or wrongly identified, simply because of lack of local expertise and—lack of scientific priority to develop the expertise. So I became involved with trying to sort out the classification. Fifteen years and 160 new species later, the incidence and effects of jellyfish bloom problems have become much more obvious all over the world. Stings to tourists. Clogging of fishing nets. Aquaculture kills. Emergency shutdowns to power and desalination plants. A great many jellyfish bloom incidents have caused companies and governments many millions of dollars, and these costly incidents appear to be increasing in frequency.

Dr. Claudia Mills of the Friday Harbor Laboratories of the University of Washington is one of the pioneers in the study of jellyfish blooms. Claudia is a tall woman with a commanding presence and a playful smile. Her science is clean and meticulous, while she conveys a sense of wonder and admiration for her subjects. Pondering jellyfish unhurriedly from her office, which overlooks the laboratory docks and off into the picturesque Puget Sound, she became concerned about what others saw as an unrelated collection of anomalous events; in Claudia's view, however, these seemingly isolated blooms looked like a growing problem. That was in 1995. Today we have much more data, far more workers, and hundreds of published papers, and yet a few still question whether there is indeed a problem. In fact, a big hoopla erupted on this very issue in early 2012 when a group of experts announced that a global increase in blooms is unsubstantiated (Condon et al. 2012). This was unfortunate wording, and the media had a field day with it. The simple truth is that jellyfish swarms come and go like blooms of flowers as a normal part of their life cycle—always have and always will—in response to environmental stimuli. We are increasingly fishing out their predators and competitors, and we are altering the physical properties of the seabed and chemical properties of the oceans to favor jellyfish. And we are more frequently using the oceans, putting our bodies and industries into the pathways of jellyfish blooms. But despite our increasing use of the sea, we have surprisingly few datasets about jellyfish.

You see, the problem is that nobody foresaw the potential for these simple creatures to wreak the havoc they are now causing all over the world, and so very few long-term datasets exist from which we can quantify the degree of change.

Wild animal populations rise and fall continuously to some degree . . . it's just the nature of nature. But long-term trends—and more importantly, huge declines or inclines over the short term—cry out to be explained.

Lucas Brotz did just that. As a master's student at the University of British Columbia, he investigated the question of jellyfish blooms. Out of 66 large marine ecosystems covering the world's coastal waters and seas, Brotz found quantifiable jellyfish trends in 45, the overwhelming majority of which showed an increase, while only 2 showed a decrease and 12 were stable (Brotz 2011). Furthermore, Brotz found significant correlations between jellyfish blooms and human activities. His thesis is available on the Internet. It's worth a read.

One must be cautious in today's political and economic world, where words can be twisted and evidence is often ignored. If you hear someone say, "jellyfish blooms are bunk," start asking questions. They are not bunk; jellyfish bloom as a normal part of their life cycle in response to environmental stimuli. The more stimuli, the more response—it's not all that complicated. If you hear someone say, "There's not enough data to show that jellyfish are increasing," this doesn't mean there's a lack of data that jellyfish are causing problems—oh, they're causing plenty of problems. While you need a trend to predict future events, you don't need a trend to detect that something is a problem now. And you don't need a PhD to take a punt that ideal conditions for mayhem might result in mayhem.

We would be astonished if a doctor told us, "We have only just noticed this growing cancerous tumor and I'm not sure what it means; let's just wait and see if it becomes untreatable." As the patient wanting to survive, we expect aggressive action to fix the problem rather than a sit-and-wait approach to collect more data.

As with many such complicated issues, there are essentially two ways to approach this: (1) from the standpoint of scientific enquiry, or (2) from the perspective of management. Science seeks to quantify a problem in order to understand its origins and predict its future. Management seeks to minimize negative effects now and in the future, regardless of origins and explanations. It's just two different ways of looking at the same thing. One is not better or worse than the other—*it is not a competition!* They simply seek different outcomes. But we confuse the two at our own peril. As a scientist, I will argue the importance of data and understanding until the cows come home. But as a practical person wanting safer, healthier, more sustainable oceans *now*—not waiting 40 or 100 years until we finally have sufficient long-term data for scientific consensus—I will argue that jellyfish blooms are a problem that needs to be dealt with, pronto.

Jellyfish as organisms may or may not be experiencing a sustained global increase, but there can be no doubt that the incidence and severity of problems they are causing globally are increasingly reported (Mills 2001; Purcell 2012). Either way, the havoc that jellyfish are causing is difficult to ignore, and our lack of long-term comparative data in most regions leaves us vulnerable to being caught by surprise as blooms cause trouble and ecosystems falter. In all likelihood, yes, jellyfish probably are on the increase worldwide, as human-impacted ecosystems change to become less favorable to some species and more favorable to jellyfish.

In addition to Brotz's thesis work, a persuasive argument is to be found in the intersection of two recent studies. Dr. Ben Halpern of the National Center for Ecological Analysis and Synthesis in Santa Barbara, California, and his colleagues created a scoring system to rank the human impacts on 232 marine ecoregions around the world (Halpern et al. 2008). Meanwhile, Dr. Jenny Purcell of Western Washington University tabulated a variety of environmental indicators and found that 6 of Halpern's top 10 rankings, plus 8 others in the top 100, coincide with regions of notable recent jellyfish blooms and high indicator values (Purcell 2012). Many of Halpern's other higher-ranking ecoregions may well be experiencing jellyfish bloom problems too, but remain undocumented for whatever reason—as the old saying goes: absence of evidence is not the same as evidence of absence.

Furthermore, harmful and costly jellyfish incidents are likely to increase in severity and frequency as our oceans become more and more stressed by manmade disturbances—"anthropogenic perturbations," they are called.

A great many books—some of which are fascinating reads—have been written on climate change, overfishing, and the pollution of our ecosystems. But there is a pattern to these books that I believe is not completely accurate, and is perhaps somewhat misleading. They leave the reader with the feeling that if we would just stop polluting, everything would be okay—that if we would just stop overfishing, the oceans would return to normal. These ideas sound good, but are not what we observe actually taking place.

It is a very egocentric view to assume that *Homo sapiens*, wise as we are, hold all the levers of change. We pollute, but we cannot so effortlessly unpollute. We overfish, but we cannot so easily restore fish populations to normal. There is another variable in the equation that is invisible to us. As invisible but as real as low pressure is to a hurricane. As invisible but as real as DNA is to the existence and daily functioning of our own bodies. As invisible but as real as a poisonous vapor. Jellyfish.

Anthropogenic disturbances and jellyfish blooms are not mere hypothetical concepts. The causes and effects are happening around us now—we don't have to wait hundreds of years to see if it's all true. We can see evidence of degradation all around. And the problems all tie in together; they are all part of a cascade of events leading to a very changed world from the one we know. You can go to a marina near you and see introduced species and lots and lots of jellyfish now. You can go snorkeling and see bleached and dying reefs now. You can go fishing and see changed relative species abundances now. You don't need to spend money to go to some exotic tropical or polar locale. You don't need a PhD to figure out the chemistry or understand the biology.

Don't take my word for it—ask your own questions. Go to any public aquarium or marine theme park and ask why they have so many species of jellyfish on display but no Yangtze River dolphins. Go to the beach—any beach, anywhere—pick up a handful of sand, let it trickle through your fingers, and notice how much of the grainy mixture contains tiny fragments of plastic. Go to your local fishmonger and ask when was the last time they carried Atlantic halibut, or Newfoundland cod, or California white abalone, or Chesapeake Bay scallops, and why the price was so high. Google terms like "jellyfish blooms" or "jellyfish climate change." Read the literature and make up your own mind.

So many bits and pieces of evidence are around us all the time, but few see them in a connected way. When we think of overfishing, we forget that the warming waters of climate change are reducing the dissolved oxygen, making it harder for fish to respire and survive, and thus further contributing to the loss of fish. When we think of pollution, we think of smelly nasty corners of marinas, or beer cans and plastic drink bottles washed up on beaches, but we don't think about the heavy metals or pesticide residues accumulating in our food supply and in our own bodies as a result . . . or about the excess nutrients flowing into estuaries and bays, creating vast dead zones . . . or about the many exotic species transported around the world in ballast water every day . . . or about the changes in ocean pH caused by carbon dioxide, leading calcium carbonate to leach out of snails' shells and corals' skeletons. And we hardly ever consider the consequences of jellyfish inheriting disturbed ecosystems— from the jellyfish perspective it is certainly a perfect Hollywood ending, but from the human perspective, it is more like Hitchcock or Poe.

This book is not meant to be an exhaustive account of the perturbations vexing marine ecosystems, nor even a complete treatise on jellyfish bloom biology. Rather, I have written this as an overview to give a feel for the depth and breadth of the problems that arise where disturbances and jellyfish intersect. Many of the stories relate directly to jellyfish blooms, while others relate indirectly through the destructive effect that they have on the marine ecosystem or its inhabitants.

I hope that as you read this book, you will come to realize two things: that the statistics and stories recounted herein are just a small sample of the total problem of ecosystem degradation, and that jellyfish blooms are the inevitable outcome of extreme changes in the marine environment.

Normally, predicting the future is the suspicious enterprise of palm readers and fortune-tellers. However, in this case, the scientific evidence is overwhelming, with a very large number of independent datasets all pointing to the same trends. The conclusions presented here are conservative, and the predictions stemming from them are scientifically reasonable. It is hard to imagine that they will not come to pass, despite being socially unfathomable and essentially apocalyptic. In the words of Professor Jeremy Jackson of the Scripps Institution of Oceanography (2010, 3772), "The question is not whether these trends will happen, but how fast they will happen, and what will be the consequences for the oceans and humanity."

In my view, anthropogenic disturbances are merely different types of stimuli that trigger a common biological response. And more often than not, they work synergistically. This is the story of the response—the role that jellyfish play in driving weakened ecosystems to a new stable state, *their state*.

> How inappropriate to call this planet Earth when it is quite clearly Ocean.
>
> —ARTHUR C. CLARKE

PART I

Jellyfish Behaving Badly

"Do we have the courage to face the realities of our time, and allow ourselves to feel deeply enough that it transforms us and our future?"
—CHRIS JORDAN, from the movie *Midway*

CHAPTER 1

At the Mercy of Jellyfish

Here we are at the dawn of a new millennium, in the age of cyberspace
and we are at the mercy of jellyfish.
—Editorial in the *Philippine Star*, December 1999

How to Cripple a Nuclear Warship
(Queensland, Australia, July 2006)

The United States Navy boasts ten Nimitz class nuclear-powered aircraft su-
percarriers, the largest of their kind. They weigh 100,000 tons, extend 1,092
feet, stand more than 77 yards high above the waterline, and cost $4.5 billion
each. The flight deck covers 1.82 hectares and houses 80 fixed-wing aircraft
and helicopters. These warships are built to operate for 20 years without re-
fueling and to withstand any threat that a military's might or nature's wrath
can hurl at them.

The ninth in its class, the USS *Ronald Reagan* was commissioned on 12 July
2003. Just three years later on its maiden deployment, the world's most modern aircraft carrier, with a crew of 6,000 and capable of taking on an entire
nation's armed forces, met its match. Brisbane, Australia, was its first foreign
port of call. On 27 July 2006, thousands of jellyfish were sucked into the condensers (which take in seawater and cool down the ship's engines) while the

ship was docked in the Port of Brisbane ("Jellyfish Take On U.S. Warship" 2006; Mancuso 2006).

The incident diminished the ship's ability to condense steam from the turbines, rendering them less efficient. To preserve power while the jellyfish were cleared from the condensers, the ship had to shut down several on-board systems and switch over to generators. Local fire crews were placed on standby while full on-board capacities were disabled. As a result, the ship cut short its Brisbane visit. Perhaps tongue-in-cheek or perhaps being somewhat economical with the truth, the commander of US Naval Air Forces called this incident an "acute case of fouling."

Jellyfish. Stings. Slime and jiggle. "Ick-factor." Stealthy. Spooky. Scary. Alien. Lethal. Jellyfish elicit fear, especially en masse. As silent as a school of sharks, as relentless as a swarm of bees. And they are on the increase.

Jellyfish blooms are nothing new. In fact, fossil evidence shows us that jellyfish have been blooming for hundreds of millions of years. Around the turn of the nineteenth century, it became fashionable for naturalists to report all sorts of odd and unusual events from the natural world. The early issues of the journal *Nature* and others like it are full of such interesting tidbits. One such report described *Aurelia* as so abundant in Kiel Bay, Germany, that an oar pushed down between the jellyfish remained standing upright (Möbius 1880). Today, just about any bay or harbor has *Aurelia* shoals so dense that one may wonder whether there is actually enough water between each jellyfish for it to obtain enough oxygen to survive.

So while the fact of jellyfish blooms is not new, what does appear to be new is the increasing frequency and duration of blooms and the similar effect they are having on dissimilar ecosystems. The most surprising part isn't the blooms, per se, but that it got "this bad" without us noticing.

Some jellyfish bloom incidents from around the world are highlighted below. In many cases there may be an obvious anthropogenic cause—something that we humans have done to disturb the ecosystem sufficiently to cause a wobble—but such a cause is not always immediately apparent. Some blooms are of such minor consequence as to almost raise a smirk or a giggle, while others cause enormous financial loss, ecosystem shifts, infrastructure problems, medical problems, or even human death.

To most people, jellyfish problems are about stings. The public health aspects of jellyfish blooms can be personally stressful and can greatly impact tourism. Increasingly, jellyfish blooms are also impacting fisheries and industrial endeavors, such as power stations and desalination plants. These problems of human inconvenience are the subject of this chapter, while more serious problems of shifting ecosystems are presented in the next chapter.

Jellyfish have an uncanny knack for getting stuck. In pipes. On nets. Against screens. Like the proverbial bull in a china shop, jellyfish simply *will* get stuck if even remotely possible. Imagine a piece of thin, flexible plastic wrapper in a pool, where it can drift almost forever without sinking, until it gets sucked against the outflow mesh. Such is the problem with jellyfish. Most are unable to swim against a current, and so, when even the gentlest flow leads them toward artificial structures, such as screens, pipes, and nets, things go badly. *Very* badly.

Jellyfish and the Military Coup
(Philippines, December 1999)

The Philippines has long had unstable governments. Military raids and hostile takeovers are an ever-present concern. After only a year in office, Philippine president Joseph Estrada was already on shaky ground, with pundits questioning whether he would complete his term to 2004.

So when on the night of 10 December 1999, some 40 million people across the northern half of the country were suddenly plunged into darkness by a power outage, many thought a coup d'état was underway. It was not: this time the enemy was . . . jellyfish.

In fact, fifty truckloads of jellyfish had been sucked into the seawater cooling system of the coal-fired Sual power station, causing a cascading blackout ("Asia: Dark Days" 1999).

Estrada was meeting with senators at the time of the power failure and remained in the dark for ten minutes before generators could restore the power. The public, however, remained in the dark until the following day when the crisis could be rectified and its cause clarified.

This incident made international headlines. Media all over the world reported it first as a coup, then with a bit of a giggle.

Jellyfish against Nuclear Power
(California, October 2008)

Of all peculiar associations that one could possibly imagine, this may well be the most bizarre. Long before jellyfish first pulsated through my fascinations, back when I was trying to "find myself," I joined an antinuclear protest. I was not quite eighteen. It was a good cause, and I still believe in it, though I must say I get arrested less often for it these days.

The 2,240-megawatt Diablo Canyon nuclear power plant supplies power to over 1.5 million homes in California. It is located on the picturesque coast at Avila Beach near San Luis Obispo, California, where it was built on two known earthquake faults, about 250 kilometers (150 miles) upwind of Los Angeles. In September 1981, the plant was due to begin low-power testing, which triggered a massive public demonstration—the largest act of civil disobedience in US antinuclear history—and I was one of about 1,900 people who were arrested in peaceful protest, along with the actor Martin Sheen. And yes, for the record, I wore my patchwork jeans, my tie-dye shirt, and my "One-Love" armband. Of course.

Skip ahead: when I first began working with jellyfish in 1992 in Southern California, the species that most caught my fancy was what I thought could be a new species of moon jellyfish. It certainly was quite different from others I was working with, *Aurelia aurita*, which was the only accepted species of *Aurelia* at the time. Drilling down into the nomenclatural history of the group, I found that this "new species" had actually been formally named and classified some 175 years earlier, as *Aurelia labiata* (see plate 2), and that the act of merging them under one species name many years earlier had been an error. Through revalidation in a technical publication (Gershwin 2001), the species again became recognized.

Now skip ahead again: in 2008, I was living in Australia, far away from *Aurelia labiata* and nuclear power demonstrations. On 21 October, news hit that a species of jellyfish—none other than *Aurelia labiata*—had been sucked into the cooling water intake racks of the Diablo Canyon plant in such huge numbers as to cover 80 percent of one rack and 40 percent of another. The incident triggered a total shutdown of containment 2 and reduction to half power of containment 1 . . . for *three* days (DiSavino 2008). At last, the jellyfish were able to do what decades of activists had failed to accomplish.

But this isn't just a story about one's scientific progeny "doing their master's bidding"; this is about jellyfish blooms out of control, causing millions of dollars of loss in a man-made system that simply cannot stop the avalanche of nature out of balance. If this were an isolated incident, perhaps it would be entertaining. But it's not.

Another nuclear power plant with similar problems is the Madras Atomic Power Station in Kalpakkam, India. Unscheduled outages due to jellyfish occurred on 30 October 1983, 4 November 1983, and 21 April 1985 (Rajagopal, Nair, and Azariah 1989). In fact, these are just a few of the many such incidents that occur regularly there. From February 1988 to April 1989, staff counted the number of jellyfish removed from the intake screens. The total for the 15-month study period was over 4 million. The heaviest infestation was in May 1988, with over 1.5 million removed (about 315 tons); the maximum quantity collected on any one day was a whopping 31 tons, on 21 July 1988. During May 1988, 12 of the 16 weld mesh screens were severely damaged by jellyfish; during other months, 16 gates were damaged.

A decade later, the problem recurred. During 1995–1996, the plant was plagued by up to 18 tons of jellyfish a month, necessitating shutdowns costing about ₹5.5 million (approximately $122,000) per day (Masilamoni et al. 2000). During that year, peaks in jellyfish arrivals coincided with the reversal of coastal water currents during the two monsoon seasons, that is, early June for the southwest monsoon and November for the northeast monsoon.

There's Just Something about Power Plants . . .

Strangely enough, power plants in general seem to be somewhat magnetic when it comes to jellyfish problems (see plate 4). Nuclear plants and nonnuclear plants. Northern Hemisphere and Southern Hemisphere. Plants drawing cooling water from the oceans, or bays, or estuaries, or lagoons.

Around 1960, Japanese authorities began to notice increasing problems with the moon jellyfish *Aurelia aurita* clogging intake pipes. By 1969, it was reported that "almost all domestic electric power companies have been suffering from the damages since 1965" (Matsueda 1969, 187). The authors of that report noted that four power stations were subject to frequent jellyfish attacks, which in 1967 alone caused eight load restrictions and six shutdowns. At one station, jellies removed from intake screens between May and August amounted to 2,000 tons, reaching as much as 150 tons a day. Just to give an idea of how pervasive this problem is, more than 30 additional major nuisance events with jellyfish vexing power plants are listed in table 1 in the appendix—these are just the ones that have been reported.

And the jellyfish ingress problem isn't just with power plants. Desalination plants have also been afflicted: some are listed in table 2 in the appendix. In fact, the ingress problem isn't limited to power plants and desalination plants, but can involve any type of intake screens in regions where jellyfish occur. Indeed, even the Monterey Bay Aquarium in California, one of the first places in the world to successfully culture and display jellyfish, was plagued by one of its "babies"—a species for which it had resolved the life cycle a decade earlier and has been cultivating since. In August 2009, millions of the brown sea nettle *Chrysaora fuscescens* bloomed in Monterey Bay (see plate 3), clogging the aquarium's seawater intake screens (Tucker 2010). These jellyfish grow to the size of basketballs, so one might imagine that millions of them might have a collectively large appetite. The aquarium blamed the bloom on ideal feeding conditions caused by high nitrate levels in the water. Of course, the curious would ask why the nitrate levels were so high (the issue of sewage and fertilizer pollution is the subject of chapter 7).

Intensive research around the world on methods to keep jellyfish away from the intakes of power stations and desalination plants has had mixed but interesting results. Chemical repellents don't work, because jellyfish drift on the current and can't respond. Electric shocks don't work for the same reason. Acoustic shocks don't work, because jellyfish, not having a brain, aren't afraid of noise. Bubble curtains don't work because the bubbles kill them and, alive or dead, they block the flow of water all the same. Biocides don't work for the same reason. Diversional nets don't work because they foul too quickly and can't be left in place. Research continues. And so do ingress problems.

It's easy to see how passively drifting jellyfish can become entrained in the powerful intake currents of desalination plants and power station cooling systems. Alas, it doesn't take much of a current to entrain a jellyfish. . . .

Particular Issues with Salmon Farms
(Estuaries and Bays Globally, since 1997)

Jellyfish seem to *really* like salmon farms. Or given the damage they cause, perhaps it would be more appropriate to say that jellyfish really *don't* like salmon farms.

New Zealand. In November 1998, a massive salmon kill occurred at Stewart Island, off the far southeast of New Zealand's South Island. A swarm of large *Aurelia* had moved into Big Glory Bay with the tide, and within 30 minutes, 56,000 3-kilogram salmon were dead. Needless to say, the salmon farm owners (and their insurance agents) were horrified at what seemed at the time like a freak event.

The salmon all swim in one direction inside the circular pens, creating a fairly strong vortex that sucks water in from the surrounding area. The *Aurelia*, being passive drifters, became entrained in the vortex. Too large to pass through the mesh, the jellyfish were pinned against the netting. As the jellyfish struggled against the current and the netting, their mucus, which is profuse and packed with stinging cells, was sucked into the cages. It appears that as the salmon inhaled the mucus, it blocked the oxygen-exchange surfaces of their gills, causing them to suffocate. The stinging cells exacerbated the problem by alarming the salmon, causing them to breathe faster, thereby serving to suffocate them faster.

Australia. Just one week after the New Zealand incident, a similar situation occurred in the Huon Estuary in southern Tasmania. This time it was 25,000 harvest-ready salmon. While these *Aurelia* were smaller and able to penetrate the mesh cages, the mechanism, speed, and outcome were like déjà vu.

Since these two events in 1998, both regions have had almost annually recurring *Aurelia* swarms vexing the salmon. Most recently, the New Zealand farm lost another 2,000 salmon in November 2010.

Besides the mass losses at the time of each incident, many fish die in the days that follow as a result of their injuries, and fish that survive these episodes often fail to grow efficiently (J. Handlinger, personal communication).

Norway. In November–December 1997, a mass occurrence of the very large and very stingy siphonophore *Apolemia uvaria* invaded coastal and offshore waters from western Sweden to northern Norway. This colonial species can grow to 30 meters (100 feet) long, looking somewhat like a long feather boa. For about six weeks, salmon farms were impacted by high rates of stock mortality and morbidity due to stinging lesions on the bodies and gills of the salmon. Previous mass fish-kill problems with no fewer than 5 other species of jellyfish, including comb jellies, which do not sting, were also reported (Båmstedt et al. 1998).

Chile. So too, Chilean salmon farms have been plagued by jellyfish blooms and algal blooms, incurring huge economic losses (Carvajal 2002). In March 2002, one salmon farm in the Quemchi area of southern Chile lost about 120,000 fish, while another lost about 45,000, both due to jellyfish "attacks." Many salmon farmers believe that jellyfish are an omen of the arrival of El Niño.

Ireland. More recently, in November 2007, a massive swarm of *Pelagia noctiluca*, the so-called mauve stinger or pink meanie, wiped out a salmon farm in northern Ireland. The densely packed swarm occupied an estimated 26 square kilometers (10 square miles) and was 10 meters (35 feet) deep. All 100,000 fish at the farm, worth $2 million, were killed. The following week, a separate farm owned by the same company was wiped out by the same jellyfish swarm. All told, about 250,000 salmon were killed (Doyle et al. 2008).

Scotland. There was even a case of a seemingly innocuous hydrozoan causing a similar problem. In August 1984, the dime-sized, so-called water jelly *Phialella quadrata* swarmed around ten salmon cages at a farm in Scotland. Over 4 days, about 1,500 fish died from "hypersensitivity to the jellyfish toxin" (Bruno and Ellis 1985).

In the *Phialella* incident, it was noted that the lion's mane jellyfish *Cyanea capillata* was also implicated in earlier mortalities of farmed salmon. The lion's mane is well known for its stinging abilities, having featured as the murder weapon in Sherlock Holmes's *Adventure of the Lion's Mane*.

In August 2002, nearly 1 million salmon were killed at 2 Scottish farms by *Solmaris*, a tiny hydromedusa. Another company lost 400 tons of salmon that were fully grown and ready for market.

A Scottish salmon industry protest group newsletter, the *Salmon Farm Monitor*, reported the following for the period 1999–2002 in Scotland alone:

- More than 4.4 million farm salmon died in their cages in ninety separate incidents, 50 percent of deaths were caused by "algal blooms," 45 percent by "jellyfish" and 5 percent by "plankton" (which could be algal blooms or jellyfish).

- In the period 1999–2002, mass mortality incidents increased more than sixfold (12 to 78), while in the same period fish deaths increased eighteen-fold, from just over 240,000 to over 4 million.
- Average mortality per incident rose from just over 1,000 dead fish in 2000, to 44,000 in 2001, to 70,000 in 2002.

While mass mortalities from suffocation and toxic overload are grist for the media mill, a seemingly smaller problem with the capacity to become far bigger appears to be brewing. Hydroids are the most prevalent organism to foul salmon cages (see plate 4), and when they are cleaned off, they break into zillions of tiny fragments, each of which carries both a damaging sting and the capacity to settle and start a whole new colony nearby (Guenther, Misimi, and Sunde 2010).

Many more problems with jellyfish and fish farming operations have occurred around the world. As we become more reliant on mariculture and as jellyfish become more abundant, it is fair to say that we can expect to see increasing problems where the two meet. It may be that our most prime aquaculture areas will soon become thickets of hydroids and chowders of jellyfish.

In addition to the acute problems of toxicity and of suffocation by mucus secretions, jellyfish have also been linked to bacterial diseases in farmed salmon. It appears that small jellyfish, such as the diminutive and "harmless" *Phialella quadrata* can carry large numbers of a type of bacteria that can be transferred to fish gills during stinging (Ferguson et al. 2010). It is thought that the centimeter-wide jellyfish easily pass through the cage mesh and into fishes' mouths during respiration, initially causing a mild stinging that is then exacerbated by the transferred bacteria. So while these smaller species are less likely to cause mass mortality within minutes, by acting as vectors of disease they can be responsible for high mortality rates through slower illness.

Recently, the larger, more virulent *Pelagia noctiluca* has been shown to carry the same bacteria (Delannoy et al. 2011). As jellyfish become fragmented by contact with the nets, small pieces are able to penetrate the mesh and thus sting the fish; it is thought that the bacteria aggravate the sting lesions.

The underlying causes of jellyfish blooms that vex salmon farms are complicated, and are likely to be the result of two quite different phenomena.

First, the extra nutrients from salmon waste and uneaten food are probably driving ecosystem changes that favor jellyfish blooms and algal blooms. On the west coast of Canada, "The 49,600 tonnes of farmed salmon produced in British Columbia in 2000 contributed as much nitrogen as the untreated sewage from 682,000 people, and as much phosphorous as the sewage from 216,000 people" (Tirado 2008, 19). The situation is even worse in Puget Sound, where 4 of about 12 salmon netpens in the state discharged 93 percent as much "total suspended solids" into the sound as did the sewage treatment plant serving more than a million people in the city of Seattle (Goldburg, Elliott, and Naylor 2001). In extreme cases, dead zones develop under salmon pens, surrounded by a ring of decreased animal diversity extending up to 150 meters (500 feet). This process, called "eutrophication"—or pollution through excessive fertilization—is the subject of chapter 7.

Second, inshore water movement patterns are probably responsible for the relatively high number of fish kills in salmon farms. Jellyfish drift with currents as a normal part of their lifestyle. As currents move in and around embayments, it is inevitable that jellyfish will pass by or through those with salmon farms. However, the strong vortex effect created by the salmon swimming at the same speed and in the same direction in their cages is often enough to pull the current toward them. Most jellyfish are unable to fight even the slightest current, so they become entrained in the flow toward the salmon. Furthermore, there is increasing discussion about the salmon farms also creating extra structures for the jellyfish larvae to settle and grow on, both with the salmon cages and with the dead zones caused by the accumulated waste below. These two issues are explored in more detail later.

Regardless of the cause, one thing seems clear: the problem is getting worse, as summed up by Emily Baxter of University College Cork in Ireland and her colleagues (2011, 1): "With aquaculture predicted to expand worldwide and evidence suggesting that jellyfish populations are increasing in some areas, this threat to aquaculture is of rising concern as significant losses due to jellyfish could be expected to increase in the future."

Many problems with jellyfish blooms relate to interference with human enterprise, such as shipping, fishing, farming, or infrastructure. Of course, another big problem with jellyfish is that they sting. And sometimes they kill people. The full scope of stinging jellyfish is not detailed here because the subject can take up volumes. However, an overview of the problem may be helpful in order to answer the often-asked question of whether the deadly species are likely to spread.

Box Jellyfish: The World's Most Venomous Animal

Chironex fleckeri and its recently described Okinawan counterpart *C. yamaguchii* are arguably the world's deadliest animals (see plate 1). They can kill a healthy adult in as little as 2 minutes—the average time to death is just 4 minutes. It is common for people to think this is caused by an allergic reaction, but that is untrue. Death is merely a consequence of "enough" jellyfish tentacle coming into contact with unprotected skin—for an adult this is about 3–5 meters (9–15 feet), for a child it is 1–2 meters (3–6 feet)—not much, considering that a mature jellyfish has approximately 120–180 meters (350–550 feet) of combined tentacle length in its arsenal.

Chironex grows to about 30 centimeters (1 foot) across the body, but it is lethal from the time it reaches about 8–10 centimeters (3–4 inches). When young, they are dangerous to only small shrimp and fish, but as they grow, they undergo a change in the potency of their venom, as well as a dramatic increase in the ratio of lethal to nonlethal stinging cells. They grow into killing machines.

The mechanism of death is poorly understood, but it involves a component of the potent venom locking the heart in a contracted state. It seems in most cases that the heart progressively contracts over a brief period, perhaps a couple of minutes. There have been successful resuscitations before the heart is locked completely. However, there have also been many cases where victims could not be resuscitated. In Australia, 76 fatalities have been recorded since 1884.

In places like Thailand, the Philippines, and Papua New Guinea, where record-keeping is often nonexistent and diagnoses inaccurate or lacking, confirmed deaths are likely to be gross underestimates of the true problem. For example, 6 fatalities and 10 potentially fatal stings have been reported in Thailand since 1996 (Fenner, Lippmann, and Gershwin 2010), and Malaysia has recorded 3 fatalities and 4 potentially fatal stings since 2000 (Lippmann et al. 2011); these reported stings all involve tourists—the rate is undoubtedly higher amongst local children, fishermen, and others who use the water.

In the Philippines, it is conservatively estimated that 20 to 50 fatalities a year are caused by box jellyfish. However, it is very difficult to get reliable data. Similarly, in Papua New Guinea, India, Myanmar, Indonesia, China, and

Korea, it has been difficult to get data, but sting-related fatalities are known to occur.

Irukandjis: Funny Name, Not So Funny Sting

The other type of highly dangerous jellyfish is the Irukandji (see plate 1). The name is from an Aboriginal tribe from the region near Cairns, Australia, where the strange illness caused by the jellyfish was first discovered. Currently, more than 10 species are known. They are generally small, peanut- to thumb-sized, although a few can be the size of a large takeaway mocha. All species of Irukandjis have only 4 tentacles, though these can be up to 100 times as long as the body of their owner. Despite their diminutive stature, there is nothing small about the severe medical complications from their stings—it's like something out of a Hollywood murder mystery.

Many people have never heard of Irukandji. At first sight most people find their petite bodies and highly dangerous nature difficult to reconcile. But it's true. The first-discovered species, *Carukia barnesi* from the Cairns region, reaches a mere 14 millimeters (half an inch) on a good day. But it can floor a healthy adult with just a brush of tentacle.

Its sting causes Irukandji syndrome, a constellation of seemingly unrelated symptoms. The sting itself is often not even felt, or is so minor that it is dismissed. After a characteristic delay of about 20–30 minutes, the lower back begins to ache, then rapidly escalates to fully debilitating, cramping, or pounding pain. Patients often describe it as akin to being hit in the kidneys with a wooden bat again and again and again, or feeling as if an electric drill is drilling into the back. Then nausea begins, along with relentless vomiting— every 1–2 minutes for up to 12 hours. The syndrome rapidly develops into full-body cramps comparable to the bends, with shooting spasms in the arms and legs and behind the eyes; difficulty breathing; profuse, drenching sweating; coughing; and muscular restlessness. Many patients feel a "creepy skin" sensation, often described as feeling like spiders crawling on or worms burrowing into the skin. Many patients feel an "impending doom," believing they are going to die. Some go so far as to beg their doctor to put them out of their misery.

The symptoms and severity of Irukandji syndrome vary with species. Some species, in addition to the above symptoms, also cause severe hypertension (high blood pressure). Two people with confirmed cases of Irukandji syndrome died from hypertension-related strokes. It is believed by medical and scientific experts that other Irukandji-related deaths have occurred—perhaps many—but that they were misinterpreted as heart attacks, strokes, or drownings. Irukandji stings typically leave no mark and nothing to test in the body—no trace.

Irukandjis and their stings became widely known following two highly publicized fatalities at the Great Barrier Reef in 2002. Possibly due to awareness of climate change increasing at the same time as publicity about dangerous jellyfish, questions have arisen about future jellyfish bloom patterns. In particular, people want to know whether jellyfish problems are likely to get worse and whether they will spread to cooler, more populated areas of Australia, such as the Gold Coast and Sydney, or up toward America. We explore these questions below in chapter 10 in the discussion on climate change.

Some may be tempted to think that the increase in reports of Irukandji stings from regions not previously reported to have them might reflect geographic spread of the species, but this is questionable. Irukandji as a name has long been regional and so was likely to be disregarded overseas. The species are usually small and cryptic—and essentially invisible in water—and are therefore often overlooked. The syndrome involves so many apparently unrelated symptoms that most doctors would not realize the significance until they had treated several cases and noticed a pattern; but even then, most don't publish, so it would be unlikely to become widely known.

However, some doctors do publish, so we now know of some remarkably distant cases that have occurred in Florida; Hawaii; Thailand; Goa, India; Perth, Western Australia; Cape Town, South Africa; and even North Wales in the UK . . . In fact, it seems that Irukandjis (as a group of species producing similar syndromes) are distributed from about 55°N to 38°S latitude, that is, most of the recreationally usable oceans and seas of the world.

CHAPTER 2

Some Astonishing Ecological Impacts

All around the world, jellyfish are behaving badly—reproducing in astonishing numbers and congregating where they've supposedly never been seen before.

—ABIGAIL TUCKER, "Jellyfish: The Next King of the Sea"

If the expanding problem of jellyfish blooms only affected human convenience, it might not be that serious. We could find ways to deal with those sorts of problems. But in fact, major global ecological changes are occurring in our oceans today—and jellyfish blooms are one of the few things they have in common as an outcome. Indeed, jellyfish blooms are visual evidence of degrading ecosystems, and, in many cases, the drivers of further decline.

It should become evident from the following case histories that a truly astounding variety of species are causing an equally impressive variety of problems—and not just in one or two locations, but across the oceans and around the world, in just about every longitude and nearly every latitude, every depth, and every habitat. This is not a problem with an easy solution—this is a problem that is fundamentally changing our oceans globally.

Long considered the last wild frontier, Antarctica now appears to be "flipping" to a jellyfish-dominated ecosystem.

Jellyfish Replacing Penguins
(Antarctica, 2010)

On 3 November 2000, a news story was carried around the world about a research project investigating reports by British military personnel on the Falk-
land Islands that penguins would look up in unison to watch aircraft fly over,
then topple over in regimental order. The whole colony. Like dominoes.

No doubt hilarious to some, toppling is not as funny as it sounds. Eggs
break. Young get crushed or exposed to the bitter cold. Domino toppling
would have serious repercussions on penguin reproduction and survival.

The reports of toppling penguins have been debunked as urban myth by a
scientist with the British Antarctic Survey. It seems that they wobble but not
topple. Hmmm. . . .

Domino wobbling isn't the only modern, industrial-era threat that Antarctic penguins face. Like many other feathery, furry, and blubbery species,
penguins feed on small crustaceans called krill. Vast swarms of krill used to
dominate the surface waters of the Southern Ocean, where they were preyed
upon by the abundant colonies of seals and albatrosses, pods of whales, and
innumerable penguins. But now, the growing industrial demand for krill for
aquaculture feed has exceeded supply in some areas. Krill are called "pink
gold" in the industry. But as men mine their millions with ever-improving
technology, life gets harder and harder for the penguins and other marine
creatures who must now expend more energy traveling greater distances in
search of food.

"Penguins in Antarctica to Be Replaced by Jellyfish due to Global Warming"—
this was the newspaper headline from a story in the *Telegraph* on 19 February
2010. Imagine my curiosity. . . .

Krill are absolutely fundamental to the Antarctic food web. Whales, seals,
seabirds, penguins—just about everything larger than a krill eats krill. And
krill eat phytoplankton. Lots of it. Phytoplankton are, quite literally, tiny drifting plants (*phyto*, meaning "plant" + *plankton*, meaning "drifting"). Two
main types of phytoplankton are fundamental to discussions about jellyfish:
diatoms and dinoflagellates, which are discussed in detail in chapter 13. The
other main type of plankton essential to discussions about jellyfish is zooplankton, or animal plankton. Copepods, a group of extremely abundant tiny
crustaceans, are a key component of the zooplankton, but the term technically

includes any animal organism that is at the mercy of currents, including larvae of most aquatic invertebrates and many types of fish, as well as jellyfish.

Huw Griffiths of the British Antarctic Survey has been researching the organisms that make up Antarctic food webs. He found that krill are being replaced by copepods, which are about 120 times smaller than krill—far too small for penguins but perfect for jellyfish, which filter small particles in the water with tentacles rather than hunt by sight.

One key to the problem is the shrinking sheets of sea-ice. Krill are concentrated in a narrow band along the edge of the sea-ice, where the young feed on algae living on the underside of the ice (Brierley et al. 2002). As the ice melts, so too does the primary grazing ground for the young krill and the length of the ice edge where the adults live. Furthermore, penguins use sea-ice sheets as breeding grounds, so any rise in temperature melts away the area available for raising young.

The other key to the problem is overfishing of krill. Never mind the politics, krill abundance has reduced at a rate of 40 percent per decade since 1976 (Atkinson et al. 2004). Already, penguins are showing signs of severe decline. According to Professor Ove Hoegh-Guldberg of the University of Queensland, the Adélie penguin population size has decreased by 70 percent since 1987 and emperor penguin numbers have declined by 50 percent since the 1970s (Hoegh-Guldberg 2005).

As penguins are replaced by jellyfish, other species that depend on krill will also face a similar fate. Consider Migaloo—the "White Fella"—Australia's beloved albino humpback whale, and all the others on their krill-fueled migrations from feeding grounds to breeding grounds. What will sustain them?

While penguins and whales are emotive and charismatic, the real problem isn't the loss of penguins and whales, per se, but the concomitant changes and losses of all the other components of the Antarctic ecosystem: the seals, the big fish, the seabirds, and the "other 99 percent" of species, that is, the invertebrates down the food chain. All are affected by these changes. Like a ratchet, click by click, each year the Antarctic ecosystem is constantly shifting toward a different community structure—one supporting jellyfish instead of penguins.

On the far side of the planet in the somewhat less chilly Arctic, a similar phenomenon is occurring. A type of seabird called a razorbill is the Northern Hemisphere equivalent of the penguin. In 2003, razorbills began stealing food from each other, a behavior that ecologists call "kleptoparasitism." In fact, more than 60 percent of razorbills have been recorded stealing, while breeding success has declined to less than 40 percent (Lavers and Jones 2007). These changes have taken place as the birds' food has shifted to less nutritious varieties, a shift thought to reflect the ecosystem and food chain effects of warming waters.

Similarly, it seems that climate change is causing polar bears to become cannibals (see plate 12). Prior to the 2000s, such events were rarely observed, but in the last eight years numerous incidents of polar bears stalking, killing, and eating other polar bears have been recorded (Amstrup et al. 2006; Mulvaney 2011). Researchers now believe that diminishing sea-ice and concomitant reduction in foraging opportunities may underlie this apparent behavioral shift.

Our frozen Arctic is thawing, and this has implications for all the organisms that depend on it. Consider high-latitude places like Hudson Bay, Baffin Island off the coast of Newfoundland, and the icy Qaanaaq in northwestern Greenland. These locales are normally frozen by late November; however, the winters have been getting milder. Data from 1979–2009 indicates that the winter "freeze-up" in Canadian and Greenlandic narwhal summering localities has been progressively later by about 1 day each year (Laidre et al. 2012). At this rate, Santa's elves may soon begin wearing shorts, T-shirts, and sandals to work, and snow globes may soon become as obsolete as the phonograph.

With animals that live on land but rely on the sea, such as penguins, razorbills, and polar bears, struggling to find food, it seems that the more intimately associated with the sea an animal is, the harder its struggle to survive. In this next vignette, we examine what happens when key components of marine ecosystems experience what may initially seem like a minor blip. Except this time, instead of penguins, it's sea otters. The mechanism is different, and the culprit jellyfish species is different, but the coupling phenomenon of jellyfish blooms with major ecosystem change is similar, as is the reverberation up and down the food chain.

The World's Richest Fishing Grounds in Peril
(The Bering Sea, since 1990)

If you enjoy McDonald's Filet-O-Fish, you can thank the Bering Sea—that's where 100 percent of McDonald's fish comes from, as well as about half of the annual US fish catch and nearly a third of the fish caught globally.

Spanning over 2 million square kilometers (almost 800,000 square miles), the Bering Sea is that huge body of frigid water cupped in by Alaska's Aleutian Islands and Siberia's Commander Islands. This vast area is divided into two quite different regions as far as the flora, fauna, and fisheries are concerned. The western Bering Sea is very deep, up to 3,500 meters (12,000 feet); the Russian fisheries from this region are worth about $600 million a year. The eastern Bering Sea is primarily a shallow shelf region (less than 150 meters, or 500 feet), a combined fishery worth about $1 billion. This extraordinarily productive ecosystem benefits from vast stores of nutrients brought up by ocean currents into shallow waters from the deep sea in a process called "upwelling."

This freezing cold, blustery region has long been known for its unlimited fish stocks, including Alaskan king crab, salmon, walleye pollock, cod, halibut, and sole. Bering Sea fish and invertebrates don't just feed us, they also support a large diversity of whales and dolphins, seals and sea lions, walruses and polar bears, and 80 percent of the seabird population in the United States.

In the geographical center of the Bering Sea is a very large fishing spot called the "Donut Hole." This area was intensively fished in the late 1980s, with 5 consecutive years each yielding in excess of 2.2 million tons of pollock. By 1992 the Donut Hole pollock fishery had completely crashed, and it has never recovered.

Around the same time, in 1990, fishermen and scientists began seeing wild and drastic changes in the jellyfish abundance, mainly just a single native species, *Chrysaora melanaster*. In fact, jellyfish biomass was so high that the area north of the Alaskan Peninsula became known as the "slime bank" and was shunned by fishermen (Brodeur et al. 2002).

Dr. Richard Brodeur is a fisheries oceanographer working at the National Oceanic and Atmospheric Administration's Northwest Fisheries Science Center in Newport, Oregon. Ric has studied the Bering Sea fisheries for over 20 years and led the team of scientists studying the jellyfish blooms. In a series of publications since 1999, Ric has investigated what may well be the most important jellyfish bloom of our time relating to climate change (Brodeur et al. 1999, 2002, 2008a, 2008b).

Ric and his colleagues adapted a model used in the North Sea (Lynam, Hay, and Brierley 2004) to explain the rise and fall of *Chrysaora melanaster* in the Bering Sea. In a nutshell:

Warm periods (e.g., before 1989 and after 1999) are characterized by unfavorable conditions, such as warm sea surface temperature and low sea-ice cover, which leads to a late-spring bloom and low zooplankton production. This in turn leads to low jellyfish larval survival, low medusa biomass, low reproduction, and reduced settlement and survival of polyps.

Cool periods (e.g., between 1989 and 1999) are characterized by favorable conditions, such as moderate sea surface temperature and sea-ice cover, which leads to an early-spring bloom and high zooplankton production. This in turn leads to high jellyfish larval survival, high medusa biomass, high reproduction, and high settlement and survival of polyps.

An important conclusion of Ric's work is that increasing ocean temperatures associated with global warming don't necessarily indicate increasing biomass of jellyfish in all ecosystems; more probably, a suite of biophysical factors is responsible for the fantastical increases in jellyfish abundance being witnessed around the world.

It appears that the Bering Sea jellyfish may have been responding not only to the climatic changes but also to changes in fish biodiversity. Long-term studies on populations of 17 species of flatfish (e.g., halibut and sole) and more than 76 species of roundfish (e.g., mackerel and rockfish) indicate that fish populations have reorganized in abundance, favoring bottom-dwelling species (Hoff 2006). The biggest shift in fish biodiversity change came in 1988 . . . right before the jellyfish began their dramatic increase.

But fish and fishermen aren't the only ones affected by this. The Aleutian Islands are essentially a wildlife preserve. No tourists disturbing the local wildlife. Nobody feeding the animals stale bread and other unnatural snacks. No fireplace smoke from B & Bs or commuting cars smogging up the air. And yet, wildlife are vanishing at an alarming rate.

Overfishing Affects the Top of the Food Chain

The overfishing problems in the Bering Sea started about 4,000 years ago, when the Aleutians were settled by aboriginal man (Simenstad, Estes, and Kenyon 1978; Jackson et al. 2001). By the time the first Europeans arrived in 1741, the once-abundant Steller's sea cow populations had dwindled to near-extinction. Europeans killed the last sea cow just 27 years later, in 1768. Meanwhile, the aboriginal Aleuts had been hunting sea otters too, beginning about 2,500 years ago.

The European fur traders in the 1800s hunted the otters almost to the point of extinction. The primary prey of the otters was sea urchins. So the urchins, enjoying reduced predation, flourished. And the hungry sea urchins voraciously grazed on kelp, stalk by stalk, blade by blade, reducing the once-lush kelp forests to moonscapes of bare rock and crustose algae. The many-stories-tall macroalgae of the kelp forests are analogous to trees in terrestrial woodlands, providing habitat for a whole ecosystem of organisms: fish darting between branches like birds, bryozoans and hydroids growing on blades like lichens on leaves, snails slowly gliding up stalks, and a myriad of gribblies living in the understory. As the kelp forests disappeared, the fishes and invertebrates that depended on them for shelter and food vanished too.

Sea otters became protected by international treaty in 1911. As they slowly came back from the brink of extinction, their sea urchin prey gradually returned to normal population numbers, and the kelp forests began to grow again. By the 1970s, the sea otters had reached their equilibrium capacity in many areas. That is, until 1990.

Professor Jim Estes of the United States Geological Survey and his colleagues have been studying sea otters in the Aleutians for the past several decades. They documented a 25 percent decrease in the Aleutian sea otter population each year throughout the 1990s, resulting in an overall decrease of 78 percent by 1997 (Estes et al. 1998). In just six years, the sea otters were again heading for oblivion. So too, the sea urchins' populations exploded . . . and the kelp forests again largely disappeared.

The sea otters' incredibly fast disappearance seemed to be a mystery. Never before had any marine mammal population experienced such a rapid rate of decline (Cone 2000). Estes and his colleagues worked within the logical framework that the population declines were limited to reduced fertility, increased mortality, or redistribution. Redistribution was immediately dismissed, because the declines were taking place over a large geographic area. Reduced fertility was examined but also dismissed, because birth rates were found to be similar to those of stable populations. What then, could be killing the otters?

Disease, toxins, and starvation were dismissed because there were no carcasses of dead otters, and no sick or emaciated ones. It was like they were being eaten. . . .

Estes and his colleagues proposed a novel explanation for the sea otters' population crash based on a chain of ecological interactions. It seems that the decline in forage fish in the late 1980s, which has been attributed to overfishing and climate fluctuations, set into motion a cascade of events that was so comprehensive, it affected not only animals up and down the food chain but also the physical habitat structure and therefore the species composition over a broad geographical region.

As forage fish stocks declined, so too did their pinniped predators, that is, the seals and sea lions. This left killer whales searching for high-calorie mammalian prey. After living side by side for millennia, all of a sudden the orcas were attacking otters, which were now the only mammal left. But the scrawny otter was no substitute for a nice, plump, meaty pinniped.

Estes and his colleagues calculated that 40,000 otters would have to have been eaten to drive the observed 78 percent decline in 6 years. By measuring the caloric value of sea otters and the metabolic rate of orcas, they estimated that an adult female killer whale would need to eat 3 male or 5 female sea otters per day, while an adult male whale would need 5 male or 7 female otters per day. At an average of 5 otters per day, a single orca would consume 1,825 otters per year, leading Estes to conclude that a mere "3.7 killer whales feeding exclusively on sea otters would be sufficient to drive the population decline" (Estes et al. 1998, 475).

But otters are "keystone predators," meaning that they control other species through predation. As the otters declined and there were fewer mouths to feed, the biomass of their urchin prey increased eightfold, but as the hungry urchins competed for kelp, the lush underwater forests declined by a factor of 12 . . . and with them the fish and invertebrates that live, eat, and breed in and on the kelp . . . and with them the seabirds. . . .

> The Aleutians offer proof that one small ecological change can move like a tsunami throughout the entire ocean realm. (Cone 2000)

That small ecological change is believed to have occurred in 1977, when the Gulf of Alaska waters experienced a sudden and slight warming—only 2°C (3 ½°F) on average. But that seemingly small change was apparently enough to set into action a cascade of events that have had a huge impact.

Climate Change Affects the Bottom of the Food Chain

Other significant changes to the Bering Sea ecosystem have taken place, including massive blooms of a microscopic single-celled alga called a coccolithophore (see plate 14).

Coccolithophores are strange little critters. They photosynthesize like plants, but they are not plants. Their bodies are covered with highly intricate limestone plates like sea urchins, but they are not sea urchins, or even animals for that matter. They are more closely related to giant kelp, but not very closely related to them either. And they are really, *really* small.

Sometimes coccolithophores bloom in truly magnificent numbers, like they did in the Bering Sea in 1997 . . . and 1998 . . . and, in fact, most summers since. Even though each individual coccolithophore is only tiny (about one-hundredth the thickness of a dime), when they occur en masse, they turn huge regions of the ocean milky whitish or turquoise because of the large amount of light they reflect. So vast and vibrantly colored, these blooms look like auroras when seen from space.

Coccolithophore blooms are usually short-lived. But these Bering Sea blooms persisted for months. And they reappeared each year. This unusual pattern was thought to signal significant changes in the environment and climate, and it was very bad news for the fish and birds. Due to the combined effects of the reduced cloud cover (which allowed the sun to warm the water) and the calmer seas (which reduced the mixing and therefore brought less nutrients to the surface), the blooms of the typical phytoplankton, called "diatoms," were far less robust than normal. But these warmer, calmer conditions favored growth of the coccolithophores.

Because the coccolithophore blooms reflect light, the light did not penetrate the surface waters sufficiently for the diatoms, kelp, and other algae to grow. With insufficient phytoplankton to graze, the populations of zooplankton, such as copepods, were reduced. And therefore, the larger zooplankton, small fish, larger fish, and so on up the food chain starved. Krill also graze on diatoms, and many types of seabirds and whales eat krill. But they too went hungry.

Jellyfish, however, feed on survival food, and even survival rations if necessary: other gelatinous zooplankton, the biofilm on surface sediments, flocculent organic matter . . . apparently even dissolved organic matter absorbed directly through the epidermis. They can eat anything, and often do: things that fish won't eat, and things that fish can't eat. And they've been doing really well in the Bering Sea.

The walleye pollock fishery in the eastern Bering Sea is the largest and most lucrative fishery in North America, and it is also said to be the most-researched and best-managed fishery in the world. However, it is also showing serious signs of collapse (Morell 2009).

Not all that long ago, pollock were considered worthless. But then the Japanese developed a method for reducing its white meat into a protein paste called *surimi*—this is the basis of those yummy crab sticks and imitation lobster snacks. After the collapse of the North Atlantic cod fishery in 1992, pollock took up much of the slack, and it is now considered the world's largest single-species fishery, averaging about 1.4 million tons per year . . . that is, until 2007. Surveys in 2007 showed that pollock numbers were declining. Reductions in quotas in 2008 and 2009 were expected to bring the numbers back up, but surveys in 2009 found instead a 30 percent reduction in abundance. In response to the stock being at its lowest level since 1980, Greenpeace has added the Bering Sea pollock fishery to its Red List of unsustainable stocks. Faced with the decision of how to manage the surprising news of the dwindling population, the North Pacific Fishery Management Council chose to reduce the quota by only 2,200 tons in 2009, drawing criticism from a broad array of scientists. Quotas have been set lower each year since, but still stocks are down. . . .

Epilogue

Right around the time the pollock started decreasing again, guess what started making a comeback? Jellyfish. Yes, the pesky sea nettle that bloomed like crazy last time the pollock stocks declined. According to Ric Brodeur, "The upshot is that the biomass has again increased the last three years after being relatively low for most of the last decade" (personal communication, 2012).

The exact relationship, if any, between the decline of pollock and the increase in jellyfish is not well understood. But even if there is no direct relationship, that is to say, if they were responding to environmental stimuli but not to each other, at the very least it is a bit of a wake-up call about how rapidly and drastically what we count on as normal can change.

On a more ominous note, this may be the fisheries equivalent of a death rattle: the final breaths of a dying soul, too late to be saved with any amount of heroic effort. Only time will tell if the down-up-down pattern of the walleye pollock will have a similar outcome to those of the Newfoundland cod, the California sardine, and so many others that have rallied before crashing completely.

Norwegian Fjords Mimicking the Deep Sea
(Western Norway, since 1973)

Major ecological change isn't only occurring in the polar regions of Antarctica and the Bering Sea. It's also occurring in the high northern latitudes of Norway. In this case, the cause isn't loss of krill or sea-ice, or overfishing, or climate change. It is murky water.

There is a fairly beautiful but common species in the deep sea called *Periphylla periphylla* (see plates 2 and 3). It looks somewhat like a Santa's hat, with the red cone pointing up and the white frill around the base (but without a white pom-pom on top). Like many deep-sea species, *Periphylla* is bioluminescent, meaning that it produces its own light. Some species flash, while others emit a longer dull glow. But *Periphylla*, mesmerizingly, twinkles. It scintillates. In the darkness its tiny pinpoint sparkling blue-green lights subtly suggest, yet at the same time camouflage, its conical shape.

Its stomach is entirely lined in dark red pigment, thought to mask any flashing signals given off by its dying, struggling prey. Yes, it's a vicious dog-eat-dog world down there in the deep sea. Many species are bioluminescent, not only jellyfish. If an organism flashes, it attracts the attention of visual predators. Many deep-sea jellyfish have developed pigmented stomachs as a means of hiding the fact that they have just caught a meal—to advertise this could mean that they in turn become another species' prey.

These deep-sea adaptations make *Periphylla* perfectly suited to its deep, dark, slow-moving environment. Maybe that's why it seems so unusual that *Periphylla* has not only colonized—but *deluged*—the shallows and even the surface waters of Lurefjorden in Norway. In the deep sea its abundance is quite low, perhaps one individual per cubic kilometer (about 4 per cubic mile). But in this fjord, *Periphylla* is found in staggering numbers of more than one per cubic foot. This small, 30-square-kilometer (12-square-mile) fjord is home to over 35 million *Periphylla* (Youngbluth and Båmstedt 2001).

Ulf Båmstedt of the University of Bergen has been studying *Periphylla* with numerous colleagues since 1992, when scientists were alerted to the mass presence of these jellyfish causing problems for nearby fisheries. Just 50 kilometers (30 miles) north of Bergen lies this unusual body of water, Lurefjorden. There is very limited water circulation in this 450-meter (1,500-foot)-deep basin because it lacks significant freshwater input from rivers and has a very narrow and shallow sill exiting to the sea.

These unusual hydrographical conditions make Lurefjorden more like a marine lake than a typical fjord. Furthermore, even though Lurefjorden is quite shallow, it has a relatively larger, darker habitat than otherwise similar fjords. In particular, it is thought that the shallow sill allows only shallow coastal water to enter the fjord, rather than the heavier, saltier waters of the deep Atlantic. Shallower coastal water tends to be less salty and is more prone to eutrophication processes (see chapter 7). The water of Lurefjorden, therefore, contains more particles, which absorb more light, meaning that less light reaches the depths of the fjord. Hence, visual predators are at a disadvantage. Quite simply, it's as if they were in the vast darkness of a deep ocean.

Enter *Periphylla*. Living in the deep sea, beyond the penetration limit for light, doesn't require eyes. *Periphylla*, like many other deep-sea species, is a tactile predator, which makes it perfectly suited to dark and murky Lurefjorden. So too does the constant temperature at most depths within the fjord, which is similar to that of the deep sea. In these conditions, it is thought that *Periphylla* lives for 10 to 30 years, performing a daily vertical migration of hundreds of meters. The medusae spend the night feeding near the surface where copepods and other zooplankton live, then the dawn drives them back to the depths, from where they migrate up again as the light fades, and so on (although medusae cannot "see," they have rudimentary light receptors with which to tell light from dark).

Periphylla has no obvious predators or competitors in Lurefjorden. Studies have shown that the abundance of fish is considerably less in Lurefjorden than in other nearby similar fjords with less light absorbance, that is, where light is reaching deeper. This affects the whole ecosystem. The copepods that fish would normally eat are larger and more numerous in Lurefjorden than in other fjords with visual predators.

Furthermore, although the reason is unclear, species of jellyfish that would typically compete with *Periphylla* are few in Lurefjorden. Hence, *Periphylla* has found itself a veritable candy store of unlimited food with no restrictions.

It wasn't always like this. According to local fishermen, Lurefjorden had long contained *Periphylla* in small numbers, at least as far back as the late 1940s, but they began "taking over" in 1973 (Fosså 1992). What, then, possibly could have happened in 1973 to cause such an incredible shift from a fish fjord to a jellyfish fjord? This question is under investigation by Tom Sørnes and his colleagues, who are working with Ulf Båmstedt. Sørnes recently reported that two other nearby fjords have experienced similar shifts from fish to jellyfish (Sørnes et al. 2007). It is thought that the shallow sill, which acts as a levee for the majority of the basin water, plays a key role in maintaining the incredibly high density of jellyfish in Lurefjorden. Small young *Periphylla* stay near the bottom, so are never advected out via the exchange of upper water. The larger, adult *Periphylla*, on the other hand, migrate up to the surface waters each night, and are thought to be advected out of the fjord. Curiously, whereas the blooms of most jellyfish species have a strongly seasonal component, this is not the case for *Periphylla* in Norway. In all study locations, *Periphylla* has a stable abundance throughout the year.

A similar phenomenon is occurring in another fjord in another country with another species of jellyfish. Hans Riisgård of the University of Southern Denmark and his colleagues have been studying the Danish Limfjorden, and found that annual fish landings have decreased from about 2,500 tons in the early 1920s to only about 20 tons in recent years, while the moon jellyfish, *Aurelia aurita*, have increased (Riisgård, Andersen, and Hoffmann 2012).

Jellyfication in the Open Ocean
(Namibia, since 1990)

Jellyfish blooms are not exclusive to high latitudes—far from it. Their effects are often much worse at mid-latitudes. Many blooms are in enclosed or semienclosed bodies of water, for example, bays, harbors, fjords, bights, and seas. The theory behind these events is that these small-ish bodies of water get out of balance, whereupon, because they are essentially their own little ecosystem, they stay that way. However, this is not the case for the persistent bloom in the Benguela Current off the west coast of Africa, which is essentially in the open ocean.

The northern Benguela Current off Namibia has historically been a highly productive region for fisheries, including those for sardines and anchovies. It is one of the most intensively fished regions of the world, where mass-schooling fish support high-energy predators like larger fish, sharks, and marine mammals. Like other continental west coasts (e.g., Peru and California), the Benguela region receives nutrient-rich waters from seasonal upwelling. But studies over the last 40–50 years paint an increasingly dismal picture.

In the 1960s, catches peaked at more than a million tons of sardines per year. The fish stocks seemed unlimited. Extensive scientific sampling did not mention the presence of large jellyfish, although small ones were noted; it therefore seems likely that large jellyfish were not present in large numbers, if at all. However, as fish catches began to rapidly decrease to very low levels, reports of two large species of jellyfish have steadily increased:

- *Chrysaora hysoscella*, the so-called European sea nettle, which occurs primarily in inshore waters along the coast, and
- *Aequorea aequorea*, a large and voracious hydromedusa, which occurs primarily in offshore waters.

These two species of jellyfish stretch like a double-walled fishing net in the upper layers of the water column and thus are able to exert a consistently strong predation pressure across the continental shelf. Jellyfish have responded to this heavy exploitation of fish by expanding phenomenally. In 2006, jellyfish biomass was estimated at 13 million tons, far surpassing the once-abundant total fish biomass, now estimated at just 3.9 million tons (Lynam et al. 2006).

A recent study on the underlying causes of jellyfication of the Benguela concluded that the answer lies in the humble sardine (Roux et al., in press). By unusual circumstance, the northern and southern Benguela provide an "accidental experiment," allowing researchers to compare fishing patterns between the two regions over the past half century. The same species occur in both regions but are separated into distinct stocks by a strong upwelling off Lüderitz in southern Namibia. Both regions were intensively fished after World War II, but strategies diverged when the sardine fisheries collapsed in the 1960s. Both regions began targeting anchovies as sardines dwindled, but South Africa applied conservative management measures in an attempt to rebuild sardine stocks and maintain high stocks of other small pelagic fish (forage fish), whereas Namibia elected instead to heavily fish anchovies, partly believing that this would reduce competition with sardines. In the early 1970s, the northern Benguela populations of small pelagic fish crashed completely and have not recovered; this coincided with early observations of growing numbers of jellyfish. The southern Benguela, in contrast, has maintained large populations of small pelagics and does not suffer the jellyfish bloom problems that now characterize the north.

Other upwelling regions with large fisheries of sardines and anchovies have seen similar collapses. In Peru, the anchoveta fishery recovered within 20 years, whereas sardine fisheries off California have not—even after almost a half century.

It seems clear from Namibian data that the fish were overexploited, after which jellyfish seized the opportunity for a population explosion and have since become the dominant organisms in the community. Intriguingly, whereas most jellyfish blooms are strongly seasonal, this one has persisted more or less continuously for decades.

The economic nuisance from the jellyfish isn't limited to drastically reduced fish landings. Like in so many other areas with jellyfish bloom problems, fishing has become vexed by burst trawl nets and spoiled catches. Power stations have experienced problems with blocked coolant intakes. And jellyfish have interfered with diamond mining operations in southern Africa by clogging vacuum tubes used to suck up sediments from the seafloor.

The causes and effects of this bloom are still hotly debated. Some think that indigenous jellyfish have responded to a decrease in competition and a resulting increase in food due to the intense harvesting of fish. Others think that the strong current has imported jellyfish along with their food. Some think that climate change has increased both the jellyfish and their food species, which then allowed the jellyfish to outcompete the fish. Others think that the ongoing bloom is part of a natural boom-and-bust cycle of jellyfish and fish populations.

By mid-2010, the once productive Benguela fishing region had become a "ghost town," or "dead zone." Dead and dying jellyfish sink to the bottom and rot. Millions of phytoplankton that were once eaten by copepods and other zooplankton, now uneaten, also die and sink to the bottom to rot. These masses of decaying carcasses create a zero-oxygen zone of hydrogen sulfide where nothing can survive. Vast expanses of the seafloor are now a moonscape, an eerie graveyard almost completely devoid of living things. Jellyfish dominate the surface waters above this dead zone. The jellyfish have excluded most other living things by partitioning their vertical and horizontal space into a stingy-slimy killing field impacting over 30,000 square nautical miles.

Strangely enough, though, one fish species called the bearded goby has capitalized on this situation. Bearded gobies can live for hours without oxygen—essentially holding their breath—while hiding from predators that cannot. By day, they feast on dying jellyfish, phytoplankton, and worms in this oxygen-free area. By consuming these dead-end organisms that most fish don't eat, the bearded goby is making this dead ecosystem once again productive. At night, the goby swims toward the surface, using the jellyfish to shield itself from predators like hake, horse mackerel, seabirds, penguins, and seals. In the oxygenated waters near the surface, the goby can start breathing again and digest its earlier meal (Marshall 2010; Moloney 2010; Pennisi 2010; Utne-Palm et al. 2010).

While scientists were busy trying to figure out how jellyfish could take over an open ocean region, another similar situation came to light elsewhere. It appears that comb jellyfish have been on a persistent and widespread increase along the northeastern US shelf, from North Carolina to Nova Scotia, for more than 25 years. Using a rather low-tech but innovative sampling method, Doctors Jason Link and Michael Ford from the National Oceanic and Atmospheric Administration tracked a two- to eightfold increase in the frequency of occurrence in the stomachs of the spiny dogfish since the early 1980s (Link and Ford 2006). The authors interpreted this sustained change as a likely indicator of major changes in the structure and function of the northeastern shelf ecosystem.

While partially or fully enclosed marine ecosystems have generally been accepted as vulnerable to runaway ecosystem changes, these drastic changes in the open system of the northeastern shelf suggest that "virtually all the world's marine ecosystems are vulnerable to such increases" (Link and Ford 2006, 158), leading Link and Ford to conclude, "Some scientists have jokingly suggested that continued ocean perturbations may leave us with marine ecosystems populated predominantly by jellyfish; we may be closer to such a situation than we suspect" (p. 158).

A Thousand Miles of Sea Tomatoes
(Western Australia, 2000)

The Southern Ocean, Bering Sea, Norway, and Namibia all share a history of heavy fishing; this is not the case with the Indian Ocean coastline of Australia. Western Australia is one of the most pristine regions of the world—it is remote, unpopulated, unexploited. And yet it too is showing signs of jellyfish bloom problems.

From April to June of 2000, arguably one of the biggest jellyfish blooms in history occurred, but with no official witness. For about 10 weeks, the tomato-sized *Crambione mastigophora* (see plate 2) bloomed "cheek by jowl" along the entire Indian Ocean coastline of Western Australia, from Derby in the north to Rottnest Island in the south, some 1,200 kilometers (750 miles). Helicopter footage showed a thick red belt along the coast for as far as the eye could see in both directions. It was quite the "talk of the town"—in every coastal town—for many months, but, as all too often happens in remote areas, especially with poorly studied taxonomic groups, nobody quantified it in a meaningful way. Nobody studied the ecological effects. Nobody documented the numbers.

Traveling through the region eight months later with my colleague Dr. Wolfgang Zeidler, the senior curator of marine invertebrates at the South Australian Museum, we were bombarded by people saying that they had never seen so many jellyfish. People's observations included the following: the bloom scared off the whale sharks, it prevented the turtles nesting for the year, fishermen couldn't catch any fish after that, the lobster fishery had a banner year because of it, and larval fish studies at the Dampier research facility were skewed by near-zero recruitment (fisheries lingo for "survivors"). The account of this bloom was rejected from publication in international journals—twice—on the basis that it was "too anecdotal" and lacked solid scientific quantification. And so, there is no record. It came and went without a trace in scientific history.

But the fact that there is no official scientific record certainly doesn't mean that it didn't happen. This is the old "if a tree falls in the woods" argument. In this case, when the tree fell, it made a huge crashing ruckus—and many people heard it—but because it wasn't studied and reported in the "right way," in essence, it didn't happen.

In fact, many events and details are lost to science and history because they are not reported, or not reported in the right way. It is good to keep this phenomenon in mind as one reads various accounts. Even assuming accurate reporting, scientific documentation is bound to be skewed toward underestimation.

CHAPTER 3

Jellyfish Completely Out of Control

It devours—even beyond its capacity to digest—huge quantities of zooplankton, small crustaceans, and the eggs and larva of fish. In short, it not only kills fish directly, but also indirectly by depriving them of food.

—JOHN TRAVIS, "Invader Threatens Black, Azov Seas"

Collapse of an Ecosystem
(The Black Sea, since 1982)

Unarguably the best-studied jellyfish bloom in history is the *Mnemiopsis* invasion of the Black Sea and the unfathomable events that followed as it ate its way through Europe.

Due to an accident of natural history, the Black Sea has always been toxic. For about the last 7,500 years, both geology and water chemistry have created a situation whereby about 87 percent of the volume of the Black Sea is anoxic (i.e., without oxygen) and contains high levels of hydrogen sulfide. Only the shallow surface and shelf waters—the upper 13 percent—can support life. *Thirteen percent*. However, recent human-induced changes are putting severe stress on even this precious 13 percent.

In recent decades, the enclosed Black Sea has become the sole drainage sump for the waste of a combined population of 165 million people in 22 industrialized and developing countries through its 5 tributary rivers—the Danube, Dniester, Dnieper, Don, and Kuban. Sewage, pesticides, fertilizers, heavy metals, and radioactive waste have all combined to make a toxic cocktail, earning the Black Sea the reputation of being one of the world's most polluted bodies of water (Land 1999). Indeed, the excess nutrients dumped into the Black Sea over the decades have caused a dead zone along its shelf areas. In 1973, this dead zone occupied about 3,500 square kilometers (1,350 square miles); by 1990, it spanned some 40,000 square kilometers (15,500 square miles) of the northwest shelf (Zaitsev and Mamaev 1997). The northwest shelf is only about 64,000 square kilometers (25,000 square miles), so the dead zone now occupies over 60 percent of this once highly productive region. At present, favorable conditions for bottom-dwelling organisms are limited to depths of less than 5–6 meters (15–20 feet).

One of the more distressing effects of the pollution is stimulation of the growth of single-celled phytoplankton into massive blooms. These blooms are often toxic. In the 1950s and 1960s, phytoplankton blooms were so rare that they merited scientific publication. But by the late 1960s and early 1970s, the blooms had become nearly continuous, except in winter. In August 1974, one such bloom was calculated to contain a mind-numbing 139 million phytoplankton per liter.

Healthy phytoplankton blooms contain a high percentage of microscopic algae called diatoms, but many blooms instead contain dinoflagellates, another type of microscopic algae, at the expense of diatoms. Diatoms support a higher energy food chain than do dinoflagellates. Phytoplankton and the implications of this trade-off phenomenon are discussed in greater depth in chapter 13. In the 1950s and 1960s, dinoflagellates comprised less than 20 percent of the phytoplankton biomass in the Black Sea. By the 1970s, their biomass had nearly tripled. When phytoplankton blooms die off, the decaying of their bodies further depletes the oxygen. When hypoxic (low-oxygen) areas become anoxic (without oxygen), hydrogen sulfide forms, further exacerbating the anoxic and toxic problem. Few organisms can survive conditions of low or no oxygen. Anoxia is estimated to cause the deaths of up to 8 million tons of bottom-living animals per year, totaling some 60 million tons from 1972 to 1990, including 5 million tons of fish.

The problem is not limited to the disappearance of valuable commercial fish but also involves enormous ecological damage. Many bottom-living animals have planktonic larvae, which are a significant component of the zooplankton food for species higher up the food chain. Furthermore, many bottom-dwelling organisms are filter-feeders, that is, they feed by actively filtering tiny particles from seawater. In doing so, they not only acquire their food but also cleanse the water of impurities and detritus. According to Zaitsev and Mamaev (1997), one square kilometer (⅓ of a square mile) of mussels can filter an average of 15–20 million cubic meters (500–700 million cubic feet) of seawater every 24 hours. Mussels, like many other seabed dwellers, cannot move to more favorable habitat, so they are doomed to die when the sea around them becomes uninhabitable.

In addition to its anoxic dead zones and the shift to dinoflagellates, the Black Sea had been severely overfished, which led to severe habitat degradation. The larger piscivorous species (i.e., those that eat other fish) had become uncommon. Tuna. Swordfish. Sturgeon. In the 1960s, 26 commercial fish species were caught in the tens or even hundreds of thousands of tons, but by the 1980s only 5 species were left (Zaitsev and Mamaev 1997). Mackerel, which also eat jellyfish, had been fished out since the late 1960s. Similarly, the bonito and bluefish fisheries had crumbled by the end of the 1970s. At first, freed from predation, catches of anchovy and sprat rose. But by 1990, the anchovies, hitherto the main diet of Black Sea dolphins and the extinct (or nearly so) monk seal, also collapsed.

Heavy bottom-trawling rubbled and denuded the seafloor habitat, but far more lethal was the resuspension of bottom sediments—silting of vast bottom areas, more than 5,000 square kilometers (2,000 square miles), typically 2–5 centimeters (1–2 inches) deep—up to 40–50 centimeters (10–15 inches) deep in some spots. Mussels on the northwest shelf declined more than 95 percent, from about 12 million tons to less than half a million—and with them the biofiltration services that they performed. Water transparency declined, causing a depression of photosynthesis. So too, vast reproductive and feeding areas for most bottom-living fish were silted beyond utility (Eremeev and Zuyev 2007).

Bottom-living fish began disappearing in the early 1970s, around the same time as mass mortalities in the dead zone. Bulgaria and Romania together caught just under a million tons of turbot in the 1950s and 1960s, but only 242 tons in the 1970s and just 12 tons in the 1980s. Similarly, Georgia's estimates of the sturgeon population were about 75,000 adults in 1973–1974, and down to about 20,000 in the 1990s.

There is little, if any, dispute that these changes in phytoplankton composition and the drastic reductions in fish stocks are the result of destructive overfishing practices and the still-developing dead zones created by excess nutrients dumped into the Black Sea over decades.

When we think of marine biodiversity, most people think of fish, corals, whales, dolphins . . . maybe penguins or polar bears, maybe seagulls or pelicans. But the fact is that the vast majority of marine species are the invertebrates— the starfish and sea urchins, the crabs and shrimps, the snails and clams, the worms, the isopods (small shrimplike crustaceans that appear to be squashed flat) and amphipods (similar to isopods, but squished side to side), and more worms, the sponges and tunicates, and still more worms. As most of these generally small, camouflaged creatures are unappetizing to *Homo sapiens*, it might seem that they would be in very little danger of exploitation or human-driven extinction.

But most of these invertebrates live on or in the sea bottom, which now-adays is a very dangerous place to be. Especially in the Black Sea. Over 300 trawlers work the Black Sea day in and day out, raking the bottom for fish. Over 300 trawlers' worth of mechanical damage. Over 300 trawlers' worth of fine sediments suspended. Over 300 trawlers' worth of bycatch. These issues are treated in detail in chapter 6.

Commercial bottom-trawling was banned in the Black Sea in the early twentieth century but was resumed in the 1970s. Biodiversity of bottom mac-rofauna, such as crabs, declined on the Romanian shelf, from 70 species in 1961 to just 15 in 1994.

The northwest shelf is the most productive area of the Black Sea, histori-cally producing more than 90 percent of its fish and algae. But the Black Sea nutrient load has increased dramatically since the 1960s and has been de-scribed as "critically eutrophic" (Mee 1992, 279). By 1992, some 60,000 tons of phosphorus per year were pouring into it—four times more than that in the Baltic—as well as some 340,000 tons of nitrogen per year—more than double the load from the Rhine.

The synergistic effects of increased phytoplankton, increased suspended microparticles, and decreased filter-feeders have reduced light penetration. Just a few decades ago, a unique and extensive meadow of red algae occupied about 100,000 square kilometers (38,000 square miles) of the center of the northwest shelf. An incredible diversity of fish and invertebrates made their home in this meadow. But with decreased light, the algae withered. By 1990, only 50 square kilometers (20 square miles) of red algal meadow were left.

The sponges, sea anemones, isopods, amphipods, shrimps, crabs, tunicates, swimming fishes, clinging fishes, and all the other inhabitants of the algal meadow simply ceased to exist (Zaitsev 1992). So too, the eelgrass meadows have decreased by 90 percent. Algae and seagrass are now confined to a very narrow strip along the shoreline about 3–5 meters (10–15 feet) in depth.

In addition to light penetration impacts, widespread hypoxia on the shelf took a heavy toll on the Black Sea's flora and fauna. Many bottom-dwelling species, such as oysters, mollusks, sponges, and sea anemones, failed to survive long enough to reproduce. Very large numbers of bottom-dwelling fishes have also died out. Gobies. Blennies. Stingrays. Flounder. Sole. Mullet. Sturgeon catches have fallen by 80 percent. Turbot catches have decreased tenfold.

After the jellyfish-eating mackerel disappeared in the 1960s, several jellyfish species bloomed in very large numbers. The first, *Rhizostoma pulmo*, a type of "blubber jelly," bloomed from the late 1960s to 1974. These beachball-sized jellyfish occurred in exceptionally crowded densities of up to 2–3 individuals per cubic meter.

As the *Rhizostoma* bloom was subsiding in the mid-1970s, the population of *Aurelia aurita*, the moon jellyfish, began to explode by several orders of magnitude. From 1949 to 1962, the average biomass of *Aurelia* in the Black Sea was estimated at 670,000 tons, but from 1976 to 1981, it dramatically increased to average 222 million tons; in the late 1980s, its total biomass had risen to 300–500 million tons. By the late 1980s, 62 percent of the annual production of all Black Sea zooplankton was being devoured by *Aurelia* alone (Zaitsev and Mamaev 1997)—62 percent of all the copepods, 62 percent of all the fish eggs and larvae, 62 percent of all the invertebrate larvae. Sixty-two percent of all the available food.

The Black Sea was already in ecological freefall when *Mnemiopsis* arrived. As in so many other cases, something that eventually becomes a disaster begins with just a whisper—a whisper that could have been choked back and silenced. *Mnemiopsis leidyi* (see plate 2), a type of comb jellyfish native to the eastern coastlines of the Americas, was first observed in the Black Sea in 1982 (Ivanov et al. 2000).

At first, it was only found in the northwest in small numbers near the coast. Between the late 1980s and 1998, *Mnemiopsis* surged to become the Black Sea's dominant planktonic species. Summertime blooms contained 300–

500 specimens per cubic meter (Zaitsev 1992); to put this into perspective, that would be about 300 clenched fists in an area no larger than the leg room under a small breakfast table. Its population was estimated at over 1 billion tons, more than the world's total annual fish landings (Zaitsev and Mamaev 1997; Ivanov et al. 2000). Then the population of the moon jellyfish *Aurelia* crashed. Then the zooplankton crashed. Then the anchovy fishery crashed. It is unfathomable how many zooplankton, fish eggs, and larvae *Mnemiopsis* must have eaten to expand so explosively—these were the zooplankton, fish eggs, and larvae that had previously supported the diverse fish fauna and once-profitable fisheries.

The *Mnemiopsis* story is one of mystery and tragedy, an astonishing tale of "the worst that can happen." It may seem inconceivable that a mucousy little jellyfish, barely bigger than a chicken egg, with no brain, no backbone, and no eyes, could cripple three national economies and wipe out an entire ecosystem. But it could. And it did. If there were a poster child for unthinkable ecological disaster, certainly *Mnemiopsis* would be it.

How *Mnemiopsis* Reached the Black Sea

The mode and date of *Mnemiopsis'* arrival may never be determined with certainty. Until recently, it was believed that *Mnemiopsis* had been transported from the Chesapeake across the Atlantic and to the Black Sea in a ship's ballast water, possibly a grain ship or oil tanker (Mills 2001; Shiganova and Malej 2009). It has only recently been demonstrated with DNA studies that Eurasian *Mnemiopsis* represent at least two separate introductions, one from the Gulf of Mexico (e.g., Texas or Florida to the Black Sea), and the other from the northeastern United States (e.g., Massachusetts or Rhode Island to the Baltic Sea or North Sea) (Reusch et al. 2010; Ghabooli et al. 2011). These multiple introductions highlight the importance of proper ballast management (IMO 2011).

Ships need to carry large amounts of water for stability on long and often tumultuous journeys across open oceans. It was standard practice for many years for ships to load up with ballast when in port at one location then dump it in port at the next location. Harbor habitats tend to be fairly similar around the world: their waters are quiet and protected from heavy surf and currents, and the "flavor" of the water tends to reflect additives from shipping and industry that often exclude sensitive species. So a species comfortable in a harbor in Tokyo is likely to find a similar level of comfort in Sydney or San

Francisco. That means that not only any animal or seaweed that is sucked into the tanks will be transported, but that *every* animal or seaweed sucked in will be transported—all the larvae, all the eggs, every opportunistic weed, and the drifting jellyfish. And harbors generally teem with all sorts of larvae and other drifting things that are easily entrained and sucked up. So it's not hard to imagine how so many species were transported around the world and introduced into nonnative habitats before we not-so-cluey hominids caught on to what was happening. What is clear, however, is that *Mnemiopsis* was alien to the Black Sea. It had never been reported there before, nor anywhere nearby.

How *Mnemiopsis* Got Out of Control

Four hypotheses have been put forth to explain how the balance shifted from robust anchovy fisheries and low numbers of jellyfish to infestations of jellyfish and crashed fisheries: (1) anchovy stocks collapsed because of heavy predation on fish eggs and larvae by *Mnemiopsis*; (2) heavy competition by *Mnemiopsis* on zooplankton, causing fish and their larvae to starve; (3) severe overfishing on anchovies made more food available for jellyfish; and (4) a climatic shift triggered both a decline in zooplankton (and therefore a decline in fish stocks) and a bloom of jellyfish (Bilio and Niermann 2004). It appears that the answer is multidimensional, with probably all of these factors at play. From available data, it seems that overfishing and climatic effects on zooplankton caused fish stocks to decline to the point where *Mnemiopsis* could successfully compete for food and decimate the anchovy's eggs and larvae.

Environmental deterioration in the Black Sea ecosystem had already led to the food web being dominated by other weedy, opportunistic species before *Mnemiopsis* arrived. *Aurelia aurita* was rarely observed in the 1950s and 1960s, but by the early 1980s had become a major gelatinous predator in the ecosystem (Oguz, Fach, and Salihoglu 2008). Other weeds like the red-tide phytoplankton *Noctiluca scintillans*, the comb jelly *Pleurobrachia rhodopsis*, and the blubber jellyfish *Rhizostoma pulmo* had also become common.

Mnemiopsis was able to exploit an ecosystem made vulnerable by pollution and overfishing—the Black Sea was an ecosystem already on the brink of collapse. Perhaps without *Mnemiopsis*, it could have righted itself in time. But *Mnemiopsis* found a supportive niche and rapidly expanded its population to over a billion individuals (Zaitsev and Mamaev 1997). Then things went from bad to worse. Between 1950 and 2001, anchovies had accounted for more than 60 percent of the total catch (Kideys et al. 2005), but the combination of overfishing by humans and intense competition from the growing *Mnemiopsis* population drove anchovy stocks down, giving *Mnemiopsis* a progressive competitive advantage of food consumption.

Mnemiopsis was able to take over with such force due to a combination of factors, some inherent to the Black Sea and others just the nature of the beast. In other words, the Black Sea was an ideal habitat for an invader like *Mnemiopsis*.

Characteristics of the Black Sea

An ailing system. As discussed above, the Black Sea was already highly polluted and overfished. Just as a person under stress is more vulnerable to disease than an unstressed person, so are ecosystems. But this was no ordinary stress. This was the marine equivalent of a patient with HIV and emphysema . . . You pray that they don't get pneumonia. Or in this case, *Mnemiopsis*.

Lower salinity. For some years before the arrival of *Mnemiopsis*, the Black Sea had grown increasingly saline due to diversion of freshwater rivers for agriculture. However, in the 1980s, the Soviet decision to increase freshwater input created a habitat far more favorable to *Mnemiopsis* than to the *Aurelia* blooms of previous years.

Lack of predators. *Mnemiopsis* is kept in check in its native habitat by three species of jellyfish: the lion's mane (*Cyanea*), the sea nettle (*Chrysaora*), and the jelly with jaws (*Beroe*). Through an accident of history, none of these three species occurs naturally in the Black Sea. Nor does *Mnemiopsis*, of course.

Lack of competitors. Three Black Sea species of fish prey predominantly on zooplankton: the anchovy, the Mediterranean horse mackerel, and the sprat. These are commonly called "forage fish"; they are the smaller fish that are eaten by bigger fish. But all three species became the main targets of commercial fisheries in the 1980s following the decline of the larger piscivorous species, such as tuna and swordfish. As *Mnemiopsis* began blooming and eating the eggs and larvae of these forage fish, as well as their plankton food, the combined fishing and predation pressure rapidly depleted their stocks too, particularly the anchovy. Soon, they too were overfished, and *Mnemiopsis* had little remaining competition for food. *Mnemiopsis* was even able to outcompete the moon jellyfish for food.

Morphometry. The Black Sea's semienclosed shape and its almost total iso-lation from other seas and oceans heightened its vulnerability. Quite simply, flushing is so limited that once something gets in, it is hard to get rid of.

Characteristics of *Mnemiopsis*

Weedy nature of the beast. *Mnemiopsis* is a weed. A marine weed. We nor-mally think of weeds as those hardy and persistent plants that somehow survive no matter how poor the conditions and no matter how many times we try to remove them from our gardens. Dandelions. Nightshade. Thistles. Soursob.

It might seem strange that an animal could be a weed, and a jellyfish at that. But like all successful weeds, *Mnemiopsis* is highly opportunistic and can thrive on any combination of numerous nutritional sources, survive a broad range of environmental conditions, and reproduce and grow rapidly whenever conditions are favorable.

Weedy feeding behavior. Not surprisingly, *Mnemiopsis* has a complex of prey preferences for different conditions (Javidpour et al. 2009a). Overall, it prefers small-sized, slow-swimming prey. During winter, it seems to pre-fer barnacle larvae. In summer, about 60–70 percent of its prey are the early swimming larvae of *Aurelia*, up to 621 larvae per *Mnemiopsis* per day have been counted. At high densities, *Mnemiopsis* cannibalizes its own larvae, com-prising up to 76 percent of its prey. And surprisingly, although copepods are generally considered to be a ubiquitous food source for just about every spe-cies big enough to eat them but not so big as to overlook them, *Mnemiopsis* only rarely preys on copepods when other food is available.

In conditions of high food densities, *Mnemiopsis* is able to eat over 10 times its own body weight and more than double its body volume per day (Reeve, Walter, and Ikeda 1978). The reason for this unbelievable growth rate is that its "overhead"—the break-even point of energy it needs to consume to sur-vive and grow—is only 16 percent of its biomass (Finenko et al. 2006). Be-cause of the high efficiency with which *Mnemiopsis* converts its prey to its own body mass, it is able to grow much faster than, and therefore outcompete, slower growing species.

One of the most significant aspects of the feeding behavior of *Mnemiopsis* is that its feeding rate is proportional to food availability. It will gorge itself to the limit of its ability to ingest and digest, then continue to collect prey before "spitting it out" again because it is unable to take more (Kremer 1979). The effect of this behavior on the food chain is the same whether or not *Mnemiopsis* ingests and digests—its prey are still removed from the system. In this way, *Mnemiopsis* not only consumes, but also *destroys*, extremely large amounts of food. In fact, several studies credit *Mnemiopsis* with over 50 percent mortality of total zooplankton populations, while some studies have found that *Mnemiopsis* ate over 30 percent of the copepod population *per day* (Bishop 1967; Kremer 1979) as well as between 38 percent and 65 percent of the anchovy eggs produced *per day* (Purcell et al. 1994).

Whereas most types of jellyfish rely on horizontal movement to capture prey, *Mnemiopsis* spends most of its time either moving vertically up and down in the water column or else hovering in a mouth-down position (Larson 1988). This behavior not only conserves energy but also requires less space per individual. Furthermore, whereas most species have a fairly narrowly defined range of prey based on tentacle size and foraging mode, *Mnemiopsis* is able to feed on a wide range of prey by catching smaller or inactive prey with its tentacles and larger or more active prey on its sticky, mucus-covered lobes (Larson 1988; Waggett and Costello 1999). Unlike species that use only their tentacles for food capture (and that therefore must stop capturing food while ingesting), *Mnemiopsis* can capture and ingest simultaneously.

While subadult and adult *Mnemiopsis* eat a smorgasbord of copepods, fish eggs and larvae, and other zooplankton, the young larval *Mnemiopsis* are voracious consumers of microplankton, including diatoms, dinoflagellates, ciliates, euglenoids, and rotifers (Sullivan and Gifford 2007)—a truly unusual mix of microscopic plants, animals, and "other." What this means is that *Mnemiopsis* can eat "anything"—and given the very large number of larvae they produce, *Mnemiopsis* begins crippling the food chain from the bottom up, from the earliest stages of its life cycle.

Weedy reproduction. Hermaphroditism, or an individual being both male and female, is relatively common in lower invertebrates and plants. It's nature's way of increasing the reproductive opportunities of an organism. Species in which hermaphroditism occurs are often "sequential hermaphrodites," meaning that the individual can be either male then female, or female then male, but not both at the same time. Some species, particularly worms, are "simultaneous hermaphrodites," meaning that the individual is both male and female at the same time; in these cases, there are generally mechanisms to keep them from being able to fertilize themselves. However, *Mnemiopsis* is a self-fertilizing simultaneous hermaphrodite. This may well be one of the keys to its weediness—it only needs one individual to grow and multiply into an entire population and a successful invasion.

Besides its extreme, if not unusual, promiscuity, *Mnemiopsis* is astonishingly fertile. While an individual's life expectancy can be up to several months, it can begin laying eggs within 13 days of its own birth (Baker and Reeve 1974). By the seventeenth day, it lays eggs daily and can lay up to 10,000 per day. Even young individuals lay over 1,000 eggs per day. The number of eggs produced increases with age, and *Mnemiopsis* is believed to lay eggs throughout its lifetime, depending on food availability. Eggs hatch within 12 to 20 hours of being laid.

Mnemiopsis is not only highly reproductive, it is also capable of rapidly regenerating lost body parts or sections. Experiments with cutting the animals into halves, thirds, or even quarters have demonstrated that *Mnemiopsis* can replace missing parts within as little as 2 to 3 days and then resume life as normal (Coonfield 1936).

Weedy salinity tolerance. *Mnemiopsis* is what is called a "euryhaline" species (*eury*, meaning "wide" or "broad" + *haline*, meaning "salt" = a species tolerant of a broad range of salinities). It has been found in salinities ranging from 0.1 to 25.6 in Chesapeake Bay; elsewhere it has been found in even higher salinities, up to 38 (Purcell et al. 2001b). To put this into perspective, the salinity of seawater is generally about 35, while a salinity of 0.5 is typical for drinking water. *Mnemiopsis*, therefore, occurs in hypersaline habitats, all the way down to almost distilled, laboratory-grade freshwater. Its salinity tolerance, therefore, enables it to find refuge from competitors, such as other jellyfish, and zooplanktivorous fishes, such as the anchovy.

Weedy temperature tolerance. In Chesapeake Bay, temperature does not appear to be a limiting factor for *Mnemiopsis*; it has been found over a range of temperatures from 10–29°C (50–90°F), throughout the year. However, cool spring temperatures are associated with smaller body size, and they are unable to survive the 4°C (39°F) winters in the Sea of Azov and parts of the Black Sea. Similarly, warm temperatures are correlated with rapid population increase, whereas the higher temperatures of the Aegean and Mediterranean appear to limit its bloom success.

The Damage It Caused

The main commercial fish in the Black Sea, the anchovy, had declined sharply by the late 1980s. *Mnemiopsis* first depleted the zooplankton that are the main food source for fish, then turned to eating anchovy eggs and larvae.

To say that *Mnemiopsis* decimated the Black Sea's ecosystem may sound sensationalistic, but it's true. In fact, it's an understatement. By 1993, it was estimated that *Mnemiopsis* comprised up to 95 percent of the total wet weight biomass (Travis 1993)—including all the copepods, all the anchovies and sardines and their eggs and larvae, all the invertebrates and bottom fish. Ninety-five percent of all living things.

To put this into perspective, Lily Whiteman (2002), in an article for *On Earth*, noted that "the total weight of the Black Sea's comb jellies was more than ten times the weight of all fish caught throughout the world in a year."

The collapse of the ecosystem cost the Black Sea fishing industry more than $350 million. The region once supported plentiful subsistence fisheries and profitable commercial fisheries, including caviar from sturgeon. Tourism in the region has, likewise, been decimated.

Epilogue

In general, jellyfish infestations are hard to manage, let alone eradicate. And few plagues are as pervasive and shocking as that of *Mnemiopsis* in the Black Sea. However, through the unexpected and fortuitous *Perfect Storm*-like timing of three independent sets of events, the *Mnemiopsis* invasion was interrupted and largely brought under control. Even more surprising, the timing of these three unlikely occurrences has helped bring about the reinvigoration of the Black Sea ecosystem.

First, as the Black Sea fish stocks collapsed, their fisheries shut down. These closures eased the intense fishing pressure on the anchovies, thus slowing the spread of *Mnemiopsis* by restoring its primary competitor. After the population peak of *Mnemiopsis* in 1989, *Mnemiopsis* and *Aurelia* alternated in peaks of abundance through the 1990s.

Second, due to economic turmoil with the breakup of the Soviet Union in 1991, fertilizer subsidies to farmers were cut, thereby slashing the amount of fertilizer applied to crops. Reduced use led to a significant decline in agricultural runoff, and with it, a decline in hypoxia. With improved water quality and dissolved oxygen levels, other species were better able to survive. By the late 1990s, both *Mnemiopsis* and *Aurelia* populations had declined somewhat.

Third, another alien species of comb jellyfish, *Beroe ovata*, a key natural predator of *Mnemiopsis* in the Chesapeake Bay, was accidentally introduced into the Black Sea. *Beroe* requires more than 20 percent of its body weight in gelatinous prey per day to grow, and it can double its size in 10 days if it consumes 75 percent of its weight. Studies have shown that when *Mnemiopsis* substantially outnumbers *Beroe*, the latter consumes about 5 percent of the *Mnemiopsis* biomass per day, but when *Beroe* outnumbers *Mnemiopsis*, it can consume the entire population in a day (Bilio and Niermann 2004). In other words, with sufficient available prey, *Beroe* will eat constantly, and as it grows, it simply eats more. *Beroe* was first observed in the Black Sea in 1997 (Shiganova et al. 2001a). Originally thought to have arrived via the Sea of Marmara from the Mediterranean, it was later shown to have been introduced from the Americas in much the same way as *Mnemiopsis* (Bayha et al. 2004). *Beroe* bloomed explosively in August 1999, and the abundance of *Mnemiopsis* suddenly declined. Already by autumn 1999, increases were observed in zooplankton biomass, density of fish eggs, and the number of species overall.

These three events together created a spectacularly effective control of the *Mnemiopsis* problem. The Black Sea has begun to rebound. The *Mnemiopsis* population has declined, and some fish are beginning to build up populations again. However, it is unlikely that the ecosystem will ever be fully resurrected to its former glory, as the missing mammals and sturgeon are unlikely to revive.

In the words of Claudia Mills, the aforementioned blooms pioneer, "This system provides the most graphic example to date of a highly productive ecosystem that has converted from supporting a number of valuable commercial fisheries to having few fishes and high numbers of 'jellyfishes'—medusae and ctenophores" (Mills 2001, 61).

Mnemiopsis Again . . . Crisis in the Caspian
(Since November 1999)

The Caspian Sea is the largest inland sea on earth, with a total coastline of some 7,000 kilometers (4,400 miles) and a surface area of about 386,400 square kilometers (150,000 square miles). In the deeper, more saline middle and southern basins, the water is about one-third as salty as the ocean. The shallower northern basin is almost freshwater. Its only outside connection is the man-made Volga-Don Canal to the Black Sea.

At the same time it was wreaking havoc in the Black Sea, *Mnemiopsis* also entered the Caspian. Early warnings that the impact of this species on the Caspian ecosystem might be much worse than in the Black Sea have proven to be true (Kideys and Moghim 2003; Finenko et al. 2006; Roohi and Sajjadi 2011). Indeed, in just over a year, *Mnemiopsis* spread across almost the entire Caspian, and by August 2001, its population had expanded over the entire middle and southern parts of the sea. Its abundance is strongly seasonal, but the species is present all year. One study reported up to 1,200 specimens per square meter (about 100 per square foot) in summer, but only about 50 specimens per square meter during the coldest months of winter. Another study found up to a smothering 2,285 specimens per square meter (more than 200 per square foot) (Kideys and Moghim 2003).

In general, the smaller, younger specimens have the biggest appetite. Over 80 percent of the summer population and about 100 percent of the winter population is under 10 millimeters (⅓ inch), that is to say, the size class capable of doing the fastest ecological damage is by far the most abundant. Indeed, during winter and spring, *Mnemiopsis* consumed the available stock of zooplankton every 3 to 8 days, but took only 1 day to clear the water column in summer and autumn.

Predation by *Mnemiopsis* is far more destructive in the Caspian than it is in all other regions affected. In the northern Caspian, zooplankton biomass declined sixfold, and in the middle and southern Caspian, biomass dropped by factors of 4 and 9, respectively. Along the Iranian coast, the zooplankton declined by half between 1996 and 2006 (Roohi et al. 2008). Particularly hard-hit were the cladocerans, a group of small planktonic crustaceans, with only 1 of the 24 species found in 1996 remaining in 2006.

As might be expected, the strong predation pressure on zooplankton has caused significant decline in the fisheries of small pelagic fish, such as the anchovy kilka, which compete with *Mnemiopsis* for food. The problem is that kilka used to support a profitable commercial fishery, and were also important prey for endangered large carnivores, such as the beluga sturgeon and the Caspian seal.

Professor Ahmet Kideys of the Institute of Marine Sciences in Erdemli, Turkey, and his colleagues have reported substantial fisheries losses in Iran, Azerbaijan, and Russia. In 1998 and 1999, Iran caught 90,000 and 91,000 tons of kilka, respectively. But by 2001, just the second season after the arrival of

Mnemiopsis, Iran's kilka landings had fallen almost 50 percent to 49,000 tons, a loss estimated at a minimum of $15 million.

Curiously, *Mnemiopsis* populations differ conspicuously between the Caspian and the Black seas. For example, in the Black Sea, they grow to 180 millimeters (7 inches), but only reach 65 millimeters (2 ½ inches) in the Caspian. It is thought that this might be a result of the lower salinity in the Caspian. Similarly, *Mnemiopsis* produces fewer eggs in the Caspian than it does in the Black Sea, possibly because of its smaller body size. Nonetheless, *Mnemiopsis* produces an average of about 900 eggs per day in the Caspian, with a maximum of about 6,200.

The final words of the report by Galina Finenko and her colleagues (2006) sum up the situation: "Such a high pressure exerted by this ctenophore would not allow zooplankton biomass levels to rise, and, as a consequence, no recovery can be foreseen with respect to the catch of planktivorous pelagic fishes until *M. leidyi* levels decrease substantially" (p. 183). Professor Henri Dumont (2001) of Ghent University considers that a catastrophe occurring in the Caspian is almost unavoidable: "*Mnemiopsis*, an indiscriminate pelagic hunter, will primarily attack the zooplankton, and because it is capable of expanding as long as it can find food, it will likely drive the zooplankton to near-extinction. Some more vulnerable species may effectively go extinct. Unfortunately, the 'naïve' Caspian endemics . . . are likely to be among them."

Facing bigger losses than those from the Black Sea fiasco, attendees of an April 2001 scientific workshop in Baku, Azerbaijan, debated whether to introduce the American butterfish or *Beroe,* both natural predators of *Mnemiopsis*. A conclusion of the meeting was that *Beroe* was likely to arrive in the Caspian on its own accord via ballast water, just as *Mnemiopsis* had. Nonetheless, fisheries representatives voted in late 2001 to introduce *Beroe* into the Caspian as a means of biocontrol (Cameron 2002). But unconfirmed reports of its presence in the Caspian surfaced in September 2000, September 2001, and summer 2003; finally specimens were captured to confirm its presence in May 2004.

At first it was a wait-and-see game; nobody was quite sure what would happen between *Beroe* and *Mnemiopsis*. Would the Caspian ecosystem be as lucky as the Black Sea? Unfortunately, it looks like the answer is "no" (Roohi and Sajjadi 2011). Even after several years, abundance and biomass of *Mnemiopsis* have remained high. Zooplankton stocks have remained low, and biodiversity is a mere third of what it was before *Mnemiopsis*. Phytoplankton have increased. Eutrophication has increased. Fisheries have collapsed. Fishing-based economies have collapsed.

About the Beluga Sturgeon, *Huso huso*

> Caviar. The very word evokes glamorous lifestyles, exotic travel, and glittering
> festivities. Yet the world's source of this luxury item, the sturgeon, is in grave
> danger. Sturgeon have survived since the days the dinosaurs roamed the earth.
> The question now is whether these "living fossils" can survive the relentless fish-
> ing pressure, pollution, and habitat destruction that have brought many species of
> sturgeon to the brink of extinction. (Speer et al. 2000, i)

These are the opening words of a report about the plight of the sturgeon published just thirteen months after *Mnemiopsis* was first detected in the Caspian
and more than a year and a half before its catastrophic impact was felt.

Let's examine the case of the beluga sturgeon in order to better appreciate
what is at stake when species go head to head over limited resources. It is one
thing to say that jellyfish outcompete other species and take over ecosystems,
but quite another to get to know the winners, the losers, and what's really at
risk. In this case, the survival of an ancient species is on the line, and prospects don't look promising.

The Caspian is home to many scientifically interesting and commercially
important species, including the beluga sturgeon. This venerable fish "can live
for a century, weigh a ton, and span the length of a pickup truck" (Benningfield 2006). Relentless overfishing to feed the world's appetite for caviar—the
fish's unfertilized eggs—had already driven the beluga to the brink of extinction well before *Mnemiopsis* arrived . . . then things got worse.

The International Union for Conservation of Nature (IUCN) noted that
beluga had been extirpated from the Adriatic Sea by the early 1970s and from
the Sea of Azov more recently as a result of overfishing and loss of spawning
sites due to damming of rivers (Gesner, Chebanov, and Freyhof 2010). Since
vanishing from the Adriatic and Azov seas, beluga stocks elsewhere have rapidly dwindled. Black Sea stocks have been almost eliminated, with the last
remaining wild population migrating up the Danube. The last remaining Caspian population migrates up the River Ural. Current populations in the Sea of
Azov and the Volga River are comprised almost entirely of stocked fish.

The beluga was once the largest species in the Caspian, and in fact, the
world's largest freshwater fish (although it spends its nonbreeding time in
saltier water). It grows throughout its lifetime and can reach up to 5 meters
(15 feet) in length and 1,000 kilograms (2,200 pounds) in weight. The massive
females can carry almost half their body weight in eggs. However, the large,

old individuals have long been fished out. Today they reach only half that size.

Beluga are very slow growing, with males becoming reproductive at 10–15 years old and females at 15–18. The species has a life expectancy exceeding 100 years, and they spawn every 3–4 years. But by 2003 the maximum age had fallen to 53 years (Gesner, Chebanov, and Freyhof 2010). Based on catch data and numbers spawning, the IUCN estimated that natural populations of the species have declined over 90 percent since 1950, and that the last wild populations will soon be globally extinct due to overfishing.

Beluga caviar is considered the finest in the world and can fetch up to $8,000 a kilogram ($4,000 per pound), depending on quality and taste. Global fisheries statistics are disheartening (Gesner, Chebanov, and Freyhof 2010):

- The catch has declined 93 percent between 1992 and 2007 (from 573 to 36 tons).
- The number of fish entering the Volga (nearly 100 percent hatchery-reared), dropped 89 percent in 33 years, from 26,000 (1961–1965) to 2,800 (1998–2002).
- A mere 2,500 beluga migrated up the Ural in 2002.
- In the Caspian, the annual catch has declined 95 percent since 1945 despite intensive restocking efforts. As of 2003, for each decade the averages (in tons) were 1,521, 1,414, 1,789, 935, 557, and 66.
- The agreed catch quota for 2007–2008 was 110 tons. It was not achieved—there simply weren't enough fish.
- In the Sea of Azov, the (stocked) population was estimated at 551,000 in 1979–1981, but only 25,000 in 1988–1993. After 1994, beluga were only caught sporadically (98 percent of those caught being juveniles), despite a commercial fishing ban in 1986. After 1986, the major threat was being caught as bycatch.
- In the Danube, annual catch dropped from an average of 25 tons in 1972–1976 to 7 tons in 1985–1989, a decline of 67 percent in 12 years. Catch for Romania dropped 60 percent in three years, from 23 tons in 2002 (85 percent of quota) to just 8 tons in 2005 (34 percent of quota). Catching of beluga in the Danube was banned in 2006.

Beluga face numerous threats. A major one is overfishing for meat and caviar, with current harvest rates of the Caspian population at 4 to 5 times the sustainable levels . . . *or more*. In 2001, Russia claimed $40 million in caviar exports, but some observers say that the figure was closer to $100 million—for *legal* exports—while Russia's Interior Ministry estimated the *illegal* trade at $400 million (Stone 2002).

Another major threat is the damming of rivers; for example, in the Volga alone, damming has decreased the available spawning grounds by about 90 percent. There is also evidence of pesticide contamination accumulating in fat, which has caused many problems, including reduced reproductive success. Bycatch is also a threat, and difficult to control. Finally, another possible threat is the "Allee effect," that is, when a population size becomes so small that the rate of population growth naturally slows due to competition between individuals of the same species; this is futher discussed in chapter 11.

The beluga was formally listed as endangered in 1996. In 2006, the United States banned all imports of beluga caviar. Citing improvements in conservation measures being undertaken by Caspian countries, trade bans on beluga caviar were lifted in 2007 by CITES (the Convention on International Trade in Endangered Species). However, continued decline of the species led the IUCN to elevate the beluga's status to "critically endangered" on 18 March 2010. In fact, all 18 species of sturgeon are now on the Red List of endangered species.

The Role of *Mnemiopsis* in Commercial Extinction of the Beluga Caviar Industry

Mnemiopsis was first noted in the middle Caspian in November 1999. By 2000, "a stunningly rapid expansion of *Mnemiopsis* had taken place" (Ivanov et al. 2000, 256). Then its population exploded in 2001, as the findings of one study indicate (Shiganova et al. 2001c). The bloom was small in June 2001 in the southern Caspian, with the highest density estimated at just over 2,000 individuals per square meter (about 200 per square foot). As the bloom expanded, so did its density. By late July, *Mnemiopsis* had spread across the whole middle Caspian. By August, it was distributed throughout the Caspian, reaching staggering densities of 3,756 individuals per square meter (about 350 per square foot). While its density doubled, its spatial distribution containing that higher density more than tripled.

A major food item in the beluga sturgeon's diet is anchovy kilka. It was estimated that sturgeon and seals consumed about 440,000 tons of kilka annually during the 1960s and 1970s (Daskalov and Mamedov 2007).

While seals and sturgeon were gorging themselves, so too was the Caspian fishing industry. In the 1970s and 1980s, improved fishing techniques led to mammoth takes of more than 400,000 tons of kilka per year. After about 1990, the stocks declined sharply to about half that, save for an exceptional year in 1999 when about 300,000 tons were taken. However, the high fishing rates prior to the introduction of *Mnemiopsis* were no longer sustainable after 2000. In 2001, just 64,000 tons were taken; by 2005, the take had dropped even further to just 59,000 tons (Daskalov and Mamedov 2007). The kilka fishery had collapsed, and with it, the primary food source of the beluga sturgeon.

Sadly, it's not just about the rich people of the world losing their beloved blini topping; it's also about changes to the whole ecosystem with some disastrous results. Like the beluga sturgeon, the Caspian seal, *Phoca caspica*, is facing death through starvation. Its plight was brought to the world's attention in a paper by Vladimir Ivanov from the Caspian Fisheries Research Institute in Russia and his colleagues (Ivanov et al. 2000).

The Caspian seal is one of the world's smallest seals. According to the Caspian Seal Project, "the Caspian seal is the only marine mammal in the Caspian Sea, and is found no where else in the world. At the start of the 20th century there were around 1 million Caspian seals. It is an iconic animal for the region, and a key indicator for the health of the Caspian, upon which the livelihoods of thousands of people depend. Today the seal population has fallen by more than 90% and continues to decline" (Caspian Seal Origin 2011).

Like many animals facing extinction, there are numerous reasons for its diminishing numbers, and the complexity of their combination makes for difficult challenges in conservation efforts.

One of the main reasons for the seal's decline was commercial hunting by the former Soviet nations through much of the twentieth century. Tens of thousands of pups and adult seals were taken each year for their blubber and fur. Since the collapse of the Soviet Union, commercial hunting has been scaled back considerably, with currently only a few thousand taken each year.

Another significant reason believed to be responsible for the dwindling seal population is female infertility. Since at least the early 1980s, poor recruitment has been observed. High levels of pesticide residues, such as DDT, have been found in seal tissue and are believed to cause the infertility (Watanabe et al. 1999). High levels of bioaccumulated heavy metals also found in seals could be contributing further to reproductive failure or shortened life span (Watanabe et al. 2002).

Furthermore, the Caspian seals have suffered from highly contagious infections, such as canine distemper and a mysterious avian flu–like illness, with mass mortalities in 1997, 2000, and 2001 (Forsyth et al. 1998; Kennedy et al. 2000; Stone 2002). It is estimated that more than 10,000 seals died along the Kazakhstan coast during April and May 2000. High death rates were reported along other coasts as well. The cause of death in these events was confirmed by DNA testing to be canine distemper, which is also thought to be the cause of the other mass die-offs.

Other threats to the seal population include bycatch (i.e., drowning in fishing nets), pollution from oil and gas mining, and deliberate killing by fishermen to keep seals from interfering with fishing operations.

To this already distressing situation, add *Mnemiopsis*. As with the beluga sturgeon, anchovy kilka is the Caspian seal's main prey. Overfishing had already depleted kilka stocks, but since 2000, when *Mnemiopsis* invaded the Caspian, kilka have become scarce. So, on top of commercial hunting, chemical sterility, viral infections, drowning, pollution, and deliberate killing, all working in combination against the seal's survival, starvation, too, has become an increasing threat.

All the above menacing ecological changes in the Caspian have been extensively documented (Nasrollahzadeh 2010). However, there doesn't seem to be an easy answer or any remediation process in sight.

Mnemiopsis Again . . . and Again . . . and Again . . . Marching toward the Mediterranean, and the Conquest of Western Europe (1988, 1990, 1992, 1993, 1999, 2006, 2009 . . .)

It reads like diary notes of Hitler's army: try the northern route, stopped dead by the winter cold—try the southern route, success—push further—keep going, almost there—reach the target, spread out—conquer. . . . *Mnemiopsis* has spread through the seas of the Mediterranean Basin like cancer—in all directions . . . aggressively . . . lethally. The list that follows is a squinty-eyed view of the rapid spread of *Mnemiopsis* through Europe. These are merely the first occurrences; the details of its proliferation and devastation are recounted below.

- 1982: Black Sea
- 1988: Sea of Azov
- May 1992: Mersin Bay, Turkey
- October 1992: Sea of Marmara

- July 1993: Aegean Sea
- October 1993: Syria
- November 1999: Caspian
- August 2005: North Sea
- October 2006: Baltic
- March 2009: Mediterranean coast of Israel
- May 2009: Ligurian Sea
- July 2009: Tyrrhenian Sea
- July 2009: Mediterranean coast of Spain
- September 2009: Ionian Sea

Few people know the *Mnemiopsis* problem as well as Dr. Tamara Shiganova, based at the Shirshov Institute of Oceanology at the Russian Academy of Sciences in Moscow. Slight of build, blond-haired, and well kempt, her gentle and lovely appearance belies her muscular scientific accomplishments and her appropriately feisty disposition. Tamara began working with *Mnemiopsis* when it first invaded the Black Sea. In the post-Soviet era of federal funding shortfalls, cut-throat politics, and testosterone-driven alliances, she not only held her own but has since emerged as a leading expert on jellyfish invasion dynamics. She has chronicled the spread of *Mnemiopsis* through the Mediterranean Basin.

August 1988: Sea of Azov

Mnemiopsis leidyi was first found in the Azov Sea in 1988. Almost immediately the problem was obvious. Shiganova wrote in 2001 that the "effect on the Azov Sea ecosystem was even stronger than in the Black Sea. During the first months of summer, *M. leidyi* consumed almost all of the zooplankton. The stocks of planktivorous fish dropped then recovered little, due to a persistent summer abundance of *M. leidyi*. The strong effect of *M. leidyi* on the Black and Azov seas also reflects an absence of predators."

The Azov anchovy and kilka fisheries declined catastrophically to just one-third of their previous levels (Ivanov et al. 2000). This collapse was from a combination of predation by *Mnemiopsis* on fish eggs and larvae and predation on zooplankton that the larvae primarily eat.

Intriguingly, *Mnemiopsis* is unable to survive the winter cold in the Azov, so it dies off in the autumn when the temperature reaches 4°C (39°F), but then reinvades each spring through the Kerch Strait (Shiganova et al. 2001b).

Other jellyfish competitors in the Sea of Azov, *Aurelia* and *Rhizostoma*, both common since the 1960s, have virtually disappeared since the invasion of *Mnemiopsis*. The predatory comb jelly *Beroe*, which was largely responsible for gaining control of *Mnemiopsis* in the Black Sea, has not invaded the Azov in large numbers, which may be because it cannot sustain a permanent population as it starves each winter when *Mnemiopsis* dies off.

October 1992: Sea of Marmara

Mnemiopsis was first confirmed in the Sea of Marmara in October 1992, but is also thought to have invaded via upper layers of the current from the Black Sea, through the Bosporus in 1989–1990. Its initial abundance was very high, but it then failed to establish itself as it did in the Black Sea, as later studies found quite low densities. However, a Turkish survey in late July and early August 1993 found that the *Mnemiopsis* population had by then spread throughout the Sea of Marmara. Its effects on the zooplankton, fish larvae, and forage fish were similar to those in the Black Sea, but less pronounced (Shiganova et al. 2001b).

Late Spring/Early Summer 1990: Aegean Sea

Mnemiopsis was originally found in Saronikos Gulf and Elefsis Bay in 1990, but these records were not published until later, after the July 1993 find off Kusadasi on the central west coast of Turkey (Shiganova et al. 2001b; Kideys and Niermann 1994). *Mnemiopsis* is transported periodically to the northern part of the Aegean on currents from the Black Sea, but does not appear to be breeding in the Aegean, where the temperature and salinity are too high (Shiganova et al. 2001b). Furthermore, in contrast to the nutrient-rich conditions in the Black and Azov seas, which have led to high prey abundances, the Aegean is nutrient-poor, leading to a paucity of prey; this too may have contributed to *Mnemiopsis* failing to establish. The effect of *Mnemiopsis* on the ecosystem and on commercial fish populations in the Aegean Sea was minimal.

May 1992: Mersin Bay, Turkish Mediterranean

Mnemiopsis was found in the coastal waters of Mersin Bay, on the southeastern coast of Turkey, in late May 1992. The mode of transport was thought to be either in ballast water of ships from the Black Sea or by riding currents through the Bosporus Strait, the Sea of Marmara, the Dardanelles, and the Aegean Sea (Kideys and Niermann 1994). *Aurelia* and *Beroe* both occur naturally in the Mediterranean, as competitors and predators of *Mnemiopsis*, respectively. *Aurelia* abundance was lower when *Mnemiopsis* was present in the same areas. It is thought that the presence of *Beroe* may be the reason why *Mnemiopsis* abundance was low in the eastern Mediterranean (Shiganova et al. 2001b), but this does not explain its later explosive increase toward the west.

August 2005: Northern Europe

Mnemiopsis spread through Northern Europe like the plague: fast, silent, and lethal. The Netherlands. Sweden. Germany. Denmark. Norway. Finland. . . . Though it was first believed that *Mnemiopsis* arrived in ballast water from the Black Sea, DNA testing later showed that the North Sea infestation was a separate introduction altogether and from different US seed-stock, this time from the northeast rather than the Gulf of Mexico, as in the Black Sea invasion (Reusch et al. 2010; Ghabooli et al. 2011). Whether *Mnemiopsis* arrived first in the North Sea or the Baltic is unclear.

    Mnemiopsis was found in very dense blooms at numerous locations along the coast of the Netherlands from August to November 2006 (Faasse and Bayha 2006). Because it was so widespread and abundant, it seems likely to have been in the North Sea for several years. Simultaneously, another group of researchers found *Mnemiopsis* in large numbers between August and November 2006 in the Skagerrak off the western coast of Sweden (Hansson 2006). They also concluded that it was likely to have arrived earlier but initially escaped detection. Still another group of researchers found *Mnemiopsis* in the North Sea near Helgoland in November and December 2006 (Boersma et al. 2007). The population density of the species was then low. Speculation at the time that it would fail to establish was based on two reasons. First, two species of the predatory comb jellyfish *Beroe*, the type that ultimately controlled *Mnemiopsis* in the Black Sea, are normally resident in the North Sea. Also, winter temperatures there fell below 4°C (39°F), the lethal limit for *Mnemiopsis* in the Sea of Azov.

    Other researchers published papers shortly after the above three, reporting that *Mnemiopsis* had been found throughout Danish waters from August 2005 through summer 2007, often in dense clouds (Tendal, Jensen, and Riisgård 2007). The species was found in every net sample at nine locations throughout Limfjorden in northwestern Denmark, and by August 2007 had reached an average density of 867 individuals per cubic meter (about 25 per cubic foot) in Skive Fjord (Riisgård et al. 2007), exceeding its bloom abundance in the Black Sea. Other reports of *Mnemiopsis* from Sweden, Norway, and Finland quickly followed.

In Norway, it was discovered that *Mnemiopsis* had been repeatedly photographed since November 2005 in Oslofjorden but had been mistaken for the native species, *Bolinopsis*, thus raising no alarm (Oliveira 2007). The Skagerrak, the Kattegat, and the Great Belt are linked straits that separate Denmark from Norway and Sweden, thereby connecting the North Sea to the Baltic. These straits have been the subject of intense focus as the likely route of introduction and reintroduction of *Mnemiopsis* into the Baltic. Because the species is 10 times more abundant in the Kattegat than in the Baltic, and produces 50 times more eggs in the Kattegat than in the Baltic, it is believed that the Kattegat is the breeding ground for a steady source of ctenophores drifting into the Baltic (Jaspers et al. 2011).

A recent study examining the predatory impact of *Mnemiopsis* in the North Sea found that it eats far more copepods than fish eggs or larvae in this habitat (Hamer, Malzah, and Boersma 2011). Curiously, while it is a strong predator on *and* competitor with fish in other habitats, in the North Sea it appears that the primary threat to fish is competition from *Mnemiopsis* for the same prey, rather than predation on fish eggs.

Furthermore, the lion's mane jellyfish *Cyanea* is common in the North Sea. *Cyanea* preys on *Mnemiopsis* where they naturally occur together in US waters; therefore, it is possible that *Cyanea* may help keep *Mnemiopsis* in the North Sea from reaching the catastrophically destructive abundance levels witnessed in the Black and Caspian seas.

October 2006: Baltic Sea

Mnemiopsis was first reported in the Baltic during routine sampling in Kiel Fjord in October 2006 (Javidpour et al. 2006). In just over a month, it tripled its abundance from an average of 29 to 92 individuals per cubic meter (from about 1 to 3 per cubic foot). These high densities were comparable with its population expansion in the early years of its explosive takeover of the Black Sea.

Following the initial Kiel Fjord report in late 2006, *Mnemiopsis* was found at numerous locations throughout the Baltic in every month of 2007 except December (Lehtiniemi et al. 2007; Huwer et al. 2008; Javidpour et al. 2009b). By summer, the species had increased its highest densities fivefold over those observed in the previous year (up to 505 individuals per cubic meter or about 14 per cubic foot). In January 2008, it was found overwintering in the northern Baltic in enormous numbers—*more than 3,800 per square meter* (350 per square foot) (Viitasalo, Lehtiniemi, and Katajisto 2008). It is clear that *Mnemiopsis* quickly became established in the Baltic, and that it can live throughout the sea. However, it is much smaller in body size in the Baltic than in the Black Sea, which may limit its potential predatory impact.

Curiously, from March to June, when the population of *Mnemiopsis* appeared to be low, it was found to be concentrated in dense aggregations in the deep layers. But from August to September the majority of the population concentrated in the upper layers (Javidpour et al. 2009b). This behavior is thought to serve two purposes. First, concentrating the population in deep layers during low temperatures is thought to act as an energy-saving strategy. Second, shifting to the upper layers in late spring allows fast population expansion at a time when reproductive capacity is enhanced by rising temperature and increasing food.

Bornholm Basin in the central Baltic is the last remaining spawning ground for the eastern Baltic cod, as well as an important spawning area for sprat. Vertical distribution of *Mnemiopsis* has been observed to substantially overlap in layers where cod eggs and sprat eggs are neutrally buoyant and so are concentrated in the water column (Haslob et al. 2007). *Mnemiopsis* was observed in this region with fish eggs in its stomach, which suggests it could pose a serious predation threat to the early developmental stages of cod and sprat.

Concern over the effect of *Mnemiopsis* on the Baltic cod and sprat is justified, given the havoc that *Mnemiopsis* caused to the Black Sea ecosystem. Like the Black Sea, the Baltic was already unstable due to heavy overfishing and nutrient pollution before *Mnemiopsis* arrived. During the twentieth century, the Baltic shifted from its preperturbed state (with seals and cod as the top predators) to a state dominated by small schooling fish (Javidpour et al. 2009b).

Introduced species seem to love the Baltic apparently attracted or enabled by this instability—perhaps they are even in part driving it. Even before *Mnemiopsis* arrived in the Baltic, researchers had noted that it was a highly invaded body of water (Leppäkoski et al. 2002). As of 2002, about 100 alien species had been recorded in the Baltic, mostly from unintentional introduction via ballast water or hull fouling. Of these, about 70 had established breeding populations. One may wonder what makes semienclosed brackish water bodies so vulnerable to bioinvasions.

An asymmetric abundance of living things is found in brackish water bodies, such as the Baltic, Black, Azov, and Caspian seas, where native species richness is at a natural minimum at salinity levels between 5 and 8 (Paavola, Olenin, and Leppäkoski 2005). The explanation for this is that freshwater species find it increasingly difficult to survive above that range, but marine species find it increasingly difficult to survive below that range. Therefore, this phenomenon leaves an ecological gap for species with broad salinity tolerances to exploit.

Indeed, the Baltic appears to be an inherently unstable ecosystem characterized by ecological transitions (Österblom et al. 2007). Once dominated by seals and porpoises, the food chain shifted after the 1930s as these predators disappeared due to hunting, toxic pollutants, and incidental fishing bycatch. As the marine mammals dwindled, so did the predation pressure on their cod prey, which flourished . . . but not for long.

Extensive draining of wetlands and lakes, combined with increasing use of agricultural fertilizers, made the Baltic less able to withstand a large pulse of saline water and stagnation in 1951, which led to pervasive eutrophication, a type of nutrient pollution described in detail in chapter 7. A large anoxic zone now covers some 40,000 square kilometers (25,000 square miles). These comprehensive changes in bottom-water oxygenation made the region inhospitable to cod. Intensive fishing of cod since the 1970s hastened their decline, and as they vanished, their herring and sprat prey boomed. And now, with the explosive population growth of *Mnemiopsis*, it appears that the Baltic may be undergoing yet another regime shift, with jellyfish becoming the top predators.

2009: *Mnemiopsis* across the Mediterranean

After more than a decade of appearances around the eastern Mediterranean, *Mnemiopsis* was found in October 2005 in the Gulf of Trieste in the northern Adriatic (Shiganova and Malej 2009). Then, almost simultaneously, *Mnemiopsis* bloomed across the Mediterranean (Fuentes et al. 2010). In March 2009 at the coast of Israel . . . in May 2009 in the Ligurian Sea . . . in July 2009 in the Tyrrhenian Sea . . . in July 2009 off the Catalan coast of Spain . . . in September 2009 in the Ionian Sea. . . . Uh-oh.

In Israeli waters, *Mnemiopsis* wreaked havoc, announcing its presence on 3 March with large swarms blocking the intake pipes of a desalination plant (Galil, Kress, and Shiganova 2009). Within just 3 months, *Mnemiopsis* was observed along the entire Israeli Mediterranean coast, inside ports and along open shores, from the surface to a depth of 20 meters (60 feet).

In Italian waters, many sightings were made in the Ligurian Sea throughout the summer of 2009, ranging from a single specimen to thousands (Boero et al. 2009). When found at Ponza Island in the Tyrrhenian Sea, the species was already occurring in thick patches. One can guess that the single specimen sighted in the Ionian Sea was probably not alone, or at least not for long. And in Spanish waters, *Mnemiopsis* was found in many locations throughout the summer of 2009, often in very large numbers (Fuentes et al. 2009).

Because of the sudden appearance of dense aggregations across such a large area, it is probable that *Mnemiopsis* had been present in the Mediterranean for quite some time but had not been detected. Building its population in deeper waters then migrating to the surface en masse would account for its simultaneous sudden appearance in the Mediterranean. However, this explanation also portends a potentially serious crisis.

It is difficult to predict the effect that *Mnemiopsis* may have on the Mediterranean. It is a large sea with a variety of habitats and diverse flora and fauna — features that, in theory, should protect it against destruction. However, like the Black Sea, the Mediterranean has been severely altered by human disturbance. So must we hold our breath, cross our fingers, and pray for a good outcome.

The combination of factors in the Black Sea that enabled *Mnemiopsis* to take over — overfishing, eutrophication, and other forms of pollution — is also present in the Gulf of Mexico. And like the Black Sea, native jellyfish species are exploiting the instability and an invader has become established.

Wasting Away in Margaritaville
(Gulf of Mexico, since 2000)

Like Jimmy Buffet in his famous song about slow death in paradise, so too the Gulf of Mexico is wasting away. And it's our own damn fault. Immortalized in songs and prose, the coastline from Texas to the Louisiana bayou, from the Everglades to Hemingway's Florida Keys, is legendary as a place of trophy fishing, airboating through swamps and weeping willows, haunting jazz, unlimited seafood, endless summers, and decadent relaxation. But the romantic image is slipping away to the juggernaut of commercial development.

Starting on 28 September 2003, the *Naples Daily News* in Florida ran a 15-day special feature called "Deep Trouble: The Gulf in Peril" (http://web .naplesnews.com/deeptrouble/deeptrouble.html). "Deep Trouble" summarized the constellation of threats and cascade of outcomes plaguing the Gulf of Mexico. The feature was 15 months in the making, with a team of journalists and photographers researching the issues threatening the coastlines of the five gulf-rim states and Mexico. It is a real eye-opener, and well worth a peek. For example:

- In 2002, the Environmental Protection Agency called the gulf the dirtiest coastal body of water in America.
- From Iowa to the Everglades, farm fertilizer and waste from 31 states spills into the Mississippi and other waterways, which act like a hypodermic needle, mainlining this pollution straight into the heart and soul of the gulf.
- Tourism is a $20 billion-a-year industry for the gulf's beaches. But water quality has become so unsafe that there are now government warnings to stay out of the water. Once billed as "turquoise waters," the gulf's waters are now marketed as "emerald green"—yeah, due to pollution that triggers algal blooms.
- Many gulf communities are experiencing high rates of cancer and other health problems, apparently due to decades of industrial dumping of carcinogens into the air and of mercury, arsenic, and dioxin into the water.
- Seagrass meadows are disappearing. Mangroves and marshes are disappearing. Food chains supporting fish and birds are disappearing.
- Red tides, a form of toxic algal bloom, now occur with increasing frequency. So too do mass fish mortality events associated with red tides. In 1996 alone, red tide was blamed for killing 151 manatees in southwest Florida.
- Because the gulf is shared between the United States and Mexico, US coastlines have to deal with problems arising from Mexico's lax laws, such as untreated sewage, unreported industrial discharge, collapsing seafood stocks, and a "dark legacy of environmental contamination."
- A massive dead zone the size of Massachusetts is growing off New Orleans. This hypoxic area has been estimated to cost $14 billion to clean up.
- The gulf seafood industry, once the largest in the United States, accounting for 40 percent of all US commercial landings, is collapsing. The dead zone is displacing prawns and other seafood, while those surviving contain heavy metals from pollution.

- Some chemicals, such as phosphate and nitrogen, have been diverted to the gulf in an effort to save smaller bodies of water originally used as dumping grounds.
- The once-lush coral reefs of the Florida Keys are probably beyond recovery, with 93 percent of the reef already dead.
- A massive, viscous, nonoily, odorless "black water plume" persisted for months off southern Florida and the Keys, snuffing out corals and other living things in its wake; the black water was thought to be an unusual algal bloom, but its source was never resolved.
- Development has led to the loss of more than half of the Everglades, resulting in 70 percent decline in freshwater flow-through and a 90 percent decline in wading birds. The Everglades Restoration Project will cost $8.4 billion to fix environmental damage to the once-awesome "River of Grass." And that's just a small part of the gulf.

That was in late 2003.

Dr. Monty Graham is a senior marine scientist at Dauphin Island Sea Lab near Mobile, Alabama. Soft-spoken, boyishly handsome, and easygoing, Monty was recruited by DISL straight out of college with a strong background in marine ecology and particular focus on jellyfish. Since the late 1990s, he has been working on the problem of population increases in native moon jellies and sea nettles in response to the gulf's changing conditions. Sea nettles appear to be extending offshore as the area of the dead zone expands, and moon jellies appear to be increasing with the buildup of oil and gas structures as substrate for their polyps (Arai 2001). "Moon jellies have formed a kind of gelatinous net that stretches from end to end across the Gulf," he told the *Washington Post* (Dybas 2002).

But an ecosystem so severely disturbed was vulnerable to—and you might say *asking for*—really big problems. The Australian spotted jellyfish, *Phyllorhiza punctata* (see plate 2), introduced to the Caribbean around 1955, made its way into the gulf and bloomed from May to September 2000 in unprecedented numbers (Graham et al. 2003). They grew fast, both in medusa body size and population. Astonishingly, during July alone, their average bell diameter grew by 50 percent, from 32 centimeters (1 foot) to 45 centimeters (1 ½ feet), with an average weight per medusa of a whopping 7 kilograms (15 pounds). By August, the bloom spanned about 150 square kilometers (60 square miles), spread in an ovoid aggregation with an inner core of high and medium concentrations of about 30 square kilometers (12 square miles). At the high density core were an average of 2.3 medusae per square meter (with a meter being about 10 square feet)—and these are *big* animals. In all, the population was estimated at about 10 million medusae (Bolton and Graham 2004).

In areas of highest concentration, the jellyfish had an average of 1,651 prey items in their guts, dominated by bivalve larvae (presumably the commercial oyster farmed in the area), copepods, and fish eggs (probably anchovy).

To give an idea of the effect of all these hungry jellyfish on the local ecosystem, researchers studied what is called "clearance rate," or the rate of filtering or sweeping clear of food particles from a volume of water, per day. Clearance rate is used to estimate the grazing pressure of predators on prey. Imagine a single jellyfish in, say, a cubic meter (an area of about $3 \times 3 \times 3$ feet, or about 36 cubic feet) of seawater, with, say, 1,000 fish eggs scattered through it. Now assume that the digestion by jellyfish of fish eggs is 3 hours, and that its stomach can hold up to 125 fish eggs at a time. (In fact, a large jellyfish's stomach can hold far more. But in this scenario, the jellyfish is able to completely digest what's in its stomach every three hours, leaving space to eat more.) So in every 24-hour cycle it is able to eat that quantity of food 8 times, the resulting clearance rate being 1 cubic meter per day at that prey density.

In the case of *P. punctata*, their clearance rate in this bloom was calculated at an average of 92.5 cubic meters (121 cubic yards) per day for fish eggs, and 7.7 cubic meters (10 cubic yards) per day for bivalve larvae. Never mind the mathematics of it—what this means in practical terms is that these jellyfish ate a lot, digested fast, and ate more. They exerted enormous feeding pressure on the fish eggs, larvae, and other plankton of the local ecosystem, their rate of consumption far higher than the ecosystem was accustomed to sustaining. In areas of highest jellyfish concentration, the clearance rate for fish eggs was faster than once a day—at least an order of magnitude higher than the typical clearance rate.

In addition to the direct effects of predation load on the larvae, there was also some indication of an indirect effect on zooplankton production through changes in the chemical or physical properties of the water (Graham et al. 2003). Foam streaks observed on the surface emanating from the swarms were thought to indicate high concentrations of dissolved organic matter from the jellyfish. The researchers suspected that mucus shed by the jellyfish increased the viscosity of the water, making swimming more difficult for tiny organisms, and that the water also contained toxins from discharged stinging cells. Indeed, the researchers found that copepods from the swarm area were lethargic, had increased mortality, and produced fewer eggs than copepods taken from outside the swarm, presumably from living in thick, toxic water.

Copepods, bivalves, and fish eggs were not alone in being negatively impacted by the presence of the jellyfish swarms. For about two months during the bloom, commercial shrimping was severely affected by the fouling of nets and damage to the shrimp. Fisheries suffered more than a 25 percent decline that year due to interference by jellyfish, with the damage estimated to have cost up to $10 million. Apart from the havoc for shrimp fishermen, jellyfish also feasted voraciously on shrimp larvae, one of their preferred foods, reducing the survival rate for future catches.

Then came Hurricane Katrina in August 2005, the costliest hurricane in US history—and not only in dollar terms. Katrina killed at least 1,836 people in the hurricane and the floods that followed, while property damage was estimated at $81 billion. The full extent and duration of environmental damage has taken much longer to determine.

Erosion, Sewage, and Oil

The storm itself and the storm surge floods caused massive erosion to beaches and cliffsides throughout the affected Gulf Coast area. Sandy islands shifted. Bayous and wetlands were badly damaged. Bird habitats and sanctuaries were destroyed.

Already in decline before the hurricane, commercial fishing was pummeled by Katrina (Buck 2005). The hurricane hit during the peak harvesting season for shrimp, destroying or severely damaging many commercial shrimping boats and processing facilities and forcing large quantities of stored stock to be thrown out. Oyster beds and oyster vessels along the Gulf Coast were either extensively damaged or totally destroyed.

The vast waters flooding New Orleans were pumped into Lake Pontchartrain, which sandwiches the city against the Mississippi River. Like the river, the lake drains to the gulf. These pumped-out floodwaters were contaminated with raw sewage, bacteria, pesticides and other noxious chemicals, and oil. This toxic gumbo was puked straight into the gulf.

Furthermore, 44 refineries in southeastern Louisiana suffered oil spills caused by Katrina, resulting in over 26 million liters (7 million gallons) of oil being leaked (Llanos 2005). This was equivalent to about a third of the oil spilled by the *Exxon Valdez* in Alaska in 1989—but, unlike the single spill of the tanker, these 44 spills were scattered over a broad geographical area, making containment and cleanup far more challenging.

Regions affected included the gulf, as well as inland communities of freshwater fish and other wildlife. Inland areas account for much of the US-farmed catfish production (Buck 2005). Fish throughout Louisiana and Mississippi died in massive numbers. Some pundits blamed it on the storm surge pushing saltwater far inland through rivers and tributaries. But some fish found dead were anadromous, spending part of their life in the rivers and part in the sea, like salmon, and would be accustomed to different salinities. Others blamed the fish kills on pollution from flooded water treatment plants or storm runoff. Still others blamed it on rapid drops in dissolved oxygen. The precise cause of reduced oxygen in the inland waterways is not known, but it could have been the result of several factors (Schaefer et al. 2006). First, high temperatures would enhance decay of leaf and wood debris left by the hurricane, thereby spiking oxygen demand for the rotting process. Second, anoxic sediments in the gulf or rivers may have been disturbed by the storm surge of the hurricane. Finally, pollution from a number of sources could have reduced the oxygen levels.

Two weeks after Hurricane Katrina hit Louisiana, scientists from the National Oceanic and Atmospheric Administration began testing seafood samples for elevated levels of bacteria and chemical contaminants. Results showed heightened levels of hydrocarbons due to exposure to oil in the water, but in December 2005, the US government issued a multiagency announcement declaring gulf seafood safe to eat after extensive sampling and testing. As the first anniversary of Katrina approached, Carlos M. Gutierrez, US secretary of commerce, announced that gulf seafood was safe to eat, and that the yearlong testing program was concluding, with "great news . . . for all of us who enjoy seafood."

The initial jellyfish bloom simmered down for a few years after the big bloom of 2000, with few *P. punctata* being seen each year. However, by 2007, when the next massive bloom hit, the population had spread in both directions along the southern US coast, with reports as far west as Galveston Bay, Texas, east around Florida, and as far north as North Carolina (DISL 2007).

And then came the gulf oil spill of 2010, which President Barack Obama called the "worst environmental disaster the US has faced" (Obama 2010). People once used to talk about the March 1989 *Exxon Valdez* spill in Alaska in hushed tones and with teary eyes. It was bad—*really bad*. It killed a lot of birds and marine life, and now more than 20 years later, the region *still* hasn't recovered. But the gulf spill utterly dwarfed the *Exxon Valdez* incident. In fact, the Deepwater Horizon accident in the gulf spewed at a rate of one *Exxon Valdez* disaster every five days . . . *for three months*. By the time it was capped, the gulf spill had released nearly 5 million barrels—or more than 200 million gallons, nearly a *billion* liters—of crude oil into the local environment.

Within two weeks of the spill, evidence of environmental damage began surfacing, such as one report from Venice, Louisiana. In an emotive incident that left a river looking more like a gravel road than a shipping channel (see plate 8), thousands of carcasses blocked the entrance, including numerous species of fish, crab, stingray, and eel, and even a whale. It's enough to make you nauseous. Local government officials blamed it on the oil spill; state fisheries officials blamed it on low oxygen levels due to high temperatures and low tides. Either way, the ecosystem is screwed.

Epilogue

On 3 May 2010, just 13 days after the oil spill began, huge numbers of dead jellyfish and sea turtles washed up on Mississippi's shores. According to Larry Schweiger, head of the National Wildlife Federation, "it's not uncommon to see jellyfish floating dead during high winds, but the number of dead found so far is beyond normal" (AP 2010). By mid-May, scientists had discovered a "jellyfish graveyard" with a very large number of dead *Velella* (known as by-the-wind sailors) (WLOX 2010). By July, it was reported that masses of dead jellyfish were washing ashore, some soaked in oil, along with dead birds and mats of tar ("Tens of thousands of Jelly fish are washing ashore Florida East Coast" 2010). Because jellyfish naturally shoal and become beached, trying to determine which events are attributable to the spill, and to what extent, is challenging. Likewise, dismissing it altogether is tantamount to overlooking a serious problem.

Meanwhile, *P. punctata*, the spotted jellyfish blooming like crazy in the gulf, has since reappeared off the Mediterranean coast of Israel and off southern Brazil (Haddad and Nogueira Júnior 2006; Galil, Shoval, and Goren

2009), and has also shown up lately in the western Mediterranean (Boero et al. 2009). By July 2011, blooms of *Phyllorhiza* were vexing many beaches in Spain, and stings to more than 100 people required beach closures during the height of the tourist season (NewsCore 2011). Only time will tell if it gets the same toehold that it now has in the Gulf of Mexico, but it certainly sounds like it's off to a good start.

PART II

Jellyfish, Planetary Doom, and Other Trivia

Catches of wild fish are plummeting and the researchers predict that without steps to protect biodiversity, all current commercial fish and seafood species will collapse by 2050.

—CAROLINE WILLIAMS, "Jellyfish Sushi: Seafood's Slimy Future"

CHAPTER 4

Jellyfish: The Basics

Glimmering beings appear. They seem to be constructed of spider webs, fishing line and silk, soap bubbles, glow sticks, strands of Christmas lights and pearls. Some are siphonophores and gelatinous organisms I've never seen before. Others are tiny jellyfish.

—ABIGAIL TUCKER, "Jellyfish: The Next King of the Sea"

Chrysaora achlyos, the "Largest Invertebrate Discovered in the 20th Century" (California, since July 1999)

Off Southern California, a goliath jellyfish is increasing in its frequency. *Chrysaora achlyos* (see plate 1), the so-called black sea nettle, was reported in 1926, 1965, and 1989. This giant grows to a meter (3 feet) across the bell, with oral-arms to 8 meters (25 feet) long, and is the color of fine burgundy. Formally named and classified in 1997, it was declared two years later to be the largest invertebrate discovered in the twentieth century.

After the 1989 bloom that led to its description, *C. achlyos* bloomed again in massive numbers in July 1999, June–August 2005, and July 2010. It also bloomed in smaller numbers in the summer of 2001 and in August 2007 and 2012. Because the blooms spread northward on the inshore Southern California countercurrent, the species is thought to have its home somewhere in the waters off Mexico, and only to sojourn from time to time off California. In 1999, it was sighted off Monterey for the first time, some 420 kilometers (250 miles) north of Los Angeles, where it has since become a more or less annual visitor.

The fact that a species of this conspicuous size and color was found at Los Angeles where the sea is so much used and well studied, yet was only identified and classified as late as 1997, is nothing short of remarkable. This is as good an example as anyone can find to demonstrate that the oceans still remain largely unknown. And its more frequent appearance and more northerly appearance provide a big, colorful example of shifting populations.

Weeds. Fast-growing, hardy, adaptable, tenacious . . . Dandelions. Nightshade. Ragweed. Foxtails . . . Rabbits. Pigeons. Cockroaches. Mosquitoes. Flies . . . Jellyfish. Yes, jellyfish.

We are accustomed to thinking of weeds in terms of plants, but animals can be weeds, too. Weediness is a lifestyle. It is a survival strategy in harsh or unpredictable environments. Weeds persist through changes around them, because their lifestyle rolls with the punches. Too hot or too cold? They are the last ones to wilt. Inhospitable habitat? Not for weeds. Shortage of food? No worries, they will find something, somewhere, somehow. They are the survivors. New opportunities open—a volcanic lava flow or a flooded valley or an oil spill—weeds are the first on site, the first to sprout or burrow, the fastest to colonize.

What Makes Weeds Weedy?

Certain lifestyle characteristics are shared among weedy species. Generalists. Opportunistic. Versatile. Hardy. Tolerant of a broad range of ecological conditions. Will eat just about anything. Short life cycle. Prolific. They readily disperse. Resist eradication. It's these features that give weeds their edge. It's what makes weeds weeds. One of the most important features of weeds is that they thrive in disturbed habitats. When ecosystems wobble, weeds flourish.

Dandelions as Weeds

Consider the dandelions. Who can help but break a smile with the dandelion story in David Matz's book, *Chicken Soup for the Gardener's Soul*, where the father of a young girl is laboring over a scourge of dandelions in his would-be perfect lawn. Screwdriver in hand, sweat on his brow, determined to rid his yard of the infernal pests. After several attempts to get his attention, and he stalling "just another moment" to win over the pesky weeds, the young girl comes up behind him, says, "Make a wish, Daddy," and blows a thousand baby dandelion seeds over the yard.

A thing of fascination for children, intensely hated by adults, especially gardeners, dandelions are ubiquitous. Each "flower" is actually a composite of many flowers grouped into a single head. After flowering, the fluffball of seeds becomes the secret to its weedy success. The feathery parachutes ride the wind and, when they settle, they can rapidly colonize just about anywhere. They have broad tolerances to temperature, humidity, and soil types. Many species of dandelions are able to produce seeds asexually, that is, without pollination, producing clones of the parent plant.

In truth, dandelions are "beneficial weeds." All parts of the plant are edible, the flowers have been used in traditional medicines and are good at attracting pollinating insects, and the taproot brings nutrients upward for shallower plants. And of course, dandelions bring smiles to children's faces.

Cockroaches as Weeds

Like dandelions, cockroaches are notoriously weedy (but they bring notoriously fewer smiles to children's faces). Consider the common belief that cockroaches will survive nuclear war. In truth, they will survive a bit better than we wimpy humans, but they are still total lightweights in comparison to some other insects and bacteria. It all comes down to radon units, or "rads." Humans will die from exposure to some 400–1,000 rads. "Wood-boring insects and their eggs were able to survive doses of 48,000 to 68,000 rads with no apparent ill effect . . . it took 64,000 rads to kill the fruit fly, and a colossal 180,000 rads to be sure of killing the parasitoid wasp" (Dr. Karl 2006). Most cockroaches, in comparison, will die with a dose of a mere 6,400 rads.

Okay, nuclear war aside, cockroaches are pretty hardy. Over the last 300 million years of clicking along this earth with those fiendishly fast little legs, they have evolved the ability to live more than a month without food or water, or a couple of weeks without a head, and can survive steaming hot water, being frozen, or being deprived of air for 45 minutes. And some can hiss or chirp when upset. Eew, ick.

They are typically omnivorous, and can survive on *anything*—including

the adhesive on the back of postage stamps. Some species are parthenoge-
netic, meaning that the females can reproduce without the need of a male.
Other species can mate just once, then the female carries that sperm supply
to fertilize eggs for the rest of her life. A typical female cockroach will lay hundreds of eggs in her lifetime, but some species will lay far more.

Jellyfish as Weeds

Dandelions and cockroaches are weeds because they have weedy lifestyles and
weedy biological traits. So too do jellyfish. We don't usually think of jellyfish
as "weeds," but they are. They share the same weedy qualities that are the
essence of weediness in dandelions and cockroaches and their other weedy
brethren. They are highly tolerant of a broad range of environmental conditions; they grow fast, breed early and often, and have a large number of young;
and they will eat just about anything that they can get their lips around.

Understanding Weeds

The reason that understanding weeds is so important is in what ecologists call
"r and K selection"—a fancy name for a fundamentally important concept
that explains the trade-off between quality or quantity of offspring in stable
or unstable environments. It explains who survives, and how. Species tend to
adopt one of two fairly distinct life history strategies: either they are opportunistic or they are near equilibrium. Boom and bust or stable state. Weedy or
not weedy.

Equilibrium species (i.e., K-selected species) tend to be long lived, slow
growing, and have few young at a time to which they invest parental care.
Their emphasis is on living close to carrying capacity in order to maximize
their competitive advantage in stable and predictable environments. Whales.
Lions. Penguins. Redwood trees. Stable state.

Opportunistic species (i.e., r-selected species) tend to be short lived, fast
growing, and have many young at a time to which they invest little or no parental care. Their emphasis is on high population growth rate through pulses
of sheer volume, in order to maximize survival in unstable or unpredictable
environments. Rats. Mosquitoes. Cockroaches. Jellyfish. Weeds. Boom and
bust.

The weedy features of the jellyfish lifestyle are surprisingly simple but have equipped their owners with the ability to take advantage of a vast array of ever-changing conditions. Fossil evidence tells us that the jellyfish body plan dominated the planet for countless millions of years before predators with shells or teeth evolved.

You'd be forgiven for thinking that jellyfish wouldn't fossilize very easily. Yeah, it's not quite like the fossils of a big meaty dinosaur bone. Jellyfish fossils are more like footprints, mere impressions in the sediment. But some have splendid detail. The striking thing about jellyfish fossils is that the forms and families haven't changed much. Through the eons, while trilobites and dinosaurs came and went and plants and animals moved onto land and evolved respiratory machinery and mammals evolved bigger and better brains, jellyfish stayed the same.

Jellyfish survived through all the extinctions, even the "big five" mass die-off events, and among all the tried-and-failed species and genera and families in the sea, jellyfish are still around. The jellyfish life style works. While everything changed around them, they didn't need to change to survive. They survived conditions that drove others extinct and stimulated evolution of entirely new forms. And now, in our rapidly changing oceans and climate, jellyfish are again experiencing a renaissance.

Understanding Jellyfish

In order to better understand jellyfish blooms and their growing importance, this chapter is devoted to making sense of some of the basics of jellyfish biology and ecology.

Types of Jellyfish

Roughly 1,500 species of jellyfish have been named and classified, but this is certain to be a considerable underestimation of their true biodiversity. New species are still being discovered quite frequently. Most jellyfish are in the phylum Cnidaria (pronounced nye-DARE-ee-uh), along with the corals, sea anemones, sea fans, and hydras. Their phylum name comes from the Latin root word *cnida*, meaning nettle, in reference to their stinging cells—and their stings. A lesser number of jellyfish belong to the phylum Ctenophora (pronounced teen-OFF-uh-ruh), characterized by having eight rows of cilia along the body. No heart, no gills, no brain, no bones. Members of both groups are essentially a gelatinous body with one or more mouths for ingesting food, one or more stomachs for digesting food, and usually 4 or 8 gonads for making more jellyfish. Simple but effective.

Jellyfish come in some of the wildest imaginable shapes and color patterns, as hinted by their often evocative common names, including moon jellies, comb jellies, rainbow jellies, box jellies, fire jellies, sea wasps, sea nettles, sea tomatoes, sea walnuts, sea gooseberries, Venus's girdles, lion's manes, pink meanies, purple people eaters, blubbers, snotties, agua vivas (living water), agua mala (bad water), blue bottles, blue buttons, blue stars, Portuguese man-o'-war, by-the-wind sailor, and—my favorite—the long stingy stringy thingy (see plates 1 and 2).

Okay, please pardon the momentary diversion, but the last one in the list requires some elaboration. The long stingy stringy thingy is actually a group of species, not just one. They are weird colonial creatures related to the Portuguese man-o'-war. Yes, they sting: they'll make a grown man—or woman (she says, with wincing recollection)—cry. But in the water (and eh-hem, from a distance), they are a magnificent thing to behold. They are usually pink, or sometimes yellow and red. Some can grow up to 50 meters (150 feet). When hungry, they relax their thousands of long, gossamer feeding tentacles into an invisible wall of sting. And when they bloom, the water column becomes for other species like a suicidal charge into the Battle of Gallipoli.

While most people tend to group jellyfish into one of two categories— the stingy ones and the *really* stingy ones—there are actually a few more legitimate groupings than that. One of the most easily identifiable jellyfish, the Portuguese man-o'-war, belongs to a group called the "siphonophores." Siphonophores are among the absolute strangest imaginable creatures. Not quite an individual. Not quite a colony. And so, for over 150 years, many of the greatest minds in evolutionary biology have debated the proper status of the siphonophores.

Siphonophores have a gas-filled float, or swimming bells, or both. All sting, some quite fiercely. And all are quasi-colonial, with the "individual colony" (to combine two mutually exclusive concepts) being composed of repeating groups of members properly referred to as "persons": food-catching persons, digestive persons, defensive persons, reproductive persons, and swimming persons. These persons and groups are not able to live independently, thus the argument that they comprise a colony; but they function autonomously, thus the argument that they are an individual.

Some siphonophores have been causing stinging plagues at fish farms, such as the dainty *Nanomia* in the Gulf of Maine and the much larger and stingier *Apolemia* in Norway. As for the fearsome man-o'-war, its name sounds menacing, its sting hurts like hell, and its crested blue float elicits the same dread as the "dun-dun, dun-dun" from *Jaws*. Up to 40,000 stings a year are treated in Australia from its little cousin, the blue bottle—but that's just child's play compared to the *half million* man-o'-war stings a year along the east coast of America (Gershwin et al. 2009). And the *Telegraph* reported on 12 September 2009 that unprecedented numbers of man-o'-wars were washing up on British coasts.

The so-called box jellies are another easily identifiable group. They're box-shaped, with single or groups of tentacles coming off each of the four lower corners. Most of the highly dangerous species are in this group. Sometimes they swarm. For example, a May 2005 story in the *Philippine Star* reported an incident where 127 policemen were "attacked" by box jellyfish off Sebonga, near Cebu. The officers were on a counterinsurgency exercise that involved their wading chest-deep into the sea. "Before they could escape, they came under intense attack not by insurgents, but by hundreds of jellyfish" (Perolina 2005). The stings caused severe itching and rashes, along with vomiting, dizzy spells, and high fever, requiring transfer to hospital. They had every right to worry; it has been credibly estimated that box jellyfish kill up to 50 people each year in the Philippines.

Irukandjis are also in the box jellyfish group. Their stings are the ones to fear the most. Not all that long ago, Irukandji stings were considered a strange tropical Australian nuisance. Then Hawaii . . . Florida . . . Caribbean . . . Japan . . . Thailand . . . Malaysia . . . India . . . South Africa . . . Wales . . . credible reports of similar stings started popping up all over the place. But perhaps the most intriguing are the ones in paradise. Hawaiian coastlines have long held the public imagination as a place of beauty, drama, and romance. But you don't generally expect to share the balmy beaches with amorous jellyfish. Hawaiian Irukandjis spawn close to shore on the eighth through the tenth nights after the full moon. During these days, large numbers of them swim in the shallows and wash up on the beaches . . . and sting a lot of people. Irukandji syndrome isn't usually fatal, but it can be, and recall from chapter 1 that, at the very least, it will ruin your vacation.

The so-called blubber jellies and saucer jellies are another group. These are the more "jellyfish-looking" jellyfish—that is to say, they have the shapes that we most often think of as the classic jellyfish: dome-shaped body with long tentacles. Most of the species causing bloom problems are from these groups: moon jellies, sea nettles, lion's manes, spotted blubbers, nomadic blubbers, tomato blubbers, upside-down jellies, and mauve stingers (a.k.a., the purple people eater or the pink meanie).

The so-called water jellies, or hydromedusae, are typically small and transparent, more or less invisible. They are an incredibly diverse group. In the right conditions, they can bloom to the extent that it becomes harder to swim—the water seems less like a soup and more like a thick chowder. Although they are invisible, they are easily detected by their stings, which feel like "sea lice." While unpleasant, they are rarely more than a nuisance to humans. To fish and invertebrates, however, their blooms can be just as devastating as those of larger, more visible species.

Another type of jellyfish better known for its stings than anything else is called "seabather's eruption"—Ick! It sounds pretty unpleasant, and it is. *Linuche unguiculata* occurs in shallow, tropical waters, where it farms its symbiotic algae. Occasionally they gather in the millions in natural breeding aggregations, particularly in Florida, the Caribbean, and the Gulf of Mexico. When their larvae get caught under clothing and then exposed to freshwater, they cause a ghastly arrangement of small, red, polka-dot welts that can itch intensely for a few days, or up to a couple of weeks.

Finally, there are the ctenophores, or "comb jellies," which are just plain weird. There's no dancing around that fact. Everything about them is weird. Their body forms range from those that look like a grape with two tentacles to those that look more like a set of praying hands, or the batman logo, or a long sinuous belt, or a Klingon attack vessel. No kidding. The coordinated beating of the eight cilia rows, for which the phylum is distinguished, refract available light into brilliant rainbows of color as they move the animal through the water like a tractor. Ctenophores are evolutionarily blessed with having both a very primitive sort of central nervous system and the earliest form of a through gut. But showing that even evolution may have a sense of humor, their "anus" is wrapped around their "brain" (oy, the jokes that come to mind!). *Mnemiopsis* and *Beroe*, both of Black Sea infamy, are in this group.

Do All Jellyfish Sting?

Yes and no. The ctenophores don't have stinging cells, so they aren't capable of stinging, so "no." Of the cnidarian jellyfish, which are the ones that purists would tell you are *really* the jellyfish, "yes," inasmuch as they all have stinging cells, but "no," they don't all cause us pain when their stinging cells inject their venom. Some stinging cells are too short to penetrate our skin, or some venoms are too mild to cause us discomfort. But a larval fish or small plankton would feel the effects. So it really depends on how you define "sting."

Cnidarians have stinging cells (cnidae), whereas ctenophores lack stinging cells but instead have sticky cells (called "colloblasts," in case it ever comes up in Trivial Pursuit). The cnidae and colloblasts are both used in prey capture and defense, and they work in a similar way: a long thread is coiled tightly inside a tiny capsule, and is shot out at ultra-high speed and force (about 40,000 Gs, if you can imagine it). While a poisonous harpoon shoots out of the stinging cell, the contents of the colloblast are more similar to a rope covered in honey.

Jellyfish Large and Small

Some jellyfish species can get BI-I-I-IG. *Chrysaora achlyos*—widely hailed as the largest invertebrate discovered in the twentieth century—grows to a meter (3 feet) across the bell, with its entwined fleshy oral-arms forming a thick column 8 meters (25 feet) long. *Nemopilema nomurai*—often compared to sumo wrestlers or refrigerators—grows to 2 meters (6 feet) in diameter and over 200 kilograms (450 pounds). But the king daddy of them all is *Cyanea arctica* of the North Atlantic, said to reach more than 2 meters across the body.

Some jellyfish at full maturity barely even reach a millimeter (less than the thickness of a dime). *Csiromedusa medeopolis* is the size of just a few grains of sand, but is weird as weird can be with its gonads sticking out the top of its body like a cluster of icebergs or skyscrapers. Or consider *Turritopsis lata*—virtually invisible to the naked eye if not for its bright red stomach about one-fifth the size of a grain of rice.

Where Jellyfish Live

As a group, jellyfish inhabit all depths of the oceans from the seafloor to the air-water interface, all latitudes from the poles to the tropics, and all seasons of the year. As one might imagine, different species inhabit different regions and conditions. While most jellyfish are marine, a few species are native to freshwater. Most people are surprised to learn that freshwater jellyfish exist, and indeed, they are typically an object of great interest when found.

Sex and Cloning in Jellyfish Reproduction

Jellyfish sex is like something straight out of science fiction, except that there is nothing fictional about it. The methods jellyfish use for reproduction are beyond the realms of Hollywood, and the numbers involved may seem unbelievable. Millions of moon jellyfish aggregate into a massive orgy, same time, same place, every day. For months. Tens of thousands of eggs is not uncommon . . . per jellyfish. Per day. Every day. For months. Hermaphroditism. Cloning. External fertilization. Self fertilization. Courtship and copulation. Fission. Fusion. Cannibalism. You name it, jellyfish do it while they're "doing it."

Most jellyfish have both a pelagic (drifting), sexually reproducing medusa phase and a benthic (bottom-dwelling), asexual polyp phase, as well as various larval stages, in their life cycles. These distinct elements of their life histories and the ecological interaction between the two phases are largely the reasons behind their success through the eons. Jellyfish are, in the very essence of the word, weeds.

Clonal reproduction in and of itself is not uncommon among living things, particularly plants and lower invertebrates. But jellyfish cloning is beyond most people's wildest dreams. Jellyfish are able to clone—that is, reproduce asexually—in at least 13 different ways. While some may find learning about cloning interesting in and of itself, it is particularly relevant to the understanding of jellyfish blooms. The polyp stage is essentially the seed bank of bloom potential, and the ways the polyps can proliferate is the key to their ability to bloom in such incredibly rapid fashion and shocking numbers.

Their modes of sex aren't the only parts that may seem strange to us. Their whole life cycle, commonly known as an "alternation of generations," is different from anything that most people have ever heard of. The most easily recognizable comparison would be the butterfly and caterpillar, but it is not a perfect analogy, because the caterpillar is basically a larval butterfly and actually metamorphoses its whole self into the butterfly. When a butterfly lays eggs, the butterfly continues to exist, whereas when a caterpillar metamorphoses into a butterfly, the caterpillar no longer exists. Jellyfish, on the other hand, have two adult stages, that is to say, neither "becomes" the other, but rather one "spawns" or "gives off" the other. In layman's terms, when jellyfish have babies, they don't grow up to look like jellyfish, they grow up to look like tiny coral polyps or sea anemones; and when the polyps have babies, they grow into jellyfish. The "jellyfish," that is, the medusa, is the free-swimming dispersal stage and represents the sexual part of the life cycle—males and females producing sperm and eggs. The polyp is an attached stage and represents the asexual part of the life cycle—reproduction by cloning. So each generation—the medusa and the polyp—alternate with each other.

As is the case with sea anemones and coral polyps, the perennial jellyfish polyp is stuck to the bottom, usually on rocks or shells or other hard substrates. Man-made structures, such as the undersides of docks and boats, the pilings of oil rigs and bridges, sea-based wind farms, discarded waste, and other flotsam and jetsam, all make wonderful new habitats to be colonized by whoever gets there first. Jellyfish polyps, being the weeds that they are, often get there first.

Whether the polyp stage truly represents an "adult" stage or whether it is merely a developmental stage of a larva has been debated for centuries. But like many things in science, trying to shoe-horn something with its own unique way of doing things into a familiar construct is often futile.

Strobilation is another strange feature of jellyfish sex . . . well, jellyfish "asex," actually. Strobilation is where the polyp undergoes a partially metamorphic process, whereby it elongates and differentiates into a stack of discs, like a stack of tiny dinner plates or a roll of tiny coins, through a process of transverse fission. These discs develop into tiny, daisy-shaped larval medusae, then begin pulsating and eventually break away to become free-swimming. Then they start eating. A lot. And they grow. Fast.

When jellyfish polyps aren't strobilating, they are very busy making more little clones of their polyp selves. They can build vast colonies of polyps, all waiting until the conditions are right to strobilate. And being perennial, they can wait a long time if they need to.

The size and density of the bloom is determined by some ethereal equation of factors including number of polyps, health of the polyps, percent of polyps strobilating at the same time, and chance timing of the strobilation. The number of polyps may well be the key factor in bloom dynamics. Quite simply, more polyps means the potential for more medusae. While some polyps produce only one ephyra (baby jellyfish, or more accurately, larval jellyfish) at a time, most produce more. Moon jellyfish typically produce about 7–12 per strobila, while sea nettles typically produce about 50.

Future environmental conditions are unpredictable, such that temperature, salinity, and food availability may not be conducive to ephyra and medusa survival and growth. Therefore, health of the polyps is important, because healthier polyps produce more ephyrae, with a range of features suited to different conditions, which means that at least some have a better chance of survival. The percent of polyps strobilating is an interesting phenomenon, about which very little is known. In some years almost 100 percent strobilate in one big pulse, while in other years fewer than 10 percent are triggered, or the strobilation event is drawn out over time. Oh, jellyfish sex is *so* complicated!

The short generation time of many jellyfish species allows them to take advantage of this seed bank effect of polyps waiting to strobilate, where they can often respond almost immediately to ecological changes. Even in undisturbed ecosystems, jellyfish come and go seasonally as their populations track the spring and summer phytoplankton and zooplankton blooms.

Jellyfish have perfected the art of survival to a degree far surpassing that of most other types of living things. Their delicate body form means that the "physical individual" may not survive rough weather, or a stranding event, or predation or famine, but their clonal nature means that the "genetic individual" is more likely to survive through one or more of its clonal units.

Around the world, we are increasingly witnessing jellyfish blooms—more often, bigger in geographical coverage and longer in seasonal duration. They're stinging swimmers, clogging power plant intake pipes and fishermen's nets, getting chopped up in boat propellers, and depleting fish. But these visible effects are just half of the problem. The benthic polyp—the hidden other half—is churning out vast numbers of juvenile jellyfish as we are left scratching our heads, wondering what to do about it.

In order to understand jellyfish bloom dynamics, we must consider the physiological and biological effects of the environment on both parts of the life cycle. There can be no doubt that the number of medusae in a bloom is tightly dependent on the number of polyps to begin with, but also that the number and distribution of polyps is a function of medusa survival and reproduction. Jellyfish blooms cause the problems, but polyps are how they can.

The questions of which stage is the *most* responsible for the increases in blooms around the world, and which stage requires the most management, is a bit like asking which came first, the chicken or the egg. The two stages are so closely linked that they must be studied in tandem to be truly understood.

In the words of Claudia Mills, "Knowledge about the ecology of both the medusa and the polyp phases of each life cycle is necessary if we are to understand the true causes of these increases and decreases, but in most cases where changes in medusa populations have been recognized, we know nothing about the field ecology of the polyps" (Mills 2001, 55).

Jellyfish may be thought of as the "invisible variable" in ecosystem decline. However, polyps are even more invisible. They're small. They're cryptic. They're sneaky. They're hardy. They're very weedy.

Life, Death, and Immortality

Here's where jellyfish start getting really weird. Medusae have been confirmed to live for over a decade. But the polyps, being clonal, are essentially immortal. Consider this: any given polyp can live at least 14 months, but of course, even after a polyp dies, its clone mates live on. This is demonstrated by a polyp colony started in 1935 by the naturalist Frank J. Lambert in Essex, England, that is still alive and well under the care of Dr. Dorothy Spangenberg in Virginia. Some of these original polyps' genetically identical descendants have even gone into space in scientific experiments aboard a US space shuttle (Spangenberg et al. 1994).

Okay, so perhaps simple cloning may be stretching the word "immortal" a bit thin. But consider the demure but fascinating *Turritopsis dohrnii*. When the medusa dies, the cells begin to dissociate like any normal organism—that is, it disintegrates. But then something remarkable happens. As the medusa body decomposes, the cells reaggregate and transform into new hydroid colonies (Piraino et al. 1996). The whole transformation from medusa to polyp takes place within a mere five days or so of the medusa's death. This would be roughly the equivalent of a dead butterfly's cells reforming, all on their own, into a full-grown, fully formed caterpillar. Bizarre. This weird aspect of the reproductive cycle of *Turritopsis* is normal for the species; all medusae undergo the process of transformation and never actually experience full demise. This is the first known example of true biological immortality.

How and What Jellyfish Eat

Most jellyfish eat by capturing prey on their tentacles. Cnidarian jellyfish use their stinging cells for this purpose, essentially stunning or killing their prey with venom and a harpoon injury, while ctenophores ensnare their prey in their sticky threads.

While cnidarian jellies create a big commotion with their pulsating movements as they swim, ctenophores move through the water with virtually no turbulence whatsoever, and thus are able to approach and acquire prey completely undetected . . . until it's too late (Colin et al. 2010). Stealth mode is used for approaching and ensnaring prey, but getting prey into the mouth is not so "quiet" for most ctenophores. Some species have evolved with the tentacles away from the mouth and exhibit an amusing twirling and tumbling behavior as a means of draping the tentacles across the mouth to transfer the food to the lips. Other species, such as the pocket-shaped *Beroe*, have evolved special "teeth" for biting other ctenophores into stomach-sized chunks—jellies with jaws. If only Hollywood knew. . . .

We would be horrified at the thought of being eaten by a frog or a clam. There is just something about the food chain that prevents that from happening. We are accustomed to thinking of the food chain in the traditional terms of the following rules:

- Big things eat little things.
- Faster things eat slower things.
- "Smarter" things with bigger brains eat "dumber" things with smaller brains.

These "rules" are so intuitive and pervasive that Shakespeare used this concept to illustrate his argument that rich and powerful men exploit those working for them, like the whales in the sea eating the small fish (*Pericles*, 2.1.69–70):

THIRD FISHERMAN: "Master, I marvel how the fishes live in the sea."

FIRST FISHERMAN: "Why, as men do a-land; *the great ones eat up the little ones.*"

In other words, the higher something is on the evolutionary tree, the higher it should be on the food chain. Most of the time, this is more or less true.

But jellyfish are the glaring exception to these rules. Small jellyfish eat big species of clams and crabs and hard, bony starfish. Slow jellyfish eat fast species of fish and squids. Jellyfish with no brains eat species of snails and crustaceans and fish with brains. Not only do they eat them, but they outcompete them. How, you ask? Whereas most creatures feed "down the food chain," jellyfish feed "up the food chain." Jellyfish target the eggs and larvae of species higher on the food chain than themselves. And they also target the food that the larvae would eat. This is how jellyfish take over entire ecosystems, as both predator and competitor of species bigger and faster and smarter than themselves. This "double whammy" effect is discussed in more detail later in chapter 12.

Jellyfish as Carnivores

It is easy to be fooled by jellyfish. They are low on the phylogenetic tree—"primitive," you might say. And they are basically just blobs of jelly without a brain, so you may think they don't look too fierce up against species with shells and claws and hard skeletons and sophisticated brains. But you'd be wrong.

Despite their primitive form, jellyfish are fearsome predators indeed, and can consume more than half their body weight in food per day (Larson 1987). With each pulse of their bell, each sway of their tentacles, each ripple of their oral-arms, jellyfish are catching larvae and eggs of fish, crustaceans, mollusks, echinoderms, and other creatures that make up the fauna of the sea.

Jellyfish are at the top of the food chain, despite their primitive and blobular appearance. They will eat the eggs and larvae of anything they can get their mouth around, and in some cases, even that's not necessary. Sea nettles can digest fish almost as large as themselves in the external frills of their oral-arms, the long, ruffly, fleshy structures that hang from between the tentacles around the central mouth.

Most of the invertebrates and many of the vertebrates in the sea have planktonic larvae, that is, their larvae spend some portion of their early lives drifting on the ocean currents. But the water column—that part of the water between the benthos at the bottom and the air-water interface at the top—is where the jellyfish rule. And when jellyfish occur in massive swarms, they can clear the water of all eggs, larvae, and small invertebrates, such as copepods and other teeny tiny little crustaceans that occur in vast numbers throughout the oceans. And jellyfish can clear the water fast—in many instances, in less than a day. The true effect of these huge blooms is only just beginning to be understood.

Jellyfish as Medusivores

Jellyfish not only eat and outcompete the more sophisticated organisms in the community, but many also have the unique ability to survive in their own ecosystem, devoid of fish, crustaceans, and mollusks. Many types of jellyfish are obligate or opportunistic predators on other jellyfish. This somewhat independent food web has been called "the jelly web" (Robison and Connor 1999). So most jellyfish don't *need* crustaceans and mollusks and fish to prey upon. They will survive just fine without them. But they are nice. Think of them as dessert. Or perhaps canapés.

Symbiotic Species

Some species of jellyfish don't even need to acquire food to eat. In fact, many species of jellyfish, including most of the so-called blubbers, have symbiotic algae living in their tissues, similar to the arrangement between corals and their symbionts. Think of them as farmers . . . well, farmers that sting.

Darwin's upside-down jellyfish. It sounds like some kind of Frankenstein-ian evolutionary experiment, but in truth, it's just another case of jellyfish behaving badly. A child was severely stung in the face while swimming in a reputedly safe swimming lagoon in the city of Darwin, northern Australia. The lagoon had become infested with jellyfish that often inhabit quiet, shallow waters, where they rest on the bottom upside down, farming their symbiotic algae like corals do. In this case, the species turned out to be new to science. Parks and recreation staff raked in 75,000 . . . 40,000 . . . 25,000 . . . 108,000 jellyfish *per day*, many of them the size of dinner plates. After great effort and expense, the jellyfish were brought under control, but the swimming lagoon was more or less unusable. So in 2008, another swimming lagoon was constructed, newer, more beautiful . . . and jellyfish-proof. But even before it opened, the same species found it to be a comfortable new home and was blooming out of control. "Aw, bugga!"—to borrow a common Aussie phrase.

Are Jellyfish Good for Anything?

Keeping in mind, of course, that there are no inherent laws in nature stipulating that an organism must actually be "good for something" as a requisite for existence, some jellyfish do confer benefits that we consider favorable. In China, jellyfish have been used as food for over 1,700 years, as described by the Chinese philosopher Zhang Hua (AD 232–300) in his *Natural History* during the Tsin dynasty (Omori and Nakano 2001). The use of jellyfish as food appears to have spread from China to Japan, where it is now commonplace. Prior to 1970, jellyfish fisheries remained modest, with Japan importing most of its processed jellyfish from China. Because of increasing demand, better collection and processing techniques were developed in the 1970s, opening trade opportunities with countries throughout Southeast Asia, as well as to the United States, Australia, India, Turkey, and Mexico. From 1988 to 1999, the annual worldwide catch of jellyfish was estimated at about 321,000 tons (wet weight).

To put that into perspective, that's almost half the annual global catch of Atlantic mackerel (844,164 tons), and more than the totals of Pacific herring (353,068 tons), Nile perch (283,704 tons), albacore tuna (264,570 tons), American sea scallop (309,278 tons), or the blue swimmer crab (172,189 tons) (FAO 2002). A decade ago, the global value of jellyfish fisheries was worth $80 million (Kingsford, Pitt, and Gillanders 2000); today they are worth even more.

Besides being useful as a food source, jellyfish have medical benefits. While we generally associate jellyfish stings with being harmful and unpleasant, at least one species, *Aurelia aurita*, has been used successfully to treat neuralgia and rheumatic pains (Russell 1970). At a thermal spa in Sandifjord, Norway, this jellyfish was taken by its inoffensive convex side and brushed against the affected parts of the patient, producing excellent results.

One of the most brilliant examples of jellyfish being useful is the discovery and development of the green fluorescent protein (GFP) by Osamu Shimomura, Martin Chalfie, and Roger Y. Tsien, for which they were awarded the Nobel Prize in Chemistry in 2008. GFP was originally isolated from a jellyfish species called *Aequorea victoria*, but has since been found in many others. GFP glows vibrant, neon green under ultraviolet light, allowing it to be used as a marker to study interactions of proteins inside cells without destroying them. For example, it is being used to study the cellular processes associated with cancer, HIV, Alzheimer's, and development of nerve cells in the brain. Since its original discovery, GFP technology has expanded to a kaleidoscope of colors, allowing researchers to track the behavior of multiple proteins simultaneously.

Some may say "so what?" to glowing jellyfish, but GFP is one of the truly "wow" discoveries of our time. It has been described as "having a similar effect on neuroscience as Google has on cartography" (Zimmer 2011a). Indeed.

Consider "Brainbow," a transgenic process being used on mice that individually marks their neurons from a palette of 90 distinct hues and colors, enabling scientists to map the neural circuits in the brain (Zimmer 2011b). Researchers hope that the resulting wiring diagram will help them identify the defective wiring in neurodegenerative diseases, such as Alzheimer's and Parkinson's.

Consider too the axolotl, a type of large salamander that lives in the Aztec canals of Mexico City. Axolotls have the ability to regrow injured or missing limbs, skin, organs, and parts of their brain and spinal cord. Scientists at the Max Planck Institute have created a transgenic GFP form of the salamander; they can now observe its green-glowing regeneration processes through its clear skin (Zimmer 2011c). Experts predict that the biological technique used by the axolotl will be understood and usable by humans to regrow missing limbs within 10 to 20 years.

Following from the glowing proteins and Nobel Prize, it now appears that at least one company is marketing a memory-improvement and antiaging/antistroke/anti-Alzheimer's product made from jellyfish protein (Prevagen 2011). Alas, be dubious: the product never gained approval from the Food and Drug Administration and seems to have been discontinued.

Other Peculiar Jellyfish Features

Clonal Variation to Cope with Unpredictable Conditions

The offspring of most living things take their chances in unpredictable environments. If food supply is good, the young may survive and grow. If food supply is poor, competition will be fierce, and many will starve. But jellyfish have a way of hedging their bets against this uncertainty. In many jellyfish species, "babies" are "born" in clonal groups. Recall the strobilation process, whereby the polyp divides into a stack of discs, each of which develops into a juvenile jellyfish. Each one is genetically identical to the others, but may look very different. Imagine a round pizza cut into 4 equal "slices," or 5 or 6, or 3 or 2. Jellyfish are based on a similar radial symmetry, where the standard tetra-radial (four-parted) plan has 4 equal quadrants, each with 1 gonad encircling 1 stomach pouch, 1 oral-arm that is used in feeding and reproduction, 2 sensory knobs (called rhopalia), and a set of radial canals, which act like veins and arteries to distribute nutrients around the body. Other forms keep these ratios identical but have greater or fewer than 4 "slices." Hexa-radial (six-parted) forms have 6 gonad/stomachs, 6 arms, and 12 rhopalia, while a tri-radial (three-parted) form has 3 gonad/stomachs, 3 arms, and 6 rhopalia. All variants from 1 to 8 "slices," as well as congenitally joined twins and other unexpected configurations, have been documented.

The significance of the number of radial slices is that these different forms are "born" as clone mates to each other, at the same time, from the same polyp (Gershwin 1999). But they have very different ecological roles. Those with fewer sense organs pulsate less often than those with more, but those with more also have more gonadal material to produce more sperm and eggs. When food is plentiful, many of the clone mates survive, regardless of whether they have 4 "slices," or more, or fewer, and all produce sperm and eggs to make lots of "babies." But when food is scarce, those with more "slices" use too much energy, pulsating more frequently than their counterparts with fewer "slices," and the more sluggish forms have a competitive advantage. From the clone's point of view, at least its genes get passed on, regardless of future food availability.

Growth and Degrowth

If food supply gets really scarce, many jellyfish have a backup plan to their backup plan. In times of famine, jellyfish can go without food for a very long time, switching into a process called "degrowth" (Hamner and Jenssen 1974). Starving jellyfish consume their own body mass very slowly, becoming smaller and smaller, until food is once again available. When they start eating again, they rapidly recover to their normal size—with no ill effects. Throughout the degrowth period, they remain reproductively active and look and act like normal jellyfish, despite becoming progressively smaller. This impressive process has been documented to involve animals up to 18 centimeters (7 inches) in diameter that degrow to as small as 1.4 centimeters (half an inch) in diameter, then grow back up again, the whole process taking about 120 days.

This amazing process is essentially a way of reducing overhead when times get tough. We, by contrast, are bound by our skeletal limitations: we can get thinner but not shorter. We can reduce our girth to some extent, but when our fat reserves have been exhausted, we use up protein body mass from muscles, including the heart muscle. Bad news. The jellyfish equivalent would be not becoming a thinner version of ourselves, but actually becoming child-sized again, then regrowing to adult-size, all the while remaining perfectly healthy and fertile.

Staying reproductive throughout this process is an essential feature of its success. Organisms live within a nested range of parameters, with the innermost range being a narrow niche of reproductive viability. A somewhat broader niche defines their parameters of survival, beyond which that species cannot thrive. For example, a temperate jellyfish species may thrive and reproduce in temperatures between 13–18°C (55–65°F), whereas it may survive but lack the energy to grow or reproduce in temperatures within a few degrees on either side of that range, beyond which it cannot survive at all.

This is just one of the many simple ways that jellyfish have evolved for coping with unpredictable conditions. It gives them a means to "stay in the game" as a full reproductive player while other species have to take some time out just to stay alive. In the competitive world that nature is, this can easily mean being the "last man left standing" to exploit opportunities. Or perhaps not the "last man," but the one with the biggest army and the most mouths.

How Jellyfish Persist

Jellyfish polyps can be induced to grow along two different physiological pathways via different mechanisms, with low-level exposure to toxic agents (Stebbing 1991). The first mechanism of stimulation, called hormesis, is basically an increase in the rate of asexual reproduction, that is, clonal growth. For example, a colony of polyps will accelerate its rate of cloning more polyps. In this way, a very large "seed bank" of bloom potential can be built up while waiting for the right conditions to bloom. The second mechanism of stimulation, called gonozooid production, is actually a switch in allocation of growth effort from cloning to producing medusae, that is, reproductive growth. Rou-

tine practice in public aquariums and research labs is to force polyps to strobilate or produce juvenile medusae by shocking them with changes in temperature, salinity, light, and so forth. This mechanism is likely to underlie the
massive jellyfish blooms observed around the world in response to distressed
ecosystems.

Most jellyfish bloom research focuses on food availability and other growth
parameters of the medusae. However, it appears that the fundamental trigger
of observable jellyfish blooms is a response to stress on the polyps. It may be
reasoned, therefore, that any sort of stress that elicits an escape response for
the polyp may stimulate it to switch to producing more of its dispersal phase,
the medusae. Low oxygen. Salinity changes. Warming water. Changes in pH
or other "flavors." *Any stress.*

Depending on conditions, the polyp can switch its energetic investment
into the dispersal phase, either to take advantage of abundant food or to escape local conditions unfavorable to polyps. The sexually reproducing aspect
of the dispersal phase also increases the possibility of genetic adaptation to
changing conditions. Similarly, the polyp can switch its energy into building
its seed bank of more polyps as a means of continued growth during relative
dormancy. Furthermore, by regulating the allocation of energy between the
two pathways, the polyp can essentially hedge its bets for greater advantage
in rapidly changing environmental conditions. This ability to switch growth
pathways may well be the key to jellyfish survival through the eons, while
mass extinction events have repeatedly decimated their neighbors.

[CHAPTER 5]

Overview of Ecosystem Perturbations

In recent years, the warning signs of ecological deterioration, such as algal blooms, fishery collapse, hypoxia and now jellyfish blooms have increased significantly.

—ZHIJUN DONG, Dongyan Liu, and John K. Keesing,
"Jellyfish Blooms in China"

Fools Gold in Paradise
(Great Barrier Reef, 2006–2007)

The film *Fool's Gold*, starring Kate Hudson and Matthew McConaughey, is about sunken treasure and rekindled romance . . . but like all good treasure hunt stories, there is usually a curse . . . Warner Brothers chose the idyllic Lizard Island for filming during the austral summer. Lizard Island is in the northern part of the Great Barrier Reef, offshore and close to the equator. It is balmy, scenic, remote . . . perfect. The studio employed the safety advice and local knowledge of the same person who had been the advisor for Steve Irwin on his final fateful journey in September 2006. But on 12 December 2006, paradise became pandemonium. McConaughey's stunt double was stung by a small but fierce Irukandji jellyfish and had to be airlifted to the hospital on the mainland for emergency medical treatment. As described earlier, Irukandji syndrome is an unusual complex of highly distressing and debilitating symptoms, which can be deadly. The sting was phenomenally bad luck, as Iruk-andjis are quite rare at Lizard Island.

Warner Brothers relocated the film crew farther south to get away from the danger of stinging jellyfish. They chose the equally scenic and lovely Whit-sundays, a cluster of 74 islands near the middle of the Great Barrier Reef. Filming resumed, quickly turning into the Great Barrier *Grief*. On the after-noon of 15 January 2007, the curse struck again: this time, one of the safety officers was stung by an Irukandji while rinsing his face mask. He too became violently ill and had to be rushed off to the hospital for medical care. Again, filming had to be aborted and relocated.

It is not the occasional spectacular local disaster that is significant so much as the general trends, the stealthy deterioration of environmen-tal conditions in sections of the sea of vital importance for its living resources.

—PROFESSOR JOEL HEDGPETH, "The Oceans: World Slump"

When we hear the word "perturbation," it sounds so small. A ripple in a pond. A minor annoyance. That's the whole point. The perturbations described in the following pages *are* small, compared with the effects they spark off. An-thropogenic disturbances are the ripples, and changed ecosystems are left be-hind. More often than not, these stressed ecosystems become progressively infested by jellyfish blooms.

The chapters that follow are confronting. They contain overwhelming evi-dence that we are destroying our marine ecosystems in a variety of ways—burning the proverbial candle at six ends. Each of the perturbations is de-structive . . . in combination, they are much worse than that. As each of these perturbations acts on our ecosystems, jellyfish drift in to pick up the pieces, fill in the gaps, and, where possible, take control.

Around the world, jellyfish are taking over destabilized ecosystems. The Black Sea, the Bering Sea, the North Sea, the Sea of Japan, the Benguela Current, the Gulf of Mexico, the Gulf of Maine . . . one by one, these bodies of water are "flipping" to jellyfish-dominated ecosystems. Previously healthy and productive fisheries regions have changed so dramatically that the fish are decreasing and the jellyfish are increasing, and in many cases, the jellyfish have assumed the role of top predator.

Jellyfish bloom as a normal part of their seasonal growth cycle, much like flowers bloom in the spring. Like most things in nature, these blooms are triggered by favorable conditions, such as temperature and water chemistry. And like most things in nature, there is variability in the effect. Some years the blooms occur earlier or later, or last longer, or appear richer with more individuals.

But in some species and in some regions, these blooms have apparently lost the feedback mechanism that says "stop." As a result, sometimes jellyfish blooms get out of control and cause problems. These effects are startling and a matter of great concern. Consider a meadow of daisies choking out all the other flowers and grasses, causing the cattle to starve and the bees to lose their preferred nectar, which in turn causes the farmer to suffer financial losses and the bees to die. Well, it's not a perfect analogy, but it serves to illustrate the main point: something seemingly innocuous can get out of balance and cause a cascade of negative effects to other things.

Jellyfish have spent the last 600 million years (or more) perfecting the art of survival. And they have gotten good at it. They have a suite of successful attributes that enable them to exploit favorable conditions, survive poor conditions, and switch effortlessly back and forth between the two. When conditions are favorable, they may bloom in unexpectedly high numbers; when conditions are poor, they can thrive well beyond the survivable limits of most other living things.

Jellyfish as the "Middle Man" of Destruction

It may seem unfathomable that jellyfish could drive ecosystem destruction. Consider perturbations like overfishing, climate change, and pollution. Obviously jellyfish cannot drive those . . . or can they? This middle-man effect is simply the role that jellyfish play in exacerbating an existing problem, or sometimes driving a whole new one. Jellyfish are like germs that get into an open wound and cause it to become infected. Once infected, a whole host of unpleasant things can happen. It is thought that ecosystems may operate as alternative stable states, where gradual changes may have relatively little effect until a threshold is reached, at which point a major regime shift occurs which may be difficult to reverse (Scheffer and Carpenter 2003). Just as slowly increasing the pressure on a switch will eventually flip it and turn on a light, or the slowly accumulating pressures of toxins and stresses will flip healthy cells to cancerous, so too, increasing and accumulating pressures on an ecosystem may switch it to new state.

This is because most ecosystems live in a delicate, constantly fluctuating balance where a certain number of predators are supported by a certain number of prey. But it's rarely a one-to-one balance. It's usually more like a network. When the balance is upset, other things happen to adjust for that imbalance, to compensate for the change. So when the "change" is "changed back," the gap it originally created has often already been filled by an opportunistic weed. It is very difficult to restore an original equilibrium in nature once it has been altered.

It's rather like baking a cake without a recipe by adjusting your ingredients according to taste. Let's say that you get it just about right, then decide to add more sugar, but the unlabeled bag breaks into the mixing bowl and you find that you have accidentally added a lot of salt instead. It could take you a really long time, not to mention some unpleasant taste-tests, to restore the equilibrium of the ingredients to a bakeable, flavorful cake. You would probably not be able to restore it to normal but would end up throwing out the batter and starting again. But we can't just throw out a section of the planetary ocean and start again. We are stuck with our mistakes, whether we intended to make them or not.

Ecosystem Change: Responses or Responsible?

Ecosystem perturbations come in many shapes and sizes: overfishing, pollution, thermal fluctuations, plague species, and ocean acidification, to name but a few. But jellyfish are poised to handle them all—not just to deal with them, but to exploit and in some cases even drive them, especially in combination. We live at a time when the frequency and amplitude of perturbations are at a fever pitch, probably more intense than at any other time in history. And in the rapidly changing world of today's oceans and seas, the unexpected effects of multiple pressures can make it very difficult to tease apart and accurately identify the nature of cause and effect between the different perturbations.

A lot of attention has been given recently to jellyfish as a natural indicator—a thermometer if you will—of the state of the environment. The more jellyfish, the louder the siren and the redder the flashing lights that something is out of sync: "wwhoooop . . . wwhoooop . . . problem, problem!" But it's more than that. Jellyfish blooms are not only a symptom of a problem, telling us that some crucial "tipping point" has been reached in an ecosystem, but in many cases, jellyfish are actually both the inheritors of some changes and the instigators of others.

Think of jellyfish as a broker of ecosystem change, like a stockbroker, to whom you give some money, who then (you hope) turns it into more. They inherit an ecosystem that has been perturbed in some way by overfishing, pollution, or whatever—and they exploit that relatively minor change to drive it to a very changed system indeed. Jellyfish blooms can displace fish and other predators, and in doing so, take over the role of top predator. All over the world, jellyfish blooms have been driving various damaged ecosystems toward new alternative ecosystem equilibria, in some cases so different from the original state that the original flora and fauna have been almost completely replaced.

While we tend to think of perturbations, such as those highlighted in this book, as leading to loss of biomass and biodiversity, more and more it is becoming clear that the opposite is often true with jellyfish. Overfishing depletes fish numbers, but jellyfish react to this by increasing their volume. Pollution kills off entire communities, but jellyfish emerge as if coated with Teflon. Warming waters hold less oxygen than cooler waters, so one might expect anything requiring oxygen to suffer . . . but not jellyfish. They are usually the "last man standing."

Homo industrialis. That's us. In just 200 years, we have evolved from *Homo sapiens*—wise man—to *Homo industrialis*—industrial man. We have fished out our oceans with high-tech methods to the point where whales, penguins, tuna, and cod will no longer be able to find food, but jellyfish will flourish. We have polluted our deep blue seas with agricultural and urban runoff to the point that many are now bright green and unable to sustain life over vast areas. We have dumped toxic chemicals and radioactive waste in appalling quantities—numbers that are generally only met with in astronomy. We have spewed so much carbon dioxide into the atmosphere that we are changing storm patterns and corroding corals. And we have transported species to new habitats in such large numbers that many industrial ports of the world are now taxonomically homogenized. Some may proudly boast that we have tamed nature. Others may cry that we have crushed her.

Al Gore's *An Inconvenient Truth* raised international attention about global climate change. Gore brought together well-supported facts to form the compelling conclusion that we are causing changes to our planet that are not in our best interests. But the *real truth* is even more inconvenient than Gore's message: climate change is just one of a handful of major changes we are making to our planet, making it more hostile to our current lifestyle—and possibly to our continued existence.

Homo industrialis. We are so industrial (but not so wise) that we are creating a world more like the late Precambrian than the late 1800s—a world where jellyfish ruled the seas and organisms with shells didn't exist. We are creating a world where we humans may soon be unable to survive, or want to.

Jellyfish bloom events cause economic loss. Crashed fisheries. Power outages. Spoiled fish catches. Ruined nets. Capsized boats. Medical bills. But what if I proposed to you that the jellyfish bloom problems aren't really the problem; they are just a symptom. They are just the red flag, the "please help" cry that signals oceans in distress. What if I proposed to you that this distress is caused by things we humans have done to the oceans that are now coming back to bite us in the backside. And what if I proposed to you that like a silent, invasive cancer, by the time we realize that the seas are in trouble and jellyfish are out of control, it's too late to do anything about it.

Our oceans are under threat from numerous perturbations, some that we have caused and some that we may not have caused, per se, but perhaps aggravated a bit. But because of how closely our economies and food supplies are dependent on the oceans, any threat to the oceans is a threat to our financial stability and social welfare.

Environmental disturbances come in two basic varieties: chronic and acute. Acute disturbances include the *Exxon Valdez* and Deepwater Horizon oil spills. Chronic disturbances are often more difficult to identify, and, for political and economic reasons, more difficult to stop. Fewer and fewer large fish. Fishery quotas set too high. Seafloor habitat destruction. Slow leaching of contaminants from decades of dumping. Incremental increases in ocean acidity due to rising atmospheric carbon dioxide. Increases in sewage effluent and industrial waste as "normal" outcomes of urbanization. These disturbances act like a ratchet, with each gradual "click by click" and "drip by drip" worsening the problem.

The truth is, the seas are changing, and they have been for a while, and there is no sign of letting up, only accelerating . . . We are overfishing almost every region and every commercially marketable species to the point where we progressively fish farther and farther geographically and further and further down the food chain . . . We are polluting, even with chemicals we didn't used to think of as pollution and toxic wastes we thought were being dumped so far away that they couldn't *possibly* hurt us . . . We are carelessly transporting species to places where they have no predator, no natural control . . . We are making our oceans corrosive to species with shells and skeletons so that they disintegrate, the same way osteoporosis leaches away the integrity of our own bones. . . .

And by fiddling with the fragile balance in our ecosystems—*our* ecosystems, where we get our oxygen and scrub our excess carbon—and by fiddling with our food webs—*our* food webs, where we get our food—we are making changes that we cannot undo. Changes we might not like. We do not have the technology to bring back what we have lost or to stop what we have started.

Jellyfish not only tolerate but actually flourish in changing conditions. Fewer predators, fewer competitors—what's not to like? But of course, the suites of changes that we are causing in marine ecosystems don't affect only jellyfish. Nothing in nature happens in a vacuum. Just about every species has predators and prey: tweak one, and the others become tweaked in response.

It probably sounds ludicrous if you are hearing this for the first time—how could jellyfish take over the oceans? But like the old joke, "How do you eat an elephant? . . . One bite at a time," jellyfish are taking over from fish, one bite at a time. A bite of fish larvae here, a bite of fish eggs there . . . a bite here and there of plankton that the larvae would eat.

The cumulative effects of any one of these perturbations are enough to "flip" an ecosystem to a state of jellyfish control. But in many cases around the world, these perturbations are acting in concert. And like a symphony, where each instrument plays a special note but together they produce something bigger than the sum of its parts, these perturbations in combination produce ecosystem damage that, in many cases, is well beyond the sum of the parts.

I stated in the introduction that we are led to believe that if we just stop overfishing, that the fish will come back, and if we just stop polluting, then nature will cleanse itself. Our observations in many ecosystems around the world now tell us that this is just not so. We now know that there is an invisible variable in what seemed like a simple equation, but this invisible variable gives a really different answer indeed:

stop overfishing = fish will come back

stop overfishing + (huge jellyfish blooms) = fish can't come back

I don't think we were misled on purpose, I just think that what *really* happens, now that we know, was previously unthinkable. Jellyfish inherit a weakened ecosystem, and, like spinning straw into gold, they transform something of little value to others into their own perfect world, bite by bite.

The following chapters explore the causes and effects of some of the perturbations that appear to cause the most distress on ecosystems. Each perturbation is generally considered to be an isolated signal, when in fact, ecosystems are presented with and respond to multiple stressors. Think of the ecosystem as a bank account, where multiple people are withdrawing something from it—independently. It's not hard to see how very quickly the account can become overdrawn.

Singly, any one of these perturbations reverberates through the ecosystems and affects our own relationship with the oceans. In combination, these perturbations often feed off and enhance each other. And jellyfish appear to be responding to these impacts with larger and more frequent blooms. The future of our oceans is unstable. We have never before assaulted the oceans from so many angles, and we can only speculate at the multiplicative effect.

Some of the figures and statistics numb the mind: how much pollution per year we are dumping into the oceans and atmosphere, how many fisheries are in peril and how deeply critical their situations are becoming, the rapidity of change and the severity of projected impacts of acidification, bycatch, the growth rate of dead zones around the world. . . .

Don't get caught up in the numbers, and don't let them slow you down from getting the main point. The numbers are there to be factual and accurate (and for geeks like me who want to see the evidence), but averages fluctuate over time and projections change with each new dataset. So treat the numbers as relative and appreciate what they represent.

Table 5.1. Comparison of main aquatic ecosystem perturbations that are driven by or exploited by jellyfish, with notes on potential negative effects to the jellyfish.

Perturbation	Driven by jellyfish	Exploited by jellyfish	Potential negative effects
Overfishing	Jellyfish predation and competition pressure often are not considered in sustainability estimates and fishery catches.	Reduced predation pressure on jellyfish and less competition open up bloom opportunities.	Jellyfish are quite positively affected by overfishing, except for potential for jellyfish fisheries to collapse.
Pollution	Jellyfish can become slimy toxic pollution when huge blooms die and decay.	Narrow margin of survival for most other species; jellyfish are likely to be the last man left standing.	Jellyfish are not affected by most types of pollution, except by reductions in prey availability.
Eutrophication	Heavy jellyfish predation on copepods leads to excess phytoplankton, whose mass decay causes dead zones.	Shift toward flagellate-based system; more food available for rapid bloom growth, and jellyfish create fish-free exclusion zones.	No downside for jellyfish; with overfishing, this is one of the most likely causes of jellyfish blooms.
Climate change, including El Niño and other oscillations	Jellyfish goo and poo enriched with carbon favors microbes that respire heavily, thus acting like CO_2 factories in a feedback loop.	Low dissolved oxygen suffocates other species, reducing competition for food; weedy jellyfish lifestyle makes them the last man left standing; warm waters speed up body growth and clonal proliferation.	Like all aquatic animals, one of main challenges is oxygen extraction from warmer water; unpredictable rain and drought salinity changes mean problems for polyps.

Perturbation	Driven by jellyfish	Exploited by jellyfish	Potential negative effects
Plague species	Jellyfish can create gaps in ecosystems for other species to exploit.	They themselves can be introduced, and even indigenous species easily become pestilent.	Jellyfish can be preyed upon or outcompeted by other plague species.
Ocean acidification	The CO_2 from microbes digesting jellyfish goo and poo may contribute to pH changes.	Shift toward gelatinous-based system; less predation and competition from high-energy consumers.	Less downside for jellyfish than for most other species.

CHAPTER 6

Overfishing:
A Powerful Agent of Ecosystem Change

We have forgotten what we used to have. We had oceans full of heroic fish—literally sea monsters. People used to harpoon three-meter-long swordfish in rowboats. Hemingway's *Old Man and the Sea* was for real.

—PROFESSOR JEREMY JACKSON

Where have all the flowers gone? Long time passing. Where have all the flowers gone? Long time ago. Where have all the flowers gone? Girls have picked them every one. When will they ever learn? When will they ever learn?

—PETE SEEGER

In Cod we Trust
(The Gulf of Maine, since 1975)

We have not treated the Gulf of Maine very well. We have dumped our rubbish, pesticides, industrial effluents, and other contaminants into its waters. We have zigzagged huge bottom-trawlers across the seabed, turning reefs to rubble and marine communities into moonscapes. We have hooked, netted, and speared its fish with wild abandon, as if they would never run out. And now it is kicking back.

Once upon a time, the groundfish fishery in the waters off the northeastern United States and southeastern Canada seemed to be unlimited. Cod. Flounder. Haddock. American lobster. Sea scallop. The gulfs of Maine and Newfoundland are among the world's most productive marine habitats, with a rich fishing heritage and a long pre-European history. But after more than a century of intense bottom-trawling the Grand Banks area, the groundfish community has collapsed. In his best-selling book *Cod*, Mark Kurlansky traced the New England cod fishery from Viking exploration to its dying, impotent whimper in the 1990s.

It's a familiar story by now, one that plays out again and again in our oceans, whether we recognize it or not. Young cod feed primarily on copepods, as do many other types of commercially important fish in the Gulf of Maine. But so does a jellyfish called *Nanomia cara* (see plate 2). *Nanomia*, like all siphonophores, is just plain strange. These colonial forms include the fearsome Portuguese man-o'-war, but while the man-o'-war can kill healthy people with its sting, *Nanomia* can't. But when it blooms in massive numbers, it can wipe out entire populations of fish and zooplankton by eating them and their prey.

Dr. Marsh Youngbluth, formerly of Florida's Harbor Branch Oceanographic Institution, studied *Nanomia* for many years in order to better understand its bloom dynamics and potential to harm fisheries. Vast numbers of *Nanomia* colonies comprise a persistent group of carnivores inhabiting the midwater depths along the US Atlantic coast from the Gulf of Maine south to Cape Hatteras in North Carolina. *Nanomia* range in size from 0.2–3.7 meters (8 inches to 12 feet). Siphonophores of this sort have been observed eating prey from minute copepods to small fish and a variety of fingernail-sized midwater crustaceans with tongue-twister names like amphipods, mysids, and euphausiids. It is thought that in these sorts of bloom densities, *Nanomia* could impede the success of larger fish, particularly herring, through heavy predation on their larvae (Rogers, Biggs, and Cooper 1978).

Nanomia has been clogging trawl nets and intriguing scientists since the mid-1970s. Early studies reported maximum densities of up to 7 colonies per cubic meter (about 1 per 5 cubic feet), but that density had skyrocketed to 50–100 per cubic meter (1–2 per cubic foot) in the early 1990s (Mills 2001).

The Gulf of Maine is not the only place where siphonophores like this bloom in large numbers. A similar species was reported in the waters off San Diego, California, albeit in much lesser densities (Barham 1963). Each *Nanomia* colony is constructed on a plan of a gas-filled float atop numerous rows of swimming bells, from which flows a long stem of tentacles and feeding polyps.

When *Nanomia* swarms, the gas-filled floats can act as strong sound-scatterers and give misleading sonar readings on water depth.

Another kind of jellyfish bloom problem has cropped up in the Gulf of Maine—and it's not one of the "usual suspects" plaguing other regions. Strangely enough, it is suspended fragments of hydroid colonies causing these blooms (Mills 2001). It seems that they have been blooming in unusually high numbers since 1994, possibly even 1990. Feeding experiments have demonstrated that these drifting hydroid colonies may be eating half the copepod eggs and a quarter of the larval copepods produced each day. That's half the food that would be available to fish, and a quarter of tomorrow's food.

It turns out that blooms of hydroid fragments were reported as far back as 1913. Whether they are becoming more common, or just more commonly reported, is still hard to say. Claudia Mills posed an interesting question about the origin of this phenomenon of floating hydroids. She wondered whether "these bits of usually-bottom-living animals have been broken up and become resident in the water column as a result of increased trawling activities" (Mills 2001, 59).

Jellyfish blooms are not the only problem plaguing the Gulf of Maine. As is so often the case, harmful jellyfish blooms and algal blooms take turns in abundance as if by some sort of alternating agreement, first one and then the other, and back again. Toxic algal blooms in the Gulf of Maine have gained more notoriety than those of jellyfish, because anything that kills whales is bound to grasp public attention.

When we think of overfishing, we generally think of the loss of fish. We rarely think about what that means to the other organisms in the ecosystem. But the essence of any food web or food chain is that everything in it is interconnected and interdependent. Take out one species and another will soon multiply to fill its place. Take out one link and its prey will flourish, while its predators perish.

"Animals that live in the waters, particularly in marine waters . . . are protected from destruction of their species by man. Their multiplication is so great, and the means whereby they can avoid being hunted or trapped are such that there is no likelihood that man could destroy any entire species of these animals." These words were written in 1809 by the French naturalist Jean-Baptiste de Lamarck (*Philosophie Zoologique*, 76–77). Lamarck is best known for his theory that characteristics acquired during one's lifetime, such as amputation of a leg or stretching of the neck, would be inherited by one's offspring. He was wrong about that too. Similarly, Lord Byron wrote in *Childe Harold's Pilgrimage* (1818), "Man marks the earth with ruin, his control stops with the shore, again reflecting the nineteenth-century view that the sea could take any insult, any exploitation, any abuse, and continue unscathed.

This view of the sea as perpetually resilient and forgiving, and its bounty a free-for-all, still pervades our view that fishing is our birthright and the more we take is evidence of our own mastery.

This notion was grandfathered in from a romantic age when fishermen bravely wrestled against sea monsters and "perfect storms," when the sea was seen as a hostile force that needed to be conquered to be properly used. Those rare souls who were "man enough" to face their fears of giant waves and unfathomable depths with only a small rickety boat, fishing lines, and perhaps a bailing bucket were rewarded with unlimited fish, and fish stories that grew bigger with each telling.

But those days are long gone. The man-against-beast romance of it all has transformed into a Greek tragedy where everyone dies in the end. The rod and reel have been replaced by "floating factories deploying gear of enormous proportions: 80 miles of submerged longlines with thousands of baited hooks, bag-shaped trawl nets large enough to engulf 12 jumbo jetliners and 40-mile-long drift nets (still in use by some countries). Pressure from industrial fishing is so intense that 80 to 90 percent of the fish in some populations are removed every year" (Safina 1995, 49). Fishing technologies have made it like, pardon the pun, shooting fish in a barrel: sonar, radar, LORAN, GPS, satellite weather maps, spotter aircraft . . . these military tools have given us the ability to hunt and kill fish with surgical precision and inexorable might. We now have the ability to find and extract the very last fish.

The global fishing story is one marked by sequential spatial depletion: first we exhausted the big fish near population centers, then while some turned to taking smaller fish, others ventured further afield in search of more big fish; as we fished progressively distant seas and progressively deeper waters and progressively smaller fish, we have now realized the limitations of an enclosed planet with finite resources.

One of the most astonishing examples of overfishing is California's abalone fishery. A photograph shows the huge number of shells discarded into a trash heap from an abalone processing plant in Santa Barbara, California, as evi-

dence of earlier populations (plate 6). Abalones are very slow-growing mollusks, valued for their large muscular foot that tastes a bit like scallops. Fantastically abundant prior to commercial exploitation, first one species, then
another, and then another, was fished to the point of collapse. These species
now number less than 1 percent of their baseline populations (Karpov et al.
2000; Rogers-Bennett et al. 2002).

Peter Haaker was a marine biologist with the California Department of
Fish and Game from 1968 to 2006; assigned to the abalone project, he had a
front-row seat to the fiasco. He started at a time when abalone landings were
averaging between 4 and 5 million pounds per year, and by the time he retired,
nary an ab could be found.

Like all good romantic tragedies, it didn't start in 1968: in fact, it goes
back hundreds if not thousands of years. Native Americans had used abalone
both as food and a source of inland barter since time immemorial. Abalone
populations flourished when their main predator, the sea otter, was hunted
to near oblivion in the early to mid-1800s. Prohibited from collecting in their
own country, Chinese immigrants working on the transcontinental railroad
in the 1850s capitalized on the plentiful supply: they built a thriving industry
hunting abalone from the intertidal zone in the Monterey area, and by 1879
commercial landings exceeded 4 million pounds a year (Cox 1962). So many
abalones were being taken that in 1900, shallow water collection was made
illegal to protect the species.

Japanese American hard-hat divers began collecting abalone from deeper
waters where it was still permitted. They monopolized the fishery, which netted over 2 million pounds a year from 1916 until World War II, when many
divers were relocated to internment camps. After the war, abalone fishing became a lucrative Caucasian enterprise, reaching more than 3.5 million pounds
in 1952.

By the 1970s fisheries in central California were going bust or relocating
farther and farther south and farther and farther offshore into the Santa Barbara Channel Islands. Stocks continued to decline under heavy pressure from
equally voracious recreational and commercial fishers.

But then the unthinkable happened. In 1986, abalones began falling off
rocks . . . in droves. Abalones normally clamp down hard onto the surface of
rocks to protect their delicious body parts from predators. But all of a sudden,
their muscular feet were shrinking and losing the ability to adhere to rocks. It
is called withering syndrome, and it's not good. Abalones that fall off rocks either get eaten or starve to death. Haaker estimates that more than 90 percent

of abalone were wiped out by the disease—well, 90 percent of what was left after more than a hundred years of heavy fishing.

A consortium of experts including Haaker recommended closing the fishery. Recreational divers banded together, using websites, petitions, and political pressure to close the fishery. But commercial fishermen insisted they could still make a living by taking the healthy ones. The problem is, of course, that selective taking of the healthy ones leaves the unhealthy ones as the reproductive stock for the next generation. Also not good. Ultimately, by 1997, populations of all species were in such poor shape that the California Fish and Game Commission was finally persuaded to close all fishing for abalone—commercial and recreational—south of the Farallon Islands near San Francisco.

But there is also a whole other side to the abalone story, one that you don't hear very often. You see, the otters did it. Yes, those cute, cuddly, furry larrikins that were almost driven extinct by pelt hunters in the 1800s, then were decimated again by hungry killer whales in the 1990s. The problem is that it is very, *very* politically incorrect to blame a cute furry endangered animal when it obviously must be the fishermen's fault. *But what if the otters are to blame?*

Consider the Pismo clam. Clams are humble animals, content to while away the day filtering tiny particles out of the water and occasionally sticking their muscular foot out between their bivalved shells to relocate. Don't get too excited—it's not like a clam gets up and prances away, they just sort of shift a bit. But Pismo clams have been filtering and shifting for millions of years along the shallow sandy beaches of Central California, particularly Pismo Beach, for which the clam is named.

But alas, the Pismo clam grows to about 18 centimeters (7 inches), and that's a lot of succulent clam meat for relatively little shucking effort. They are not only divine in clam chowder, but sea otters like them too. In fact, Pismo clams represent about 90–95 percent of the sea otter's diet by weight, and a single sea otter consumes about 80 clams per day. Researchers studied the clam population at one beach before and after the arrival of otters and documented a complete wipeout of the legal-size clams within 6 months of the otters' arrival (Miller, Hardwick, and Dahlstrom 1975). *Six months.* Indeed, in the 12 months from April 1974 to March 1975, in Monterey, about 60,000 Pismo clams were taken by humans, compared with an estimated half million consumed by otters.

As otters rebounded from near-extinction and they spread south in search of food, they began decimating clam beds and clamming communities, causing a wave of economic catastrophe. The locals hoped Pismo Beach would be safe because the otters wouldn't cross the sandy expanses to get there. But there's nothing like a bit of hunger to make even an otter reconsider its priorities. They crossed. They ate. And it was indeed catastrophic. Don't hold it against the otters. They are just trying to survive, just like we are. It's how life works. Besides, we can't blame everything on the otters . . . they only range so far south.

Consider the white abalone. Too far south to be affected by otter predation, it must certainly be the poster child for overfishing. *Haliotis sorenseni* was not even discovered until 1940, but is prized for having the most tender and flavorful meat. Scientific surveys in the early 1970s found the species in high concentrations of about one individual per square meter in its deep water habitat in the Channel Islands. But being a delicacy is dangerous business.

The boom-and-bust fishery saw annual commercial landings averaging 45 tons in 1971–1976, peaking abruptly at 71 tons in 1972, then plummeting to about 300 pounds in 1981 (Davis, Haaker, and Richards 1996). The species reproduces by broadcast spawning, which requires aggregations of males and females. If they are more than a meter apart, dilution of sperm and eggs makes fertilization unlikely. Haaker and his team conducted a 9-month survey in 1992–1993: searching over 30,000 square meters (about 320,000 square feet) of suitable habitat at 15 locations previously known to support white abalone, they found only 3 live individuals—just one ten-thousandth the density necessary for reproduction (Haaker 1998). Haaker then went down in a submersible to get a deeper, longer look, this time finding only five in a similar-sized area.

The white abalone are now approaching extinction because isolated surviving wild individuals are too few and far between to successfully reproduce. The last year of successful spawning occurred in the late 1960s (Carlton et al. 1999). Protection efforts were eventually put in place, but probably "too little too late." After commercial landings of less than 100 individuals in 1995, the fishery was closed in 1996. On 29 May 2001, *Haliotis sorenseni* became the first marine invertebrate in the United States to be formally protected under the Endangered Species Act.

One might think that the story ends with the white abalone fading away into extinction despite all heroic human efforts to save it. But it doesn't. Like so many wars won and lost on differing ideals and competing needs, the white abalone saga is not without its politics. It breeds well in captivity, and in fact, has been doing so for quite some time. The logical answer would be to restock the wild with captive-bred animals. But it's not that easy. Captive-bred stock may contain pathogens, such as the one that causes withering syndrome, so it is too risky to release them back into the wild, where they would pose a risk to all wild abalones. And so, with no easy answer and no practical reason to continue funding its recovery, this species appears likely to soon gain the dubious distinction of becoming the first marine invertebrate driven to extinction by overfishing (Tegner, Basch, and Dayton 1996) . . . or federal management decision.

According to Haaker, "Right now there is an issue of continued protection with the white abalone, particularly in the captive rearing program which is the basis for the recovery. If you lose the captive breeding program you lose probably 4 or 5 thousand animals that are being held, and that's probably more than you have out in the wild" (personal communication).

All species of abalone in California are now in peril. Once abundant intertidally and ignored in favor of more palatable species, the black abalone population has declined by 99 percent and was listed as endangered in 2009 (Joyce 2010). Too few and far between, they too are now waiting to die.

But there's nothing like a whiff of rarity to whet one's appetite. "The human predisposition to place exaggerated value on rarity fuels disproportionate exploitation of rare species, rendering them even rarer and thus more desirable, ultimately leading them into an extinction vortex" (Courchamp et al. 2006, 2405). Australia and New Zealand quickly became the only viable source for the strong demand in Asian countries.

However, the Australian abalone fisheries are not without problems, ranging from bacterial and parasitic infections to habitat destruction by introduced species to politics. In Victoria, the fishery—which is the state's most valuable—is worth about $20 million per year, but it suffered major losses in 2005 with the spread of a devastating herpes virus that originated in mariculture (Hooper, Hardy-Smith, and Handlinger 2007). In South Australia, the abalone fishery has been continuously plagued by the protozoan parasite *Perkinsus*, which causes small nodules on the foot, making the animal both highly contagious and nonsaleable. Currently, only Western Australia and Tasmania have profitable abalone fisheries, with Tasmania supplying a whopping 25 percent of the global wild catch. However, in 2008, the abalone viral ganglioneuritis disease was found in Tasmanian processing plants and on an abalone farm. Because of fear that it would spread to the wild, the stock at that farm was destroyed. However, infected stock have repeatedly been found since, at another farm in 2009, and again (twice) in 2011.

Since 1978, the black longspined sea urchin, *Centrostephanus rodgersii*, has been expanding through abalone habitat like a black plague (Johnson et al. 2005). The urchins denude the habitat of algae, creating "urchin barrens" in which abalone, lobsters, and other reef animals cannot survive (see plate 11). Like a slow and unstoppable flow of lava, *Centrostephanus* proceeds down the east coast of Tasmania, transforming teeming reefs and kelp forests into moonscapes.

While abalone and jellyfish may not seem connected in any way, except perhaps side-by-side on a seafood platter or in a zoology exam, these areas where abalone are declining are also areas of rapidly expanding jellyfish populations. Off California, moon jellies and sea nettles have been expanding their bloom densities and dimensions up and down the coast. It seems that these responses may be correlated as visible effects of large-scale ecosystem disturbances.

The Smaller the Fish, the Bigger the Problem

In 1998, Professor Daniel Pauly of the University of British Columbia Fisheries Centre and his colleagues published a landmark paper titled, "Fishing Down Marine Food Webs" (1998). This refers to the progressive removal of large, long-lived, piscivorous fish, such as swordfish and tuna, then targeting smaller, shorter-lived, invertebrates and planktivorous fish in their place. Pauly and his colleagues demonstrated that the average size of landings dropped by 1 meter (3 feet) between 1950 and 2000 in the North Atlantic and the Arabian Sea, and off Patagonia, Antarctica, and parts of Africa and Australia. "It is likely that continuation of present trends will lead to widespread fisheries collapses" (p. 863), they wrote.

As dire as that may seem, research by other groups has suggested that Pauly's findings have underestimated the seriousness of the effects of overfishing. Professor Jeremy Jackson and his colleagues have shown that massive declines in populations of marine mammals, turtles, and large fishes occurred all along coastlines where people lived long before the post–World War II period (Pauly and Watson 2003).

Moreover, Dr. Tony Smith of the Commonwealth Scientific and Industrial Research Organisation (CSIRO) in Australia and his colleagues conducted a study using different models to examine the effects of fishing at lower trophic (food web) levels, that is, sardines, anchovies, menhaden, capelin, and krill (Smith et al. 2011). Today these lower trophic levels account for a whopping

30 percent of global fisheries' production and are important to global food security. Recall the story of the Namibian fisheries collapse: the ecosystem lost its resilience when it lost its lower trophic level fisheries. Smith's models predict that the current rate of fishing based on what is called "maximum sustainable yield" will have substantial impacts on marine mammals, seabirds, and other commercially important fish, whereas halving the exploitation rates would still achieve high yields but with much lower impacts on the ecosystem as a whole.

The "fishing down" concept has been so widely accepted that it underlies the now most widely used model for predicting the ecosystem effects of fishing. However, Trevor Branch of the University of Washington and his colleagues questioned the validity of this model, arguing that it did not reliably predict ecosystem changes observed from catches, surveys, and assessments. Nonetheless, they did conclude that the current overall increasing catch trends can intensify fishery collapses even when fishing of different levels on the food chain is stable (Branch et al. 2010).

Regardless of our methods and ability to predict future trends, persistent overfishing issues have led to a very sad state of affairs. Despite constantly increasing global fishing efforts, cumulative yields have continued to decline since peaking in 1994 (Worm et al. 2006). According to the United Nations Environmental Programme, "In 2002, 72 percent of the world's marine fish stocks were being harvested faster than they could reproduce" (UNEP 2004, 1). The current trend is confronting and impolite, projecting the commercial extinction of all currently fished species by the year 2048 (Worm et al. 2006).

> An implied endpoint of this "fishing down marine food webs" is a proliferation of previously suppressed gelatinous plankton (jellyfish) thriving on the food no longer consumed by fish. (Lynam et al. 2006)

When we think of overfishing, some of us probably think of the recent shocking movie, *The End of the Line: The World Without Fish*, or the sad saga of the Newfoundland cod or the orange roughy off southern Australia, or the history of the overexploitation of whales. But overfishing is about a lot more than simply "taking too many fish"—it's also about habitat destruction and biodiversity depletion through wanton waste. And it's about misjudging sustainability calculations through ignorance of the invisible variable of jellyfish.

We can't imagine a world without fish or seafood. Well, I can't. My taste buds would suffer desperately if they never again savored the sheer bliss of broiled lobster with lemon and melted butter, or pan-fried salmon with Dijonnaise, or coconut prawns—or prawns with XO sauce, or for that matter, shrimp cocktail or shrimp scampi. Fresh oysters, neat.

Sushi without fish . . . If this sounds alarmist and extremist, think again. A compelling paper in 1994 by Dr. Carl Safina (founder and director of the National Audubon Society's Living Oceans Program in New York, and founder of the Marine Fish Conservation Network), aptly titled "Where Have All the Fish Gone?," highlighted the following:

- The Grand Banks were described by the explorer John Cabot as "swarming with fish [that they] could be taken not only with a net but in baskets let down with a stone" (p. 38), but the groundfish are now decimated (see plate 6). The cod are at their lowest level ever, and commercially extinct.
- The haddock fishery had to be shut down on Georges Bank after landings plunged 66 percent in a year.
- Of every 100 yellowtail flounder alive at the beginning of the year, only eight survive the year; the breeding population declined 94 percent between 1989 and 1992. Many juvenile flounder are discarded dead because they are undersized.
- The Canadian cod fishery had to be closed in 1993, costing 42,000 jobs and $1.8 billion in unemployment payments.
- In 1991, a Massachusetts task force estimated that lost landings cost the region $350 million annually, with 14,000 jobs lost.
- In 1986, 12 factory trawlers targeted walleye pollock in the waters off Alaska. By 1992, 65 factory trawlers landed 3 billion pounds, making this the largest single-species fishery in the world. By 1994, the pollock were already showing signs of overfishing, and the sea lions and seabirds that rely on them for food had declined by 50–90 percent.
- Lobsters in the Gulf of Maine are officially overexploited, and sea scallops are at or near all-time lows.
- The Alaskan halibut fishery has now turned into a sport-fishing circus: with 5,500 boats fishing, the season had to be reduced to just 2 days per year, resulting in fatalities, sinkings, and enormous quantities of spoiled fish.
- In just a decade, several important Atlantic shark species have declined 85–90 percent due to exports to China for shark-fin soup and the lack of a

management plan. When an emergency plan was finally put into place in 1993, the six-month quota was taken in less than a month.

- Bycatch is the main source of adult mortality for sea turtles and albatrosses.
- Bycatch has contributed to an 85 percent decline in the Gulf of Mexico population of bottom fish, such as snappers and groupers, from 1970–1990.
- The breeding population of western Atlantic bluefin tuna declined from 250,000 in 1975 to 20,000 by 1994. Swordfish and marlin also declined 50–90 percent in this time (see plates 6 and 7).

Recall Claudia Mills, who wrote in 2005 after attending a fisheries conference, "Over and over, the audience was shown what one speaker finally called a general 'fish graph,' each showing high stocks (of any fish, nearly anywhere in the world) in the 1960s and 1970s, followed by a precipitous decline to very low and stable stock levels today. The overfishing problem is global. The situation is especially desperate in areas of great poverty where the fish are really needed for food and income. I did not hear a single speaker say that things were improving in the world's oceans, at any level" (Mills 2005).

In a recent review of patterns of historical changes in large marine mammals, birds, reptiles, and fish, it was found that across 256 estimates from 95 studies, the average decline from historical abundance levels was an astonishing 89 percent (Lotze and Worm 2009). Further, a study of 90 fish stocks around the world found that most had little or no population recovery 15 years after collapse, raising a red flag about their extinction risk (Hutchings 2000).

A controversial 2003 paper by Myers and Worm about overfishing has been described as "a shot heard around the world" (Jackson 2008, 11461). They asserted that 90 percent of all large open ocean piscivorous fish, such as tuna, billfishes, and sharks, are gone. Other scientists attacked every aspect of the paper from the publicity the authors received to their use of highly complex assessment methods. Some even resorted to name-calling; however, despite nit-picking and outrage within the scientific community, the report was credible and was largely adopted by the National Research Council (Jackson 2008).

Scientists debate these issues back and forth, partly because of our own bias in learning about marine productivity in terms of availability of resources (bottom-up control) rather than predation (top-down control), partly because of the strong implication of long-term fisheries mismanagement, of which we have collectively been a part (Jackson 2008), and partly because we are human and it's hard to accept that we got it wrong. Also, robust debate is how science progresses. However, the scientific squabbling makes it easier for society and governments to focus on details to distrust or to relegate the whole messy subject to the "too-hard basket"—but then we miss the big picture of decline. Alas, such is the nature of politics. . . .

"Overfishing" calls to mind factory ships with massive nets taking thousands upon thousands of fish. We think of commercial fishermen's lobbyists exerting political pressure to keep quotas high. But we don't think of our own vacation last year, when we whiled away some time with a line in the water, reeling in a few beauties for the family dinner. We may blame commercial fishermen for depleting the resources, because we believe that unlimited access is our birthright and we are angry that they are, in essence, taking away our fish.

Formerly only about 2 percent of the total US fish catch, the recreational fishing effort has risen substantially in the past 20 years, and now rivals commercial catch for many major fish stocks (Coleman et al. 2004). Excluding pollock and menhaden, recreational landings comprised about 10 percent of the total in 2002, but the figures are much higher for many species of concern: 93 percent of red drum off the South Atlantic United States are taken recreationally, as are 87 percent of bocaccio on the Pacific Coast, and 59 percent of red snapper in the Gulf of Mexico.

In fact, as recreational fishing is on the rise worldwide, some scientists and fishers are arguing that its growing pressure on fish stocks must be considered, along with commercial impacts (McPhee, Leadbitter, and Skilleter 2002). In many places around the world, recreation-only fishing zones have been put in place, based on the public and political belief that recreational fishing is more benign than commercial fishing. However, close scrutiny demonstrates that recreational fishing can have severe impacts on local biodiversity and long-term sustainability of ecosystems. High percentages of juvenile fish under the legal size limits and collection of high numbers of worms, crustaceans, and mollusks as bait are peculiar to recreational fisheries, yet are poorly regulated. Although the catch-per-person is considerably higher for commercial fishermen than it is in the recreational sector, the sheer number of recreational fishers makes the impact significant—and harder to control.

Another hard-to-control aspect of both recreational and commercial fisheries is human behavior. In particular, differing interpretations and commitments to regulatory controls present a largely under-recognized problem for fisheries management. Dr. Beth Fulton of CSIRO has examined this issue and concluded that the uncertainty inherent in assessing and managing the status of exploited fisheries resources is compounded by the "consistent outcome that resource users behave in a manner that is often unintended by the designers of the management system" (Fulton et al. 2011, 2). It is therefore not sufficient to simply make policies: they have to be implemented and enforced in such a way as to gain wide compliance.

Half a century ago, it was estimated that some 200 to 240 million tons of fish are naturally produced in the sea each year (Ryther 1969). If we use this as a guideline to calculate potential harvest, we have to also consider the take by top-level carnivores, such as seabirds, marine mammals, tuna, and squid, as well as a large enough fraction to keep the resource breeding at a useful level.

Ryther estimated the maximum sustainable yield for world fisheries was unlikely to exceed 100 million tons per year. He further noted that the world fish landings in 1967 were just over 60 million tons, which at that time had been increasing at a rate of about 8 percent per year for the previous 25 years. Ryther concluded that at that rate, the fishing industry could continue to expand for no more than a decade. According to the Food and Agriculture Organization of the United Nations, the global fishery harvest for 2008 was 156 million tons: 99 million from capture production plus another 57 million from aquaculture production.

At face value, it may seem like 99 million is comfortably below the 100-million-ton ceiling forecasted by Ryther. But the seas are different today than they were in 1969, and their capacity to produce has been reduced by habitat degradation and altered predator/prey balances.

Ryther recommended that expansion of world fisheries should consider exploitation of Antarctic krill, for which no harvesting technology or market existed at that time. Now just 40 years later, we are staring down the barrel of a serious krill crisis due to overfishing.

Eco-Logical or Zoo-Illogical Fishing Practices

By-kill can produce extraordinary waste and even overfish species. Estimates of discarded bycatch in Alaskan fisheries in 1990 range to well over half a million tons annually. Ten pounds of unwanted fish are killed for every pound of shrimp caught in the southern United States. Total discard in the US shrimp fishery is estimated at 175,000 tons of juvenile fish a year—fish that would otherwise grow to support other important fisheries. (Safina 1994)

When we hear the term "sustainable fisheries," most of us think it means "the opposite of overfishing," that is to say, that overfishing is unsustainable, and that sustainable fisheries are not overfished. But taking too many fish is only part of the problem. Sustainable fisheries practice isn't just about limiting the number of fish caught per annum or area, it is also about limiting the negative side effects of fishing on the ecosystems which support those fish.

The secretary general of the United Nations wrote in a report on over-fishing:

> A significant reduction of biomass is unavoidable and even necessary to obtain food and livelihood, but a large number of stocks have been reduced below sustainable levels. There are ample data to suggest that fisheries exploitation not only affects target stocks and other fish species, but also communities of organisms, ecological processes and entire ecosystems by causing cascading effects down food webs that decrease diversity or productivity. It also affects directly vulnerable habitats like reef ecosystems when gears are in contact with the reef substratum, or indirectly by altering the relationships between those communities of plants, invertebrates and fish species that determine rates of reef accretion and bio-erosion. (UN 2006, 9)

The methods used to catch many types of fish are highly destructive not only to the target species but also to other species and to the habitat as a whole, so the collateral impacts of fishing methods also raise cause for concern. Among the most destructive aspects of fishing are overfishing, bycatch, habitat destruction, and a choking fog that is often stirred up by trawling.

Overfishing, or taking more fish than a population can naturally replace, seems pretty straightforward. Take a lot of fish and stocks go down. Take fewer fish and stocks come back. But it's not always that simple. Like many things in nature, there are direct and indirect effects. The direct effects are fairly obvious: mortality among the fish taken and therefore loss of their reproductive and predator/prey potential. But the indirect effects can have far more profound implications for the ecosystem in the long run. For example, even sustainable fishing of one species may lead to the starvation of another.

Fishing affects predator/prey relationships and food chain dynamics, leading to changes in the structure of whole communities that are unlikely to revert to their original prefished condition once fishing pressure on a given species is reduced or ended.

Recall the old saying, "nature abhors a vacuum." Well, it's true. The ecosystem doesn't just leave a gap where the fished species used to fit, waiting for its stocks to return to fill the vacancy. Other species that used to compete with it will exploit the opportunity to feed without competition. Other species that it preyed upon are likely to flourish, now that they are freed from predation pressure, and this puts additional pressure on their prey. And other species that preyed upon it are likely to suffer badly if they cannot find other suitable food. It has been said that nature is as delicate as a silken spider web: if you pull one strand, you may unravel the whole structure. Hence, overfishing depletes stocks directly, but it can also upset the balance of the food chain so that niches open for others to fill.

Bycatch is the sum of unwanted or unintended fish, birds, turtles, dolphins, invertebrates, and other marine organisms that are incidentally killed, damaged, or discarded in the process of catching a target species. Bycatch are not only dead, but generally are so mangled from being tumbled about in the trawl then emptied on deck that they are usually crushed, amputated, stabbed, and largely unrecognizable. Many of these are prohibited species or young fish that are too small to sell, which by law must be thrown back, even if dead or dying. Bycatch adds to the cost of fishing by crushing the catch under the weight of other marine life, as well as adding man-hours to sort the desired catch from the unwanted. But the real cost is in the wanton waste of the living components of the ecosystem.

Consider the barndoor skate, so named because of its size. Once it was one of the most numerous skates from Cape Hatteras, North Carolina, to Grand Banks, Newfoundland, but its population has become bycatch of cod trawlers and thus has been pushed to the brink of extinction (Casey and Myers 1998). It has been estimated that on St. Pierre Bank alone, the average number of barndoor skates in the 1950s would have been about 600,000, decreasing to about 200,000 in the 1960s, and to less than 500 individuals in the 1970s. The total number of barndoor skates caught here since the 1970s is zero. Similar patterns exist throughout its range.

Because the huge barndoor skate is kind of hard to overlook as bycatch, we are able to track its demise. However, without doubt, many less conspicuous species are just as vulnerable to the same effects of trawling, but will slip into oblivion without notice.

One might find comfort in thinking that discarded bycatch specimens will be eaten by other creatures, so therefore not wasted. But in fact, eaten or not eaten, worldwide bycatch represents 40 to 90 percent of trawl contents, totaling some 41 million tons *each year*, raising the true global annual fishing catch to some 165 million tons (Pauly 1995). To put this into perspective, imagine if a paddock full of cattle were rounded up and shot, and 40–90 percent were thrown back for the wolves and vultures to eat rather than being used for human consumption. We would be outraged.

Bycatch of bony and squishy invertebrates may seem unrelated to dwindling seafood stocks, but it's not. The larvae of those dead starfish and snails and worms could have been the food of the young of target fish species, while the unwanted bycatch species, now dead, normally would help keep jellyfish numbers under control. These are fish and invertebrates that will not contribute to the renewal of tomorrow's habitat, nor to any future generation.

Next time you hear Jo Stafford's song "Shrimp Boats," consider this: shrimp are a canonical example for bycatch. According to the Greenpeace International Seafood Red List, "Tropical shrimp are fished using bottom-trawling, which results in a high unintentional capture (bycatch) of other species. The fisheries are responsible for taking 27 percent of the world's bycatch. For every kilogram of shrimps captured as much as 10 kilograms of other marine life is thrown back into the sea, dead or dying. The bycatch includes endangered sea turtles" (Greenpeace 2012b). According to an article in *Scientific American*, Bering Sea fishermen discarded 16 million red king crabs in 1992, keeping only about 3 million (Safina 1995). Furthermore, "Trawling for shrimp produces more bykill than any other type of fishing and accounts for more than a third of the global total. Discarded creatures outnumber shrimp taken by anywhere from 125 to 830 percent. In the Gulf of Mexico shrimp fishery 12 million juvenile snappers and 3,000 tons of sharks are discarded annually. Worldwide, fishers dispose of about six million sharks every year—half of those caught. And these statistics probably underestimate the magnitude of the waste: much bycatch goes unreported" (Safina 1995, 52). The National Research Council has estimated that up to 55,000 adult turtles died as shrimp bycatch each year before "exclusion devices" were mandated—they are one of the feel-good stories with a happy outcome. Go ahead and cry.

Bycatch is not limited to bottom-trawling. Longlines catch sharks, sea turtles, and seabirds in addition as their target fish species. In fact, one of the greatest threats to seabirds like albatrosses, petrels, and shearwaters is longline fisheries, where they may be victims of bycatch. Experts estimate that "in the Southern Hemisphere more than 40,000 albatross are hooked and drowned every year after grabbing at squid used as bait on longlines being set for bluefin tuna" (Safina 1995, 52). So too, loggerhead and leatherback turtles comprise significant longline bycatch along the Atlantic seaboard.

Drift nets are notoriously among the worst causes of bycatch problems. In 1990 alone, high-seas drift nets tangled 42 million animals that were not among the targeted species, including diving seabirds and marine mammals (Safina 1995).

Purse seine nets killed up to 400,000 dolphins per year in pursuit of Pacific yellowfin tuna, until "dolphin-safe" methods were instituted in 1990 (Safina 1995). Now, instead of netting around pods of dolphins, fishermen net around logs and other floating objects. By 1993, dolphin kills had plummeted to just 4,000 per year. However, there is still bycatch: when netting around dolphin pods, every 1,000 nets catch an average of 500 dolphins, 52 billfish, and 10 sea turtles, but no sharks. By contrast, every 1,000 nets set around flotsam catch, on average, only 2 dolphins, but also 654 billfish, 102 sea turtles, and 13,958 sharks. One must wonder whether *25 times* more sharks taken than the total number of dolphins that originally riled the public will garner the same support.

It now appears that the protections in place to reduce bycatch on marine megafauna are not necessarily working. Most tuna caught today do not involve the killing of dolphins as bycatch. However, dolphin populations are not recovering. It is thought that the chronic, sublethal effects of prolonged chase and frequent capture may be to blame (Lewison et al. 2004).

Another aspect of fishing that is rarely discussed or researched is ghost netting, the effects of fishing nets and other equipment being left behind or lost in the process of fishing. These nets continue to fish, often for years, and it's all wasted because they drift lifelessly and aimlessly through the oceans. This is not simply a hook here or there or a missing yard or two of netting—but many thousands of miles of netting; it is estimated that 20–40 percent of fishing gear is lost annually (Dayton et al. 1998). Deepwater gill nets. Monofilament nets. Lobster traps. The true quantity of ghost nets is impossible to calculate accurately due to lack of reporting. The United Nations enacted a global ban on drift nets longer than 2.5 kilometers (1.5 miles), but many countries, including Italy, France, and Ireland, have continued to use them.

If civilian women and children are the "collateral damage" of war, then bycatch is certainly the collateral damage of fishing. Many fishing vessels now carry independent observers in an effort to reduce the incidental catch of protected species, such as albatrosses and sea turtles. It's a good start, but it doesn't address the problem of the "other 99 percent," that is, the invertebrates and other unprotected species that comprise the majority of life in the oceans, the majority of organisms in a habitat, the majority of larvae and links in the food chain, and the majority of food for other species . . . and the majority of bycatch.

Habitat destruction occurs when large, heavy bottom-trawls are raked along the seafloor to catch bottom-dwelling species, such as sole, halibut, cod, haddock, plaice, rockfish, rays, skates, prawns, and others, in the process turning their three-dimensional habitats built up by corals, sponges, and bryozoans, into vast plains of mud and rubble.

Imagine a huge net weighing thousands of tons, with the mouth held open by a pair of huge separators called "otter boards," each weighing thousands of pounds, and the bottom lip bearing a formidable array of heavy steel weights, 40-centimeter (10-inch)-wide rubber discs called "rockhoppers" for riding over obstructions, and "tickler chains" to scare the fish and prawns to swim upward off the seabed. Now imagine the constricted posterior end of the trawl net, called the "cod end," filled with thousands of tons of fish, shrimp, mud, rocks, coral rubble, and lots and lots of bycatch. As the front parts and back parts of this heavy net—the huge gouging otter boards, the banging steel ticklers, the rolling rockhoppers, and the thousands of tons of bycatch and rocks—are dragged along mile after mile of seafloor, what doesn't get scooped up gets crushed. Long, deep furrows are commonly left by trawling activity, and it can take up to many years for these furrows to regrow over. Scraps of fishing nets, tow lines, chains, and tires are often found in trawling areas as well, presumably lost or discarded by fishermen.

Professor Paul Dayton of the Scripps Institution of Oceanography explains this the best:

Most ground fisheries destroy natural benthic communities by killing many small individuals of the target species and most epibenthic species (many of which are important nurseries or food sources for other species), and by disturbing the substratum enough that the community becomes colonized by species that can live in such heavily disturbed habitats. Note that the benthos does not become azoic; rather, it develops a completely different association of deposit feeders that have high turnover potential and may resist recovery of the natural populations. In addition to mortality of the bycaught species, there are sometimes important secondary effects of their discard, such as large-scale oxygen depletion, diseases, and the enhancement of selected species of seabirds that in turn impact other populations. (Dayton et al. 1998, 319)

Each pass of the trawl destroys the biogenic landscape, the physical structure of the environment: the large sponges, bryozoans, corals, worms, echinoderms, and bivalves and other mollusks that provide important food and hiding places for each other and others, including commercially important fishes; many of the large species that form these habitats grow so slowly that it will take them decades to centuries to recover (Jackson 2001). Most areas are dredged several times each year by different fishermen targeting different species. Globally, an area equal to all the world's continental shelves is trawled every two years (Carlton et al. 1999).

Trawling has been likened to clear-cutting of old-growth terrestrial forests (Watling and Norse 1998). However, whereas terrestrial clear-cutting has received a great deal of conservation attention, trawling, which levels about 150 times more area per year, has not. But some might say that trawling is even worse than terrestrial clear-cutting. Logged forests are an eyesore and their biodiversity plummets, but then they are at least left alone for a few decades to recover before being razed again. Trawled areas, in contrast, are typically trawled numerous times per year, every year, with no time to recover. This would be the equivalent of a forest being bulldozed six times a year in order to collect the lizards.

Most people have heard of or seen tropical coral reefs, but another ecosystem, possibly far more ecologically informative, has been largely ignored. Consider the cold-water corals that form vast reefs in the deep sea off Norway, Scotland, Tasmania, and elsewhere. These reef-builders bear poetic names like *Lophelia, Oculina, Madrepora, Desmophyllum,* and even the happy-sounding *Solenosmilia.* Far from warm, balmy tropical paradises, these poorly known corals are found mostly in the dark depths at 100–1,000 meters (300–3,000 feet) or more.

It is believed that these reefs may rival the coverage of the more familiar tropical corals. And like their warm and toasty cousins, they provide accommodation for such a glorious array of organisms that they may be thought of as the rainforests of the oceans. Despite their importance to ecosystems over such vast spans of the ocean abyss, we are only just beginning to learn about them. You see, these reefs were discovered by accident in July of 1982, while a Norwegian energy company was exploring for oil and gas. Fishermen and scientists had known for centuries that deep water corals existed, but it wasn't until spectacular video images emerged that the importance of these coral thickets began to be appreciated.

Australia's Great Barrier Reef is well known for its tremendous biodiversity and for being the only living structure visible from space, but one cold-water coral reef is 2 ½ times as long, stretching from Norway to Africa (about 4,500 kilometers, or 2,800 miles). Cold-water coral reefs appear to be as biodiverse as tropical corals. And yet, unprotected, these slower-growing corals are being destroyed by deepwater trawling.

Dr. Jason Hall-Spencer from the University Marine Biological Station in the UK and his colleagues studied the trawling damage to cold-water coral reefs at depths of about 1,000 meters along the West Ireland continental shelf break. Large chunks of coral (up to several feet across) were broken from reefs and trawled up as bycatch by commercial fishermen. Hall-Spencer and his team used carbon-14 dating to conclude that these trawled specimens were at least 4,550 years old. Trawlers sweep about 33 square kilometers (13 square miles) in a typical 15-day fishing trip. You do the math.

Vast tracks of fragile cold-water coral reefs and seamount habitats are turned to rubble by trawling, and we have no idea whether these incredibly slow-growing ecosystems are capable of recovering (Koslow et al. 2000; Koslow et al. 2001). The UN secretary general reported in 2006 on actions taken to address the impacts of fishing on vulnerable marine ecosystems:

> Comparative studies have shown clear differences in benthic community structure in trawled vs. untrawled areas. A coral bycatch of 3,000 kilogrammes [6,600 pounds] was documented from six trawls on seamounts off Australia that had not previously been fished for orange roughy, whereas the bycatch levels at heavily fished seamounts amounted to about 5 kilogrammes [10 pounds] for 13 trawl hauls. The bycatch of coral in the first two years (1997–1998) of bottom-trawling for orange roughy over the South Tasman Rise reached 1,762 tons but was quickly reduced to only 181 tons in 1999 to 2000. It also was reported that the most heavily fished seamount containing reef-building coral, where fishing for both orange roughy and oreos took place, eventually consisted of over 90 percent bare rock at most depths. Biomass and species richness were both drastically reduced and it was anticipated that should community recovery occur, it would likely be a lengthy process. (UN 2006, 15–16)

So in a nutshell, we are not only extracting fish faster than nature can replenish them, we are also destroying the habitat that the young would live in
and killing off the food that the young would eat. According to the fisheries
biologist Daniel Pauly, "Allowing trawling in coral forests is the worst thing
we are doing in the ocean today . . . Nothing could be dumber than destroying the habitats that depleted fish populations need to recover. Governments
should stop pussyfooting around and do something useful" (Pickrell 2004).
Indeed.

Other means of habitat destruction leading to reduced fish stocks include
when rivers required for spawning are dammed. Indeed, for species, such
as salmon, white perch, and shad, which ascend rivers to spawn, habitat destruction, rather than overfishing, is the primary cause of population depletion (Safina 1994). Habitat destruction also occurs when nursery grounds like
wetlands, estuaries, reefs, seagrass meadows, and mangroves are developed
in the name of progress. Profuse sedimentation following deforestation can
kill corals by blocking their sunlight, and filter-feeders, such as clams and scallops, by clogging their gills. Finally, tropical Indo-Pacific fish captured for the
aquarium trade are often stunned with cyanide, which kills the corals that
fish use for habitat, and blast fishing, or dynamite fishing, stuns or kills fish to
make them easier to collect but is devastating to the coral reefs.

Pelitic fog. The so-called pelitic fraction is the portion of marine sediment
that has the finest grain size, like talcum powder. Seafloor trawling stirs up
vast amounts of this fine sediment (see plate 9), which can in turn have a
major impact on the local ecosystem. These particles bring organic and toxic
substances back up into the water column, where they can add to eutrophication (see chapter 7) or be inhaled or ingested into the food chain. Pelitic fog
can travel up to 200 kilometers (125 miles) and take days to settle, acting as a
silty toxic cloud blanketing bottom-living and filter-feeding organisms, such
as clams, scallops, sponges, bryozoans, and tunicates. The silt and clay from
the pelitic fog can change the habitat in its outfall area, from sandy bottoms
or algal bottoms to essentially muddy bottoms; affected organisms unaccus-
tomed to mud are either displaced or simply die.

Water turbidity. One of the main impacts of pelitic fog is an increase in turbidity, or cloudiness of the water, which has many negative effects on marine
ecosystems. Tiny particles settle out, just as fine powdery dust settles out of

a breeze, eventually, slowly, and chokingly. Many marine animals, including corals, are harmed by this fine sedimentation. Sedimentation can smother or bury organisms that are stuck to the bottom or cause tissue necrosis. Algal symbionts are negatively impacted by the reduced light penetration in the water.

Coral reefs are remarkable hotspots of biodiversity: because of the lack of nutrients in the water, rates of evolution and niche specialization are higher than anywhere else in the world. Therefore, any threat to the corals is an ecosystem-wide threat. Recent studies reveal that corals in disturbed areas are becoming increasingly skewed toward large colonies; this is believed to be a result of the more profoundly destructive effects of turbidity on small colonies.

Intriguingly, many types of jellyfish also have algal symbionts, similar to those of corals, but the effects of light reduction have not been examined. Even more intriguingly, *Phyllorhiza punctata*, the spotted jellyfish, normally has algal symbionts in its tissues, but the population that has invaded the Gulf of Mexico lacks such symbionts. One must wonder whether this lack of symbionts predisposed it to finding its comfortable niche in the gulf's murky dead zones, or whether it lost its symbionts as a coping mechanism for the turbid conditions.

Macro-algae and kelps are also affected by turbidity. Because light penetration declines rapidly in turbid water, only those species adapted to dim sunlight can survive below about 20 meters (60 feet). The result is not only a substantial reduction in algal biodiversity but also a corresponding decline in the biodiversity of fish and invertebrates that live in association with certain types of algae.

The Effect of Overfishing on Ecosystems (and Politics)

The repeated patterns of overfishing, bycatch mortality, and habitat damage are so transparent that additional science adds only incrementally to further documentation of immediate effect. Although it is always possible to find exceptions to these patterns, the weight of evidence overwhelmingly indicates that the unintended consequences of fishing on marine ecosystems are severe, dramatic, and in some cases irreversible. (Dayton, Thrush, and Coleman 2002, 1)

The UN secretary general wrote in a 2006 report on overfishing, "By and large, a dominant human-caused direct effect on marine ecosystems is fishing. While fisheries are vitally important to the global economy as a source of food, employment and support for coastal communities, the impact of over-fishing on the health and productivity of marine ecosystems has grown to be a concern for the international community. Even if target species are not being overfished, fishing affects marine habitats and has the potential to alter the functioning, state and biodiversity of marine ecosystems, particularly vulner-able ecosystems" (UN 2006, 9).

Fishing unquestionably can cause changes in the population structure of marine communities. This applies to changes in biodiversity as well as in the relationships between species up and down the food chain. It is worth keeping in mind that the fish that fetch the highest prices are generally the top preda-tors: salmon, tuna, swordfish . . . big fish.

But we have fished out the large adults and are now fishing the juveniles. The trophy specimens of Atlantic swordfish once mounted and hung from the walls of seafood restaurants used to grow to more than 1,000 pounds. But by 1995, the average swordfish caught weighed only 90 pounds and was approxi-mately 2 years short of maturing to its first spawning (Hallowell 1998).

So too the collapse of the bluefin tuna is a tragedy to rival Shakespeare's finest—the protagonist is inevitably doomed despite its virtues, or perhaps because of them. This is an extraordinary fish (Safina and Klinger 2008). Capable of regular transoceanic migrations and using its elevated body tem-perature to hunt in high latitudes, despite all its might and agility it is unable to survive the politics and profits of its value to the sushi market. Western Atlantic landings peaked in 1964 at more than 18,000 tons, declining to just 1,500 tons in 2005. Despite significant scientific data indicating impending extinction, quotas still allow the fishery to legally capture all of the adult fish (see plate 7; MacKenzie, Mosegaard, and Rosenberg 2009).

Removal of substantial portions of any link in the food chain disrupts some predator/prey relationships and truncates others. Changes in structure and relationships can be so pervasive as to alter the function, productivity, and resilience of the ecosystem as a whole. The pathways of population restruc-turing include:

- Some species grow faster or reproduce more than other species. For example, fishing of slow-growing species or late-maturing species may lead to declines in their abundance, as species with faster life cycles use up limited resources, thus essentially "blocking" slower growers from replenishing.

- Bottom-trawling physically damages seabed communities and can reduce habitat complexity by destroying key structural or biological components.
- Lost or discarded fishing gear (ghost nets) can continue to catch fish well after the intended harvest. Equipment often targets certain types or sizes of fish, thereby exerting additional and uncalculated fishing pressure on those species.
- Fishing can lead to extinction of local stocks or, by intensively removing certain species with certain traits and characteristics, lead to a "directed" and essentially faster mode of evolution.
- Bycatch can lead to unintended impacts on nontarget species, such as dolphins, seabirds, other fish, and invertebrates.
- In fishing, we take on the role of competitor against higher-order predators, such as marine mammals, sea turtles, and seabirds. "Most marine mammals and all sea turtles are severely threatened or on the verge of extinction" (Jackson 2010, 3769).

We are blind to many of the effects of overfishing because they happen far away from our view and under the cloak of water. Some changes, such as changes in species fished, we see on our plates, but even then, we don't necessarily know whether these new names are smaller species or simply ones from newly opened fishing regions. In particular, we are blind to bycatch. There is nothing about the prawns on our plate that suggests the number of turtles lost incidentally, nor does the tuna in our sandwich tell us about the drowned albatrosses.

A landmark paper by Paul Dayton and his colleagues (1995), however, brings the problem of bycatch into sharp focus. The article should be required reading for all people who want to know about the human footprint on our natural world. Many who have not previously heard of the bycatch problem simply won't believe it. But alas, like gravity or evolution or the spherical nature of this planet, the truth does not depend on whether you believe it or not. Some of the outrageous stories they cover are summarized here briefly.

Net entanglements of marine mammals are confronting. More than 6 million porpoises were taken by the Pacific tuna purse seine fishery prior to 1987. Almost 600 humpback, fin, and minke whales were taken between 1969 and 1986 off Newfoundland. Almost 3,000 small cetaceans were killed by a small number of Filipino fishermen in a single year. Between 5 and 20 percent of the Bering and Western North Pacific populations of Dall's porpoises were taken by the Japanese salmon drift-net fishery before 1987. Small coastal species, such as the harbor porpoise and New Zealand's Hector's dolphins, are taken by gill nets at an estimated rate of 1 porpoise per 10 kilometers of gill net per day, and 500–1,000 porpoises are taken annually in Danish waters. The vaquita in the Gulf of California has been driven to the brink of extinction by gill nets. Highly endangered Mediterranean and Hawaiian monk seals become entangled with fishing gear. Sea otters are killed by gill nets—fur seals and sea lions, too. It seems that mammals are attracted to the fish that are captured by nets, and they become entangled in the process of trying to feed.

Sea turtles too are taken as bycatch in stomach-churning numbers. Loggerhead and Kemp's ridley turtles prey preferentially on shrimp and crabs and are often found in close association with them. But here they are captured in shrimp trawls, or killed or injured when hit by heavy fishing gear. Like mammals, turtles are attracted to fish caught in nets and on longlines: "incidental take rates of up to several tens of thousands of sea turtles are possible in the Hawaiian-based longline fishery alone" (Dayton et al. 1995, 209). Glow sticks used by swordfish longliners and marker buoys used for pot-traps attract leatherbacks, resulting in high mortality.

So too, bycatch of seabirds is gobsmacking. In one study in St. Ives Bay in southwestern England, some 900 razorbills and diving birds were killed in just 8 days. Longline tuna fishermen in the Southern Hemisphere have been implicated in a massive annual kill of at least 44,000 wandering albatrosses, though some argue that the actual annual take is twice as high. Forty-four thousand . . . per year . . . and that's just one species. . . .

More recently, a report submitted to the International Whaling Commission by a team from Duke University in the United States and the University of St. Andrews in the UK estimated that about 800 cetaceans (whales, dolphins, and porpoises) and a slightly higher number of pinnipeds (seals and sea lions) are killed each day as bycatch in fishing nets (Read, Drinker, and Northridge 2003; Kirby 2003). *Each day*. More than 300,000 a year for each group. And this is with all the legislation in place to protect them and even after the dolphin-safe practices were put in place decades ago. *Eight hundred a day*.

Overfishing doesn't just affect marine mammals and turtles and seabirds. In order to fully understand how ecosystems shift to a "new normal," we must first understand ecosystem degradation. More often than not, jellyfish don't just "take over"—they fill a vacancy or exploit an instability. Overfishing is one of the main causes of shifts toward jellyfish because it causes both: the extracted fish leave uneaten food, and fish extraction, bycatch, and habitat destruction cause all sorts of stresses on the whole ecosystem.

One of the best examples of the ecosystem-wide effects of overfishing is found in Caribbean coral reefs. "Western Atlantic reef corals suffered catastrophic mortality in the 1980s. Live coral abundance declined to 1–2% cover from values of 50% or more" (Jackson 2001, 5412). The principal cause of coral mortality is overgrowth of macroalgae. Fleshy algae and algal turfs are generally kept in check by herbivorous (vegetarian) grazers like sea urchins and many species of algae-eating fish. Overfishing of Caribbean reef fish in the nineteenth century allowed their urchin competitors to flourish, still effectively keeping the algae under control, but then an unidentified pathogen wiped out the enormously abundant sea urchins in 1983–1984. With the fish already gone, this left no herbivores to graze down the algae. The take-home lesson here is that the effects of overfishing were obscured for over a century by the ecological redundancy of the herbivores. The ecosystem was already shifting, but we didn't notice.

Krill

Dr. Angus Atkinson of the British Antarctic Survey, a top organization in Antarctic research, led a team that studied data from nearly 12,000 net hauls from 9 countries, spanning the years 1926–2003 (Atkinson et al. 2004). In particular, they looked at the distribution and abundance of krill across time and space, and also at the same criteria for salps. Salps are a type of strange gelatinous zooplankton that are in our phylum (Chordata) and therefore are more closely related to us than to jellyfish. But they look and act more like jellyfish. Salps are characterized by explosive population growth: imagine growing 10 percent of your body length *per hour* and completing an entire generation from birth to giving birth in just one to two *days* (Heron 1972).

Both salps and krill are voracious consumers of phytoplankton. Atkinson and his team found that 58–71 percent of krill are located in the highly productive southwest Atlantic sector, while vast blooms of salps occupy the less productive "other regions" of the Southern Ocean. There is a see-saw effect between krill density and salp blooms that is strongly dependant on sea-ice cover. Krill live for several years and feed on algae growing on the underside of the ice during the wintertime, whereas salps live less than a year and are inhibited by ice; therefore, winters with extensive sea-ice cover promote early spawning and good survival of krill larvae, whereas open-water conditions promote extensive salp blooms and poor krill survival (Loeb et al. 1997).

Atkinson and his colleagues went on to note that "the western Antarctic Peninsula is one of the world's fastest warming areas, and (atypically for the Southern Ocean) winter sea-ice duration in this sector is shortening" (p. 102). Therefore, the part of the globe where the krill are in their highest densities is the very same area that is warming the most quickly—that is to say, the sea-ice that drove their former abundance is disappearing.

In fact, krill densities have declined significantly since 1976 in the southwest Atlantic sector, from over 100 to fewer than 5 per square meter. During the same period, salps increased by more than 50 percent. These changes are profound, especially for animals that prey largely or entirely on krill: penguins, albatrosses, seals, and whales, to name a few.

But declines in krill aren't only about loss of a food source. Krill also help sequester carbon dioxide in the same way that copepods do, but on a much larger scale in view of their far larger bodies. At the sea surface, they consume vast amounts of phytoplankton, which are essentially tiny carbon factories. Krill dive to great depths, releasing carbon-rich waste into the deep sea. According to one study, the amount of carbon dioxide that millions of krill transfer to the ocean floor is equal to the annual emissions of 35 million cars (KrillCount 2011).

The penguin vignette in chapter 2 raises the issue of copepods capitalizing on the decrease in krill. Krill predators, such as whales and penguins cannot see copepods, and so they are starving. But jellyfish, which eat copepods, are flourishing. And salps.

Sharks

Many people, as a knee-jerk reaction, have very little sympathy for sharks. It's easy to see why. They are big and pointy-faced, and they have funny eyes. All those gills are a bit weird. And that sandpapery skin, not very cuddly. Get real—who are we kidding?—it's the teeth. Too many teeth. It's the thought of getting ripped into shreds. And that creepy nictating membrane folding over the eyes for protection just before the attack. C'mon, I saw *Jaws*: it's not a lack of sympathy—it's F.E.A.R. But despite justifiable apprehension about swimming with sharks, killing them off, whether on purpose or accidentally, is a very bad idea, at least for the ecosystems that they inhabit. As top predators vanish, instability creates a perfect context for jellyfish.

Shark Depletion in the Western Atlantic

The pelagic longline fisheries targeting tuna and swordfish in the Northwest Atlantic land a lot of sharks as bycatch—about three times more than swordfish, to be exact (Dayton et al. 1995). On top of that, coastal and open ocean sharks are hunted directly (Baum et al. 2003). It has been estimated that scalloped hammerhead sharks, white sharks, and thresher sharks have all declined by more than 75 percent between the mid-1980s and 2000. These figures do not account for declines that took place before the mid-1980s.

Julia Baum and Ransom Myers from Dalhousie University in Nova Scotia studied a longer-term dataset in the nearby Gulf of Mexico, and their results suggest an even greater problem. By comparing shark bycatch from the yellowfin tuna fisheries in the 1950s with commercial fishing for sharks during the 1990s, they found huge declines in the number of oceanic whitetip and silky sharks caught (Baum and Myers 2004). The study period 1954–1957 used almost 83,000 hooks and caught 397 whitetips and 158 silkies, whereas the study period 1995–1999 used more than 219,000 hooks and caught only 5 whitetips and 24 silkies. This represents a decline of 99.3 percent for whitetips and 91.2 percent for silkies.

Just a smidgen to the southwest in the Caribbean and Gulf of Mexico, another study tells a similar story (Ward-Paige et al. 2010). In the 1500s, a large group of sharks surrounded Columbus' ships off the east coast of Panama to the extent that the sailors were frightened by their number and ferocity. It was common for sharks to swarm and they were so abundant throughout the greater Caribbean as to be "expected anywhere at anytime" (p. 9), even through the 1950s. In the Atlantic and Gulf of Mexico, recreational fishers took up to 1.5 million coastal and oceanic sharks annually for 1974–1975 alone. But landings of sharks tripled in the Gulf of Mexico in the 1980s and peaked at more than 99 million tons in 1990. A study from 1993 to 2008 found that nurse sharks were sighted during 10 percent of dives, but other types of sharks were collectively sighted on only 3 percent of dives, leading the researchers to conclude that sharks were "expected anytime almost nowhere" (p. 9).

Shark Depletion in the Sea of Cortez, Mexico

On the Pacific side of Mexico in the Sea of Cortez, 20 million sharks and their relatives are killed each year by the shark-fin fishery, 80 percent of which are immature and many of the rest pregnant females (Bahnsen 2006). In a study of 83 artisanal fishing sites during 1998–1999, sharks dominated the landings in all seasons and made up over 71 percent of the catch, with the scalloped hammerhead as the primary species (Bizzarro et al. 2009). Of 97 hammerhead specimens measured and sexed by the researchers, all were juveniles and the majority were under 95 centimeters (3 feet) in length. Large schools of hammerheads used to seasonally frequent seamounts in the area, but have since vanished.

Depletion of Mediterranean Sharks

Direct fishing for their fins isn't the only threat to sharks. Bycatch has depleted shark populations in the Mediterranean to the precipice of extinction. A recent study examined the population abundance of 20 species of large sharks in the northwestern Mediterranean in the nineteenth and twentieth centuries (Ferretti et al. 2008). Of these, only five species had enough data for analysis—but the picture painted by these five was glum.

Consider the hammerhead shark. In the early 1900s catches and sightings were regular. But after 1963 no hammerheads were caught or seen in coastal areas. Hammerheads in open waters took a bit longer to deplete, but eventually caught up—catches were already low by 1978, but there were no more records of hammerhead sharks after 1995. The estimated decline is said to be "greater than 99.99 percent" (Ferretti et al. 2008, 8).

The decline of the blue shark is similar. During the nineteenth century this species was "very abundant," often forming near-shore aggregations off the Tuscan Archipelago. It was the most common shark species caught in open water. Since the mid-twentieth century, blue sharks in coastal waters declined by "more than 99.99 percent," whereas open-water blue sharks declined by about 80 percent in just 21 years in Italian waters and by 99.78 percent in 25 years in Spanish waters.

Repeating the statistics on the remaining species almost seems like overkill to make the point: mackerel sharks have declined more than 99.99 percent . . . thresher sharks have declined more than 99.99 percent . . . At these low levels large sharks are functionally extinct in the Mediterranean. It seems that sharks that prefer coastal habitats were killed off long ago, while other species sought refuge in open water. However, when open-water fishing for tuna and swordfish expanded in the 1970s, sharks became regular bycatch of driftnets and longlines. Prior to their ban in 2002, about 700 Mediterranean fishing boats were using driftnets, and it appears that some 1,000 to 2,000 boats are still fishing with longlines.

As an aside, other large predators, such as whales and dolphins, seals and sea lions, turtles, and all sorts of large bony fishes, have declined by similar rates in the Mediterranean. Big species often grow slowly and have few young per year, so it doesn't take much hunting pressure to reach a tipping point.

Shark Depletion in Australia

Besides direct fishing and bycatch, sharks in Australia face yet another threat. Many popular swimming beaches in the Great Barrier Reef Marine Park and throughout Queensland and New South Wales are culled for sharks for safety reasons. Australia's shark control program uses large nets and baited drumlines off ocean beaches with the aim of attracting and culling sharks to reduce their populations. But that's not all they cull. For example, according to Humane Society International:

> In NSW between 1950 and 2008, 577 great white sharks and 352 tiger sharks were caught in shark control nets. Over the same period 15,135 other marine animals were caught and killed in nets, including turtles, whales, dolphins, rays, dugongs, and harmless species of sharks. This figure includes 377 of the now critically endangered and harmless grey nurse shark, a number which threatens their future survival.
>
> In QLD, during the first 15 years of the shark control program 14,328 marine animals other than sharks were caught in the nets and drumlines. Between 1975 and 2001, 11,899 great white sharks, tiger sharks and bull sharks were killed in nets and drumlines. Over the same period 53,098 other marine animals were killed. In 2008, 578 sharks were caught in shark control equipment in Queensland and 505 sharks were caught between January and 20th November 2009. Less than half of those sharks caught were considered the dangerous or target species. (HSI 2010)

> In the last 30 years, Queensland's shark nets have killed approximately 8644 rays, 3127 turtles, 837 dugongs and 362 cetaceans. (AMCS 2011)

So that we don't misunderstand these statistics and think the problem has lessened over time, in Queensland in 2009, 16 dolphins, 6 whales, 1 dugong, and 30 turtles were caught in shark nets, and in New South Wales in November 2009 alone, a great white shark and a dugong (both threatened species), were trapped and killed in the nets. In 2004, 2 large breeding females (of critically endangered grey nurse sharks) were killed in the nets (HSI 2010).

According to the Australian Marine Conservation Society, shark-control nets give swimmers a false sense of security, allowing them to believe that the nets are, in essence, "shark-proof impenetrable barriers." In fact, "Unknown to most people, shark nets do not enclose entire beaches to protect swimmers. Instead, they are set parallel and sometimes diagonal or even perpendicular to the beach. Statistics show that 40 percent of sharks caught in beach nets are caught on the beach side of the nets" (AMCS 2011). Yikes.

Nonetheless, it could be worse. Sharks are also a problem along the KwaZulu-Natal coast (Kearney 2010). Like Australia, many popular bathing beaches are netted for protection—these too are not a continuous barrier but rather, staggered rows. On average, about 591 sharks were caught each year from 2005 to 2009, with just 13 percent released alive. The 4 most common species caught are considered nonthreatening to humans. There were other types of bycatch; from 2005 to 2009, an average of 43 dolphins per year were caught, with only 1 live release, plus 201 rays and guitarfish, 60 turtles, 5 whales, and 5 birds.

It may be tempting to think "good riddance" about the demise of sharks. But it is not that straightforward. As we shall see again and again, a decline in top predators almost always leads to a flourishing of their prey, which then impact heavily on *their* prey, and so on . . . And yep, you guessed it, these sorts of ecosystem disturbances are often exploited by jellyfish.

> Something is horribly wrong when a group of animals that has dominated the seas for 60 million years begins to disappear within a decade, despite the existence of a law that contains the words conservation and management in its title.
>
> —CARL SAFINA, founder and director of the National
> Audubon Society's Living Oceans Program and
> founder of the Marine Fish Conservation Network

Horseshoe Crabs

If sharks are hard to find sympathy for, then horseshoe crabs are really pushing their luck. They're a bit weird. Well, they look weird, for one thing, and they are related not to crustaceans but to spiders. Aaaarrrrgghhhh! They have eyes in all kinds of wrong places: on top of the shell, on the tail, and under the shell near the mouth . . . Weird, bordering on creepy. Furthermore, they are the only chelicerate (spiders, scorpions, and their kin) with compound eyes like those of flies and bees, and they can even see ultraviolet light.

Horseshoe crabs are living fossils. Their earliest fossils date back some 445 million years, and they have remained virtually unchanged for 250 million years. Hardy spiders. But despite all their peculiarities, their blood is the basis of one of the most often used and medically useful procedures in saving human lives. Bacterial residues on needles and implants can be fatal, and some can survive the heat sterilization process. But a component isolated from the blood of horseshoe crabs clots in the presence of these bacteria; a testing procedure was developed in the 1960s and is still in use today.

When most of us look at a horseshoe crab, we see a strange little flat scurrying creature, but the Japanese see something quite different. Japanese legend holds that brave warriors who die in battle are reincarnated as horseshoe crabs—their shells being the helmets of samurai warriors—brought back to earth to march eternally across the seabed.

Not only do horseshoe crabs look and act weird—and they do that weird blood clotting thing—but they're not even edible. Well, not to humans anyway. So it's asking a lot to beg your sympathy for the lowly, funny-looking horseshoe crab. But we are losing them, and with them, the vast number of birds and other creatures that rely on them . . . and the test materials needed to keep those needles safe.

Hundreds of thousands of horseshoe crabs and up to 1.5 million shorebirds congregate on Delaware Bay each spring. Droves of crabs migrate inshore to spawn—up to 80,000 eggs being laid in the sand by each female. The birds, particularly a species with the unfortunate common name of the red knot, come to gorge on the eggs to ready themselves for the long flight to their nesting grounds in Canada. Having already flown north from Patagonia on the southern tip of South America, their small bodies rely on the rich food of the crab eggs to help them survive the rest of their 18,000-kilometer (11,000-mile) migration.

But millions of hungry birds aren't the only threat to the crabs. Fishermen catch them to sell as bait for catching eels and whelks. In fact, between 1990 and 1994, at least 500,000 horseshoe crabs a year were taken by intense offshore trawling in New Jersey, Delaware, and Maryland. By 1997, over 2 million crabs a year were being harvested. Not surprisingly, by the mid-1990s, scientists began noticing dramatic declines in horseshoe crabs and shorebirds, down by 90 percent at some beaches. Restrictions to harvesting were put in place in 1997, but "despite restrictions, the 2007 horseshoe crab harvest was still well above that of 1990, and no recovery of knots was detectable," said Lawrence Niles, a biologist with the Conserve Wildlife Foundation of New Jersey (USGS 2009).

Cascading effects of the decline in horseshoe crabs and shorebirds include threats to the multimillion-dollar ecotourism industry revolving around the annual bird migrations, as well as declines in Atlantic loggerhead sea turtles blamed on the shortage of their primary food source, the humble horseshoe crab. And meanwhile, the jellyfish in these areas continue to increase. . . .

We've all heard the "fish story" about the big one that got away. Indeed, fish stories are always about large size—overexaggerated catches, or the one that got away, or the one that's simply got to be there. We connect our own prowess with conquest of the big one. However, it appears that massage of our own egos comes at a great cost to other species (Birkeland and Dayton 2005).

While many human cultures and popular media tend to favor slim, young ladies, it seems that the opposite is true in the marine environment. Indeed, the take of big old fat female fish ("BOFFFs," they're called), is contributing to vanishing fish stocks ("Big Old Fat Fecund Female Fish" 2007).

Recent experimental evidence has shown that older fish produce larvae with significantly better survival potential than those from younger fish (Berkeley, Chapman, and Sogard 2004; Berkeley et al. 2004; Bobko and Berkeley 2004). Furthermore, larger fish produce vastly more eggs than smaller fish, because egg production is proportional to fish volume, which is proportional to the cube of its length (Palumbi 2004). So if we have rationalized our predatory urge to conquer "the big one" by thinking that it is more ecologically sound to take one large fish rather than several smaller individuals, we may want to rethink our logic. As Stanford University's Stephen Palumbi quipped in his paper *Why Mothers Matter* (2004, 621), "If Momma ain't happy, ain't nobody happy."

Besides the obvious risks of crashed fisheries, extinction of species, and rearrangement of food webs, another key impact of overfishing is in its evolutionary effects, dubbed the "Darwinian debt." In net fisheries, everything larger than the mesh size is taken, leaving behind the smaller ones. In recreational hook-and-line fisheries, we throw back the undersized or the ones that don't fit our image of a good fish—healthy weight, the right color, the right proportions—the same features that natural selection and, in some cases, sexual selection act on. In visual fisheries, such as those for abalone, beche de mer, pearling, or spear-gun fishing, we target the largest, healthiest individuals of a species, leaving behind those that are smaller, scrawnier, or more slowly growing. These become the breeding stock.

However, scientists worry that age truncation, which is an inevitable outcome of overfishing, is changing the course of evolution. One such scientist is Dr. Ulf Dieckmann, who leads the Adaptive Dynamics Network Program at the International Institute for Applied Systems Analysis in Laxenburg, Austria. In a recent paper on fishery-induced evolution, Dieckmann and his colleagues present compelling evidence from three different sources (long-term field data, controlled laboratory experiments, and model-based studies) to demonstrate the changes to fish stocks (Dieckmann, Heino, and Rijnsdorp 2009). In a nutshell, the evolutionary "damage" is occurring at a faster rate than it can be repaired.

Back to cod for a moment. Newfoundland cod was one of the world's most abundant cod stocks. In the early 1960s its northern stock numbered almost 2 billion breeding individuals (Hutchings and Reynolds 2004). However, its collapse was extraordinary and swift: since 1983, northern cod stocks have declined by 99.9 percent. It was given endangered status in 2003; however, cod continue to be caught as bycatch in other fisheries, slowing any potential recovery. Many collapsed stocks do not recover.

Consider the evolutionary pressure too, for example, that occurs following an annual 2-day sport diving lobster extravaganza in the Florida Keys. During this "mini-season," approximately 50,000 people don scuba or snorkeling gear to catch Caribbean spiny lobsters. This event alone—not including commercial fishing pressure the rest of the year—reduces the population density of lobsters by over 95 percent on patch reefs and 79 percent at patch heads (Eggleston et al. 2003).

We may try to take comfort in thinking that the evolutionary effects from these activities are merely natural selection at work. However, "recent studies suggest that fisheries-induced selection acts against and tends to swamp natural selection, leading to changes in populations that may take a long time, if ever, to reverse" (Pandolfi 2009, 43).

We lull ourselves into a false sense of security by thinking that some species recover, therefore all can. It's true: there are some success stories (Dayton et al. 2002). Some of the whales that were overhunted to the precipice of extinction have since come back to robust populations . . . but others have not. Anchovies and sardines—and even cod—have undergone boom and bust cycles . . . well, some have boomed while others have just busted. The ability of a species to recover depends on some ethereal mix of rapid reproduction, habitat stability, food supply, and protection from direct fishing or bycatch. It also depends on how it was fished, and to some extent, on luck.

Dr. Elliott Norse is the chief scientist of the Marine Conservation Institute in Washington. In a recent paper, he and his colleagues (2012) used as a good analogy for overfishing the very unwise banking practice of withdrawing capital and leaving the interest. With such a strategy, one could live a pretty exciting lifestyle up until the point that you run out of capital, and then, to use a very technical term, you're screwed. Same with fishing: catch rates could be quite high right up until the point that fish become scarce.

Trawl fishing is particularly vulnerable to this problem of removing capital, because, of course, the desired large fish are removed, but the young are also lost as bycatch and even the habitat is damaged, somewhat akin to burning down the bank building in order to access your cash.

The Australian orange roughy fishery is one of the worst cases of rapid unrestrained depletion, followed by aggressive efforts toward protection. While it is often used as a canonical overfishing example, it may also go down in history as being a case of tremendously bad luck and, paradoxically, tremendously good luck. The story begins in 1986. Gemfish were in decline, leaving many fishing trawlers out of business and many fishermen looking for something else to catch. The New Zealand orange roughy fishery was robust and lucrative, leading Australia to explore this species as an option. Pre-1986 surveys found low numbers, but the following years, huge spawning and non-spawning aggregations were found on the seamounts off southeastern Australia. Without baseline data, quotas, or sustainable harvest strategies, it was essentially a free-for-all. In Tasmanian waters, more than 40,000 tons were taken in 1989 and more than 63,000 tons in 1990 (Bax et al. 2003; see plate 7). Within 4 years of beginning, the fishery was now depleted down to less than 20 percent of its prefished biomass. The orange roughy is a slow-growing, late-maturing, long-lived species, making it particularly vulnerable to rapid decline and slow, if any, recovery. Living over 100 years, it does not begin reproducing until in its thirties. Greenpeace says that "orange roughy occurs in 'pockets' of the deep oceans worldwide, with commercial fisheries off New Zealand, Australia, Namibia and the Northeast Atlantic. It has been severely overfished and has undergone dramatic population declines in some areas. It is fished over seamounts, steep continental slopes and ocean ridges using bottom trawling gear which has caused considerable damage to sensitive seafloor habitats including corals" (Greenpeace 2012). It also forms large spawning aggregations, making it a gift from the gods for fishermen but exceptionally vulnerable to overexploitation. The orange roughy is the first commercially fished species to be listed on Australia's Endangered Species register due to overfishing. The Australian fishery for orange roughy only began in the 1990s, but in just a decade its population had fallen to just 7 percent of its unfished biomass (Darby 2006).

Now, more than 20 years on, the orange roughy is showing some signs of recovery, but also many indications of a population still in distress. A CSIRO report by Rudy Kloser, Caroline Sutton, and Kyne Krusig-Golub (2012), presents the following patterns of indicators since 1990: the average age of the population has decreased by 15–25 years (variable between sex and regions); the reproduction potential of the stock is estimated to be just 32 percent of virgin biomass; and fish are now maturing younger and smaller than before— whether this is an artefact of the data because of removal of the larger individuals, or a biological response to reduced density of stock, is not yet understood. Moreover, the spawning biomass on one key seamount has decreased by 73 percent; the population appears to be recovering, but at a rate of a mere 1,900 tons per year.

Dr. Rudy Kloser of CSIRO is an expert on orange roughy. Asked why the orange roughy was overfished, why the population is recovering, and what can be learned by these events, he highlighted three things that are now being done differently. First, the precautionary principle recommended by the Food and Agriculture Organization of the United Nations has been implemented, whereby the degree of precaution exercised is proportionate to the degree of uncertainty in the system. Second, the burden of proof has shifted from the scientists having to prove depletion, to the fishing industry having to prove sustainability. Third, stock management and harvest strategies are now based on sound science, routine monitoring, and good compliance (Rudy Kloser, personal communication).

One would instinctively think that the severely depleted orange roughy,

being a slow-growing species fished by trawling, would be perhaps more vulnerable than most to the removal of capital in Norse's banking analogy. And
one might be right . . . or maybe not. By a quirk of natural history, the orange
roughy aggregations are comprised entirely of adults. The whereabouts of
the juveniles remains a mystery. There is no denying that the population has
been grossly decimated, but because of the natural geographical separation
between the juveniles and adults, only the latter have been fished out. Tomorrow's generation is still intact and growing up. So instead of fishing out
the capital and leaving the interest, the case of the orange roughy is precisely
the opposite: the spawning and non-spawning adults have been removed, but
tomorrow's generation remains. Therefore, the full reproductive potential of
these young is like a dowry in the bank. Whether this is enough to save the
Australian population, nobody knows. But it's an interesting slant to what
otherwise appears to be a worrying and bleak situation.

While we often comfort ourselves with overly optimistic beliefs that species
will recover, we are further reassured international agreements and regulatory
bodies to protect our fisheries resources. Yes, such management tools are in
place, but they are only as strong as the weakest link in the chain of those who
are expected to honor them.

> If a country objects to the restrictions of a particular agreement, it just ignores
> them. In 1991, for instance, several countries arranged to reduce their catches of
> swordfish from the Atlantic; Spain and the United States complied with the limi-
> tations (set at 15 percent less than 1988 levels), but Japan's catch rose 70 percent,
> Portugal's landings increased by 120 percent and Canada's take nearly tripled.
> Norway has decided unilaterally to resume hunting minke whales despite an in-
> ternational moratorium. Japan's hunting of minke whales, ostensibly for scientific
> purposes, supplies meat that is sold for food and maintains a market that supports
> illegal whaling worldwide. (Safina 1995, 51)

Even some of the fisheries certified as "sustainable" do not appear to be,
indeed, sustainable. The Marine Stewardship Council, based in London, is
the world's most established fisheries certifier. The pollock fishery in the
eastern Bering Sea was certified as sustainable by the council in 2005, and
this same fishery was recertified in 2010, despite a 64 percent decline in its
spawning biomass between 2004 and 2009. In another example, the Pacific

hake fishery has declined 89 percent since its peak in the late 1980s, but was nonetheless certified as sustainable in 2009.

According to a recent publication by a group of highly respected fisheries biologists, the Marine Stewardship Council grants certification to depleted fisheries that merely hope to become sustainable, whereas these experts argue that certification of sustainability should be granted on the basis of actually being sustainable (Jacquet et al. 2010).

Furthermore, any system that relies on goodwill in reporting and intent is vulnerable to shenanigans. Sometimes people and organizations try to fool themselves, and sometimes they try to fool others. A representative of the French fishing industry ridiculously asserted that "the stocks are not declining, they are changing location" (Bigot 2002, translated in Pauly, Watson, and Alder 2005). In another example, it appears that China overreported fishery landings to the Food and Agriculture Organization of the United Nations, the only institution that maintains global fisheries statistics (Watson and Pauly 2001); because of China's huge contribution to global fish landings, these misrepresentations resulted in an overinflation of fish stock estimates. It just seems that if you have to lie about it, something is really seriously wrong.

The Effect of Overfishing on Jellyfish

Overfishing may well be the single most important factor behind the largest, most destructive jellyfish blooms, and the gelatinous future that we appear to be facing.

In a philosophical sense, species in ecosystems try to survive and grow but are kept in check by others also trying to survive and grow. Predators and prey. Competitors for limited resources. This life and death struggle against each other creates a delicate dynamic balance, where any miniscule change can tip the balance in one species' favor over another. Ecosystems have therefore evolved in a relative sort of dynamic equilibrium, according to the carrying capacity of each region. If one species fluctuates, it comes at the cost of another. Increases in one species lead to increased demand on its prey and more food for its predators. Decreases in one species lead to starvation of its predators and relief from predation on its prey.

Homo sapiens, being as wise as we are, generally prefer high-energy, meaty sorts of prey. Tuna. Swordfish. Salmon. Shark. Big fish. More bang for the buck. But removing these top predators for our use comes at the cost of something else in the food chain. Prey that are no longer being preyed upon can flourish, which has upstream and downstream effects, and so on.

The mechanism by which jellyfish exploit this perturbation is simple: fishing removes the predators and competitors of jellyfish → nature abhors a vacuum → jellyfish quietly fill this vacuum.

Loss of Predators

One complication of overfishing is the loss of predators. The few predators of jellyfish have long been targeted and are now all but gone: sea turtles, mackerel, salmon, butterfish, sunfish, trevalla, albatross, and auklets.

Blue swimmer crabs are not typically listed as jellyfish predators, but they are often observed by scuba divers swimming up and snatching a jellyfish out of the water column, then settling back down to a quiet dining spot, and tearing their prey apart, transferring clawful by clawful of the crunchy-juicy shreds to their fast-moving mouth parts. But we fish a lot of them too.

Few studies exist on the direct effect between overfishing of jellyfish predators and the increase in jellyfish blooms. However, to risk stating the obvious, this may be a contributing factor.

Loss of Competitors

Another complication of overfishing is loss of competitors: mostly the oily fish, such as anchovies, sardines, pilchards, and their kin—the so-called forage fish that live in vast schools in the upper parts of the water column. These are the primary competitors of jellyfish. The primary prey of forage fish is zooplankton, the same as for most jellyfish. This dietary overlap is the key to much of the problem. Like a see-saw, when one goes down, the other goes up.

Humans prey on these vast schools in enormous quantities. We like sardines kippered in brine and packed in oil or cooked in tomato sauce topped with gooey melting cheese. We like anchovies salted as a pizza topping or as the vibrant "zing" in a Caesar salad. Some brave souls even like pickled herring. We also harvest these fish to grind up for fertilizer, pet food, and aquaculture feed. In 2000, the catch of forage fish or small pelagic fish totaled some 42 million tons—a whopping 44 percent of global marine landings (Pauly, Watson, and Alder 2005). Yeah, that's a lot. But that's also a lot of fish that aren't available to their normal predators—the sorts of fish we like to eat like cod, snapper, tuna, and halibut.

Perhaps the most celebrated story of forage fish is one woven into the history of Monterey, a once-sleepy town on California's central coast. Monterey is nestled on a jagged promontory at the southern end of a broad arching bay, south of the metropoli of San Francisco and Santa Cruz; farther south is the breathtaking and photogenic Hundred Mile Drive of the Big Sur coastline. Monterey was made famous by the novelist John Steinbeck in his book *Cannery Row*, which began, "Cannery Row . . . is a poem, a stink, a grating noise, a quality of light, a tone, a habit, a nostalgia, a dream. Cannery Row is the gathered and scattered."

Steinbeck's Monterey was a thriving fishing community, made rich in stories and wealth by fantastically abundant sardines. The fishery was started around 1900, and by World War I was at full throttle. The corrugated iron canneries lining the shore exploded in productivity from 75,000 cases in 1915 to 1,400,000 cases in 1918. And then it collapsed. Similar heavy fishing led to a second collapse after World War II.

Today, Monterey is a vibrant tourist mecca. Good food, good wine, entertainment, history, shopping, and the world-class Monterey Bay Aquarium. No more stink. No more grating noise.

Nostalgia. But still not very many sardines.

Many other places around the world have suffered similar population crashes in their sardines, anchovies, menhaden, and the like—places without famous novels glorifying the romance and stink. Peru. Namibia. Brazil. Spain. Japan. Russia. China. India. And so on. But in all cases, whether immortalized in print or not, these forage fish were the competitors of jellyfish as well as the food for charismatic megafauna, such as seals, albatrosses, and dolphins.

These devastating changes to the structure of food webs are exactly what jellyfish need to be able to stick their metaphorical foot in the door. . . . Anchovy landings in the Yellow Sea have decreased twentyfold in the last five years, while in 2009 a research cruise there found that jellyfish constituted 95 percent of hauls (Stone 2010). Research is currently underway to try to determine how many jellyfish it takes to reach the tipping point of, in essence, "doomsday" for fish.

It's tempting to picture fishermen as big, bad, burly men who wantonly pillage the oceans' wealth. But that is a simplistic and misleading view. It may well be that some fishermen don't get the concept of overfishing in a big-picture way, but not most. It would be illogical and financially suicidal to purposely exterminate one's own source of income. And perhaps it's us, the consumer, and the government regulators who have got it wrong. As consumers, we want certain types of fish, so fishermen respond to market demand by bringing us the fish that sell for the highest price, while regulatory bodies respond by issuing licenses and quotas that work against sustainability. Our refusal to consider eating other fish and the regulations prohibiting the take of other fish mean that an enormous volume of fish is taken that have no market and no legal basis for any attempt to market them.

We may choose to view fishermen as refusing to mend their profligate ways, racing one another to catch the last fish, or we can realize that we, the consumer, have backed them into a corner where their livelihood is vanishing. Nobody wants to be the first to go broke. In such a case, it is always better to be the last to go broke.

Creating Jellyfish Nurseries

While the removal of fish that act as predators and competitors of jellyfish has obvious implications, the overfishing problem may be far more complex and the outcomes far more serious in supporting jellyfish blooms. It appears that intense trawling activity may be contributing to the expansion of jellyfish populations (Thrush and Dayton 2010). In particular, trawling opens up habitat for opportunistic settlers and kills off filter-feeding species, such as clams, scallops, and sponges, that prey on jellyfish larvae and keep them from settling. Thus, trawling not only flattens the three-dimensional habitat for fish nurseries, it simultaneously creates new micro-reefs that are a perfect habitat for jellyfish polyps. The photograph at the bottom of plate 9 shows the seabed of a regularly trawled region: part of the photo shows where a trawl has just come through and raked the sediments; the large number of brittlestar arms will act as micro-reefs for small opportunists, such as jellyfish polyps. It seems plausible that jellyfish problems in the Bering Sea and Gulf of Mexico, plus many other regions, may be suffering this effect.

Aquaculture and Mariculture

It might seem that the answer to overfishing is aquaculture, that is, raising fish and shellfish in captivity. Indeed, about half the sales of both salmon and shrimp come from aquaculture; overall, about a fifth of the fish consumed worldwide comes from aquaculture. However, it appears that mariculture, or marine aquaculture, comes at a great cost in terms of collateral damage to coastal ecosystems and creating favorable conditions for jellyfish blooms. In trying to save the ecosystem, we may well be causing it more harm.

Many cultivated species require fish meal as food, which adds to the over-fishing problem. Formerly worthless catch is now used to feed shrimp, creating a demand for essentially everything, including, often, juveniles of valuable species too small to sell (Safina 1995). Because of predator/prey food efficiency principles, it can take three pounds of fish meal to produce one pound of salmon, which is far from efficient in the bigger scheme of dwindling food resources and a growing human population.

Furthermore, a constellation of local environmental impacts of aquaculture can include clear-cutting of mangroves to provide room for coastal fish pens, large volumes of organic and inorganic wastes, including antibiotics and biocides, introduction of alien species, and escapees into the local ecosystem (CIESM 2007). Often the poor of developing nations are left to struggle with dwindling fish resources used for subsistence, while the rich develop the coastlands to raise shrimp and other valuable species for export to wealthy countries.

The PEW Oceans Commission lists the following in its May 2003 Final Report (PEW 2003):

- A salmon farm of 200,000 fish releases an amount of nitrogen, phosphorus, and fecal matter roughly equivalent to the nutrient waste in the untreated sewage from 20,000, 25,000, and 65,000 people, respectively.
- A December 2000 storm resulted in the escape of 100,000 salmon from a single farm in Maine, about 1,000 times the number of documented wild adult salmon in Maine.
- Over the past decade, nearly 1 million nonnative Atlantic salmon have escaped from fish farms and established themselves in streams in the Pacific Northwest.

Many shrimp farms are unable to afford high-tech, eco-friendly methods and equipment, so they resort instead to dumping or flushing accumulated sludge. For example, China's annual effluent of 43 billion tons of wastewater from shrimp farms utterly dwarfs the 4 billion tons of industrial wastewater (Stokstad 2010). Furthermore, consider the amount of forage fish needed for food—indeed, nearly a third of the aquaculture share of fish meal is consumed by farmed shrimp. Perhaps "consumed" isn't the right word, as a lot of it goes to waste, adding to the nutrient load in the water pumped back into the ocean, resulting in severe eutrophication. According to one expert, they are polluting themselves out of business.

An interesting study led by Wen-Tseng Lo of Taiwan's National Sun Yat-Sen University elegantly demonstrated the effects that aquaculture can have on jellyfish populations (Lo et al. 2008). Tapong Bay, a small, nearly enclosed tropical lagoon in southwestern Taiwan, has been used for decades for extensive aquaculture of oysters and fish. Over time, the bay had become highly polluted due to poor circulation and organic input from aquaculture and urban sources. During the second half of 2002, all the 19,000 oyster rafts and 3,800 fish pens were removed in order to clean the bay. Monthly to bimonthly sampling for 3 years before removal and 19 months after removal provided an excellent comparative dataset.

The remarkable result of the Tapong Bay study was that after removal of the oyster rafts and fish pens, jellyfish abundance dropped to zero. This was attributed to several factors. First, water circulation in the bay improved from a residence time of 10 days to 6 as a result of removing the large number of physical objects obstructing flow. Second, the concentration of nutrients decreased, possibly as a result of improved flushing or perhaps because of increased phytoplankton, which were now not being filtered by the oysters. Third, removal of the structures not only physically took away the settling surfaces holding the existing jellyfish polyps, but also left no shady places for new larvae to settle.

The study's authors concluded that increasing demand for aquaculture, given current fishing trends and future population growth, will expand the favorable habitats for jellyfish, and therefore probably increase jellyfish populations.

The results of the Tapong Bay study are interesting in view of the role that the oyster rafts and fish pens played as favorable habitats for the polyps to colonize. To make matters worse, researchers in Norway have recently found that hydroids employ several types of entwining strategies with the net mesh fibers, making cleaning more difficult (Carl, Günther, and Sunde 2011); indeed, the hydroids colonize faster following cleaning than they did originally, suggesting that tiny bits of tissue are left behind, which are sufficient to aid rapid recolonization. Furthermore, regrowth is denser (i.e., containing more polyps) than in undisturbed colonies (Guenther, Misimi, and Sunde 2010), suggesting that cleaning of the cages acts as a stimulus for more growth.

Besides aquaculture cages, many other types of man-made structures provide suitable jellyfish habitat and are thought to contribute to blooms; however, little work has been done yet to link cause and effect in these cases. The northern Gulf of Mexico contains nearly 4,000 oil and gas platforms, around which jellyfish blooms have occurred, although a direct link has not yet been demonstrated. Finally, the likelihood of the refrigerator-sized jellyfish *Nemopilema nomurai* breeding on artificial substrates in Chinese waters has been proposed but has not yet been demonstrated.

As part of her graduate work, Sabine Holst of the Universität Hamburg came up with a simple but clever experiment to examine the settlement preferences of jellyfish larvae. Holst and her colleague Gerhard Jarms tested larvae of five types of jellyfish that are common in German waters, along with five types of materials commonly found in marine environments: polyethylene, glass, concrete, wood, and natural shells. The results were astonishing and prophetic: the polyps of all five species overwhelmingly preferred to settle on and colonize artificial substrates (Holst and Jarms 2007). A similar experiment on yet another species (Hoover and Purcell 2009) experienced the same results. It therefore appears that the human propensity to build structures on or in the water facilitates the propensity of at least some types of jellyfish to exploit these types of structures. For example:

Marinas, docks, and pontoons—these provide ideal substrate because they are generally in quiet, protected areas and are often colonized by *Aurelia*, whose larvae float at the surface and attach themselves to the undersides of structures.

Aquaculture racks and pens, and shells from bivalve aquaculture—set in the middle of flow-through currents, these receive a constant supply of food; they also generally contribute to a heightened nutrient concentration due to fish waste and unconsumed food, which triggers more frequent phytoplankton blooms, which in turn lead to rich zooplankton blooms, ideal food for jellyfish.

Artificial reefs, breakwaters, and dams—these contain many different surfaces at different angles for polyps to settle on, and different species can colonize their own most suitable habitat, for example, top, sides, undersurfaces.

Oil platforms and sea-based wind farms—also set in currents, these provide a constant flow-through of food and oxygenated water; weedy species in particular do well in these environments.

Undersides of hulls and inside holding tanks of boats—This ideal substrate for many species of polyps and hydroids enables species to spread to new areas.

Long-distance pipes and cables—these act like fresh reefs on which polyps can settle.

Industrial waste and general sea-going rubbish—flotsam and jetsam also act as fresh reefs on which polyps can settle and can transport species long distances.

Intake pipes, sumps, and screens—these are an ideal settling area for many species due to continuous flow-through of current, which delivers a constant food supply.

The impacts of jellyfish aquaculture on ecosystems have not been studied, but it is an interesting question. Jing Dong from the Liaoning Ocean and Fisheries Research Science Institute in Dalian, China, and her colleagues examined the commercial viability of enhancing the stock of a valuable species (Dong et al. 2009). The edible jellyfish *Rhopilema esculentum* is an important fishery species, but it is characterized by unstable abundance. In an effort to stabilize stock, 416 million young jellyfish were released into Liaodong Bay in 2005 and 2006. The recapture rate was about 3 percent, equaling some 5 million jellyfish in 2005 and 80 million jellyfish in 2006. With a 1:18 ratio of culturing cost to value of sales, the enhancement of stock was determined to be a very successful enterprise. One may therefore expect that stock enhancement will continue, if not increase, for this species, and perhaps become used for other species in other regions as well. However, one must wonder about the ecological effects, not only of the extra predation pressure in areas of release, but also the long-term effects of expanding feral stock.

Similarly, and significantly, no studies have been made on the effect of feral stock escaping from public aquariums. Since the early 1990s, jellyfish have become fashionable to display at public aquariums around the world; indeed, they are among the most popular exhibition at institutions that display them. But the global trade in culturable species means that the possibility for non-native species to be accidentally introduced into the wild increases with each shipment. Many public aquariums have a flow-through or semi–flow-through water supply. Accidents aside, not all full-time, part-time, or volunteer staff are equally aware of the taxonomic subtleties that separate native from non-native stock. Similarly, there is a fast-growing market for jellyfish as pets in home aquaria, further compounding the risk of jellyfish becoming feral pests in nonnative environments. Flushed in haste or freed in mercy, they still end up in the sea.

While there are as yet few other studies examining the impacts of aquaculture on jellyfish populations, the fact that increased nutrients must drive the expansion of phytoplankton and zooplankton, resulting in increased jellyfish, is indisputable. Some of the more obvious effects of aquaculture may be expected to include the following:

- Some types of aquaculture release large amounts of "jellyfish food" into the ecosystem via production of larvae, for example, oyster farming.
- Many types of aquaculture release excess nutrients in waste food or biological waste, for example, salmon farming. These can stimulate phytoplankton and zooplankton blooms, providing large amounts of food for jellyfish.
- Most types of aquaculture create new substrate for polyps and hydroids to settle on, as reported by Lo and colleagues at Tapong Bay (2008), for example, salmon cages, oyster and scallop racks, and aquarium and laboratory plumbing.
- Most types of aquaculture change the physical properties of the habitat to become more favorable to jellyfish, including by slowing water circulation and shading or reducing light penetration.
- Changes in water chemistry, such as hypoxia or temperature/salinity shifts, can make a habitat unfavorable for other, more sensitive species, thus limiting competition for jellyfish.
- Fishes and invertebrates harvested for aquaculture feed represent a significant removal of biomass from the ecosystem, thus removing competition for jellyfish, for example, zooplanktivorous fish and krill.
- Finally, aquaculture of jellyfish species can lead to the release of large amounts of jellyfish larvae into the ocean, for example, edible jellyfish, aquarium display and feed stock, and laboratory discharge water.

While it seems clear that many jellyfish species benefit from the effects of overfishing, this is not universally the case. The cannonball jellyfish or cabbagehead jellyfish, *Stomolophus meleagris*, used to be so common throughout the southeastern United States that it was a nuisance to fisheries until harvesters began exporting it to China. Estimated in 1981 at more than 2 million *per hour* drifting through the channel at Port Aransas, Texas (Huang 1986), by 2004, *Stomolophus* had all but disappeared in South Carolina (Griffin and Murphy 2012), with similar declines in other regions. Whether these declines are the result of direct take, or trawling-related habitat disturbance, or some other factor or a combination of factors is unclear.

Marine Blooms Affecting Terrestrial Ecosystems

Curiously, a similar phenomenon to the jellyfish blooms has been noted in the Galápagos, but with sea anemones (*Aiptasia* sp.) as the middleman of disturbance instead of jellyfish.

A series of seafloor surveys over the past 35 years and as recently as 1993 did not note the presence of a large number of anemones, but a survey in 1995 did note the presence of anemone patches. Apparently the problem was just starting, but was not recognized at the time for what it was to become. By 2000, the anemones had cloned to occupy "continuous carpets . . . on vast areas of shallow reef platforms . . . The *Aiptasia* sp. carpets have replaced diverse assemblages of algae, invertebrates, and fishes that once characterized these platforms" (Okey, Shepherd, and Martinez 2003, 17). It is thought that the anemones grew out of control in response to niche opportunities created by overfishing of local sea cucumbers, or by the 1997–1998 El Niño, or both.

These anemone carpets completely cover much of the eastern shore of Fernandina Island, especially at depths between 2 and 10 meters (6–30 feet), where lush algal beds existed just decades ago. The authors called these anemone-covered areas "anemone barrens," because of their similarity to the "urchin barren" moonscapes resulting from intense grazing and kelp deforestation by sea urchins (Johnston 2002; Pearse 2006).

The loss of algal cover raises particular concerns for the survival of the Galápagos marine iguanas, which feed on algae. The great naturalist Charles Darwin was not very kind to these scaly beasts, describing them in his diary as "disgusting clumsy lizards" and calling them "imps of darkness." Granted, they may have a face only a mother could love: if their wide-set eyes, flattened beady frown, and spiky dorsal scales aren't enough to give you pause, perhaps their salt-and-snot-encrusted heads will do it. You see, they ingest a lot of seawater while scraping algae off the rocks to feed, and their bodies need to get rid of this extra salt. So they have an unusual adaptation of frequent sneezing from special salt glands, and the salt often lands on their heads, hence the distinctive white wig.

But, love 'em or hate 'em, they are in peril. Because of the vast anemone barrens that have taken over the shallow reef areas, the large population of marine iguanas currently relies on only a small strip of green algae shallower than 2 meters (6 feet). They may seem simply like lizards with a sneezing problem, but they may be a prophetic example of bloom disturbances in the marine environment rippling through to terrestrial ecosystems.

CHAPTER 7

Eutrophication Almost Always
Leads to Jellyfish

Most creatures, such as fish, shrimp, and crabs, either flee the Dead
Zone or suffocate. But jellyfish, equipped with their built-in oxygen
supplies, thrive in the Dead Zone and banquet on the ubiquitous
plankton.

—LILY WHITEMAN, *On Earth*

Size Matters in Asia: *Nemopilema nomurai*, the Giant Jellyfish
(Sea of Japan, since 2002)

One's reaction goes well beyond the "ick factor" to the "wow factor" when
a jellyfish grows to 2 meters (6 feet) in diameter and over 200 kilograms
(450 pounds). In the case of *Nemopilema nomurai*, it's WOW FACTOR, in
capital letters (see plate 5). It has been described as sounding like a monster
from the trashier reaches of Japanese science fiction. It's pink. It's slimy. It's
lethal. And it is *huge*, both in body size and bloom size. And the females carry
up to a billion eggs.

The species was named and classified in 1922 in Japan (Kishinouye 1922).
It is not normally abundant: in nonbloom years, only 1 or 2 medusae per week
may be caught in a fishing net. But in the autumn of 1958, there was an "unprecedented flourishing" of *Nemopilema* (Shimomura 1959, 85). The medusae were restricted to the Tsushima Warm Current from early August to Janu-

ary and occurred in a density of 1 every few meters or yards, causing much damage to fishermen's nets. During October and November, 20,000–30,000 jellyfish were trapped in yellowtail nets every day. The species occurred every year, but big blooms were rare, the last being about 20 years earlier. Interestingly, this one coincided with the crash ending of a several-year-long bonanza in the sardine fishery.

The medusae swim in the layer from the surface to about 200 meters, migrating deeper in the daytime and shallower at night; in late autumn or winter they wash ashore or sink to the bottom and die. The heavy bloom was preceded by a long summer drought with higher temperatures, low rainfall, and shorter duration of sunshine—which was thought might be favorable to development of the larvae. In 1950, a similar drought occurred, with heavy growth of *Aurelia aurita* that year throughout the Sea of Japan, and a heavy flourishing of an unidentified small jellyfish in a brackish lake in the Akita Prefecture.

A large bloom was noted again in 1995. Then, upon entering the twenty-first century, it seemed like all hell was breaking loose. *Nemopilema nomurai* bloomed enormously . . . *again and again* in 2002, 2003, 2005, 2006, 2007, and 2009. And like a sleeping giant awakened, it was grumpy. An estimated half a billion jellyfish *per day* drifted into the Sea of Japan from near China— and they were huge.

Blooms of *Nemopilema* have become not only more frequent but also larger. Damage to Japanese fishing operations is costly—$20 million in one prefecture alone in 2003, with damage being reported in at least 17 prefectures (Kawahara et al. 2006). Clogging and bursting of nets. Increased labor costs to remove jellyfish from nets. Reduced catch of finfish. High mortality and reduced value of fish due to envenomation. Stings to fishermen. . . .

The stings are annoying but usually not lethal. A strong burning pain is initially felt on the skin, which becomes reddened and blistered. In most cases the pain subsides after about 30 minutes. However, at least 8 fatalities from *Nemopilema* stings have been reported in China (Mingliang, Shide, and Ming 1993).

In November 2009, like something out of a Hollywood movie, a 10-ton Japanese fishing trawler capsized and sank while hauling in a netful of these sumo-wrestler–sized jellyfish. The 3-man crew was thrown into the sea when the boat overturned; all were rescued. The weather was clear and the sea calm when the incident occurred; it was blamed solely on the weight of the *Nemopilema* in the net.

Authorities in China, Japan, and Korea have convened multinational meetings in an attempt to investigate and mitigate the jellyfish problem. A key focus point of research is to identify the conditions that cause these massive blooms to occur and drift into Japanese waters. In addition to the most obvious causes of overfishing, pollution, and climate change, it is also thought that aggressive construction along Chinese coastlines has created new habitat for polyps, and that damming of the Yangtze River at the Three Gorges has changed the coastal water chemistry to favor jellyfish blooms.

Indeed, the Yangtze discharges into Hangzhou Bay. And so do three other rivers. Together, some 2 billion tons of sewage, heavy metals, and petroleum residues are discharged by these rivers into Hangzhou Bay each year, making it the most polluted body of water in eastern China (Liu, Yu, and Liu 1991). Hangzhou Bay also contains the Zhoushan fishing ground, China's most valuable and heavily fished. In 1974, fishermen were landing 133,000 tons of yellow croaker per year, but by 1986 the catch had dropped to just 1,100 tons due to overfishing. Oysters, once abundant near the islands, have vanished. Butterfish, Spanish mackerel, and anchovies have decreased. These fish were the predators and competitors of jellyfish, such as *Nemopilema nomurai*. Even the edible jellyfish are overfished: the once robust catch of 800,000 tons—a third of China's annual total—had dropped by 50 percent by 1991.

Currently the blooms of *Nemopilema* appear to arise in Chinese waters then drift into Japanese waters on ocean currents. However, the threat of permanent breeding populations becoming established in the Sea of Japan, and hence becoming an annual nuisance, seems inevitable. Indeed, it would be remarkable if local settlement did not occur, given the many billions of medusae and the weedy nature of jellyfish polyps. More likely, billions of polyps are currently multiplying into astronomical numbers, waiting for the perfect conditions to strobilate into a new "unprecedented flourishing." Such a scenario would spell disaster for Japan's fisheries. This isn't the only example of jellyfish vexing trawling operations—far from it. Many more all around the world are listed in table 3 in the appendix.

Eutrophication is a special type of pollution. The root of the word, *eutrophic*, is translated from Greek as "true or good nourishment." The word has been used in different contexts to mean quite different things, but for the purpose of marine ecology, it means "excess nutrients," or, "excess fertilizer," to be exact. In order for phytoplankton to grow, they need sunlight from above and nutrients from below. But when enormous overloads of nitrogen and phosphorus flow into coastal waters from farm fertilizers, sewage effluent, and urban runoff, the water becomes stratified into healthy and unhealthy layers, and a self-enhancing feedback process triggers spectacular blooms of phytoplankton and a cascade of unwelcome events.

- One million menhaden died from algal toxins in North Carolina's Pamlico Sound in 1991.
- Over 150 endangered Florida manatees were killed by algal intoxication in 1996, following 37 that died in 1982.
- Over 400 sea lions were killed by algal toxins in 1998 along the central California coast; that same year 1,600 rare Hooker's sea lions died in mysterious circumstances at the subantarctic Auckland Islands.
- Algal intoxication was blamed for mass mortalities of hundreds of millions of fish in mid-Atlantic estuaries in the 1990s.
- The highly endangered Mediterranean monk seal population off the Sahara coast dropped by over 50 percent due to algal-related fatalities in 1997.
- Fourteen humpback whales died off Cape Cod in the northeastern United States because of algal intoxication in 1988–1989.

The above events, along with many others, were summarized in a heart-tugging—and stomach-churning—article on ecological disturbances by Bruce McKay of SeaWeb and his colleague Kieran Mulvaney, author, activist, and editor of SeaWeb's monthly *Ocean Update* (McKay and Mulvaney 2001).

So where are all these toxic algae coming from?

The majority of plants in the ocean are microscopic phytoplankton. Phytoplankton are essentially the "meadows of the sea." Just as meadows on land provide lush grass for herbivores, such as cows and lambs, which in turn provide us carnivores with meat, phytoplankton provide marine herbivores with food, and they in turn are eaten by the carnivores. The most abundant marine herbivores are tiny crustaceans called copepods, which are the food for just about every small- to medium-sized larval and adult form in the water column, from jellyfish to fish.

We humans have historically liked to settle in locations where rivers meet the coast. Sheltered bays. Deep estuaries. Up rivers. These areas are accessible by ship, protected from storms, and provide plenty of freshwater. New York. London. Venice. Copenhagen. Los Angeles. Sydney. Tokyo. Amsterdam. San Francisco. . . .

But our collective attitude that "the solution to pollution is dilution," means that the waste of our largest cities, even in the most environmentally conscious of times, gets mainlined into the oceans. Dog poo from sidewalks. Bird poo from picnic tables. Phosphates in detergents. Motor oil on the roads. Miracle-Gro from grandma's garden. Fish guts at the pier. Cigarette butts. Into the storm drains with every trickle of garden hoses, water sprinklers, and rainstorms . . . but all that's just a fraction. Agricultural runoff. Sewage effluent. Industrial waste. Nitrogen. Phosphorus. Overenrichment.

Eutrophication, an excess of nutrients in the ecosystem, occurs when these hypernutrified waters draining off the land are warmer and less salty than the seawater they are discharging into: a photograph showing this moment, the conception of a dead zone, that is, the meeting of the two bodies of water, may be found in plate 8. This meeting creates stratification in the water column, where the cooler, saltier, denser water stays on the bottom, and the warmer, fresher water floats on top. At first, the hypernutrified upper water stimulates hyperacceleration of aquatic productivity. Phytoplankton. Zooplankton. Fish. Whales and dolphins, seals and seabirds. All that fertilizer stimulates all those plants that feed all those mouths in the food chain. Everything benefits.

But when those animals and plants start dying, things go pear-shaped. "Phytoplankton not incorporated into the food web and fecal material generated by the food web sink into bottom-waters where they are decomposed by aerobic bacteria, causing oxygen depletion" (Rabalais, Turner, and Wiseman 2002, 238). But because the saltier layer is trapped below the surface layer and unable to touch air, the oxygen dissolving from the atmosphere cannot reach the bottom layer. The combination of stratification and decomposing organic matter create a zone of hypoxia (low oxygen) or anoxia (no oxygen) in the water just above the seabed, which can stay that way for months. Those creatures that can leave—and those that can't leave, suffocate. Fish flee. Crabs scurry. Clams die. Sponges die. Corals die. Tunicates die. Jellyfish flourish. Only a few worms and microbes survive in the sediments. It happens the same way in freshwater bodies too.

As it worsens, the hypoxic area becomes one of the notorious "black bottoms" with foul-smelling sediments. The surface water over the dead zone is teeming with plankton, with lots of jellyfish and some pelagic fish. But no shrimp. No crabs. No clams or scallops. No sponges or urchins or starfish. No skates or rays. No flounder or cod. No groupers. Eventually even the worms die.

Human activity, particularly extensive agricultural activity, has led to flu-

vial drainage of phosphorus and nitrogen at many times greater than preindustrial levels. Think of all the fertile farming regions that are well irrigated
by nearby rivers and seasonal rainfall: the Mississippi draining into the Gulf of
Mexico, the Yangtze draining into the East China Sea, the Los Angeles River
draining into Santa Monica Bay, the Nile draining into the Mediterranean,
the multitude of rivers draining into the Baltic, the Kattegat, the North Sea,
Black Sea, Caspian Sea . . . the list goes on and on. These seas are all major
fishery areas.

Professor Robert Diaz of the Virginia Institute of Marine Science and
his collaborator, Rutger Rosenberg of the University of Gothenburg in Sweden, recently published a thought-provoking paper on the spread of dead
zones (Diaz and Rosenberg 2008). Dead zones have spread exponentially
since the 1960s, say Diaz and Rosenberg. "Currently, hypoxia and anoxia
are among the most widespread deleterious anthropogenic influences on
estuarine and marine environments, and now rank with overfishing, habitat
loss, and harmful algal blooms as major global environmental problems"
(p. 929).

Currently, the global trend in nutrient disposal into rivers that feed into
the ocean is rising sharply. And as one might expect, the eutrophic response
is tracking that rise. In the 1990s there were about 125 known dead zones;
by 2008, more than 400 dead zones had been described around the world;
by January 2011, the official tally had risen to more than 530, plus another
228 sites exhibiting signs of eutrophication (VIMS 2011). Indeed, "There is
no other variable of such ecological importance to coastal marine ecosystems
that has changed so drastically over such a short time as [dissolved oxygen]"
(Diaz and Rosenberg 2008, 929).

Most of these areas are clustered particularly densely off the east coast
of the United States and throughout the waters of northern Europe. But just
about every nation with a coastline has at least one dead zone. There are no
known examples of recovery of large ecosystems from persistent hypoxia or
anoxia (Diaz and Rosenberg 1995). Once initiated, the low-oxygen condition
appears to be permanent, even in the seasonal cases.

Eutrophication, along with the algal blooms and hypoxia that it creates,
occurs through natural processes, but it's the rate of increase in volume and
frequency that is gaining so much attention. At this rate, "the number and
size of dead zones will increase to form continuous swaths for thousands of
kilometres along continental coastlines within the century. Toxic blooms will
also increase in size and frequency, and primary production will be increas-

ingly dominated by the microbial loop, with catastrophic effects on fisheries and aquaculture" (Jackson 2010, 3773).

Where the Effluent Meets the Affluent

Humans behave a bit strangely when it comes to the ocean. On the one hand, it is where we celebrate honeymoons and spend summer vacations, where we long to live, where the rich and famous play. We find peace in the salt air, solace in the rhythm of the waves, joy in the feel of the water and the sand between our toes. The sea features prominently in sonnets and watercolors, and sometimes as the final resting place for those we love.

However, the sea is not just for romantics and the carefree. It generates the oxygen that keeps us alive, it is a source of food and recreation, and it acts like a sponge to moderate the effects of global warming by absorbing excess carbon.

And yet, we treat the sea with a brutal ignorance, almost closer to contempt. We dump. We take. We deface. We turn a blind eye. We abuse without care of consequence. The effects of pollution and eutrophication are evident in coastal waters of much of the world. Stories of overfishing and stock depletion flood in from every corner of every sea and ocean, while their waters are becoming corrosive, from pole to pole, and to the deepest depths.

Coastal problems often begin many, many miles inland. Coral reef problems often begin with upriver sedimentation or agricultural runoff. Problems for whales and dolphins and manatees often begin with industrial effluent. Overfishing and jellyfication of the seas often reflect human appetites, tastes, and incomes, leading us to import our favorite flavors from across the planet and dump our waste into the nearest water body to convey it away. The sandy beaches, the rocky tidepools, the coastal dunes are where society's liquid, solid, and aerial wastes meet our smorgasbord of food, oxygen, carbon sequestration, recreation, pharmaceuticals, and wonder. Where the effluent meets the affluent. Today, after half a century, the words of President John F. Kennedy ring even truer:

> "Knowledge of the oceans is more than a matter of curiosity.
> Our very survival may hinge upon it."

Of course, so do the words from Joni Mitchell's song "Big Yellow Taxi":

> "You don't know what you've got 'til it's gone."

Where Do the Excess Nutrients Come From?

Sewage effluent. Fertilizer runoff. Animal waste. Urban waste. Detergents. Emissions from vehicles and factories. Lightning. Different regions and habitats have different dominant sources. Agricultural sources like fertilizers and animal waste tend to dominate river inputs. Sewage, detergents, and urban waste tend to dominate the input from major cities near gulfs and bays. Atmospheric fallout tends to dominate input in the open sea. Essentially, excess nutrients come from we wise humans being too clean with ourselves and not clean enough with our waste.

Regardless of the source, excess nutrients trigger overgrowth of algae that is utterly beyond the ability of copepods and other grazers to consume. Uneaten plankton die and sink to the seafloor by the zillions, whereupon microbial decay uses up the available oxygen, leaving the bottom-water hypoxic or anoxic.

Phosphates in Detergents

The making and use of soap dates back to the ancient Babylonians and Phoenicians, and even the Egyptians, who combined animal and vegetable oils with alkaline salts—for example, goat's fat and wood ash—to form a soaplike product.

In today's jargon, a "detergent" is a chemical substance that breaks up grease and grime, and the word "soap" is often used interchangeably; technically, however, soaps are made from natural fats and oils whereas detergents are synthetic and are based on highly acidic, highly alkaline, or neutral chemicals. Synthetic detergents began to replace soaps in about 1950, with phosphate accounting for over 30 percent of detergent by volume in laundry and dishwashing products. Phosphates act to reduce water hardness to enhance the cleaning efficiency of surfactants, or the ingredients that lower the surface tension between water and grime; to emulsify oily and greasy deposits; and to help keep dirt from redepositing during the washing process.

The same properties that make soaps such excellent cleaners of the bacteria and persistent stains on our dishes and clothing also make them hazardous in aquatic environments.

The United States banned the use of phosphates in laundry detergents in 1994, and high-phosphate dishwashing detergents were banned in 16 states in 2010. Elsewhere, phosphate detergents were banned in England in 2005 and Canada in 2011, France began its ban on phosphates in 2012, and the European Union is considering a ban beginning in 2015. In Australia, the Aldi supermarket chain announced in April 2011 that it would sell only phosphate-free detergents after 2013, but the rest of the country seems oblivious to the ecosystem effects.

Many countries are replacing detergent phosphates with more environmentally friendly alternatives, such as citric acid. Furthermore, tertiary treatment sewage plants remove phosphates as one of the final stages in cleansing the influent; however, tertiary treatment is not used at most sewage plants, particularly in developing countries. In India, phosphate detergents are still heavily used, particularly by the *Dhobis* (washer women; Toxics Link 2002), and as a result, ponds, lakes, and rivers are choking with algae. These water bodies are the primary water supply for a large section of the population.

Nitrates in Sewage

Nitrogen is an element required in building amino acids and proteins and therefore is necessary to all living plants and animals. Animals obtain their nitrogen through the process of eating plants or other animals. Plants obtain it through some of its forms in the water.

Elemental nitrogen is far more abundant than phosphorus, making up about 79 percent of the air we breathe. However, in this form, it is useless to most types of phytoplankton. Cyanobacteria, or blue-green algae, are able to convert nitrogen to ammonium and nitrates, which are usable to other aquatic plants (but don't get too excited about cyanobacteria—they also produce extremely potent toxins).

The more common natural sources of usable nitrogen are decomposition of dead plants and animals, and excretions of living animals. These sources are high in ammonium, which is oxidized (combined with oxygen) by special types of bacteria, converting the ammonium to nitrites and nitrates. The nitrates are taken up by phytoplankton as a fertilizer.

The primary "unnatural" sources of usable nitrogen are sewage and agriculture. Inadequately treated effluent from urban sewage treatment plants, poorly functioning rural septic systems, excessive fertilizer use on crops, feedlots and grazing lands. . . .

As stated in the earlier section on the Gulf of Mexico blooms, there is a massive area of hypoxia—a dead zone—just off the Mississippi Delta and New Orleans, stretching west along the coast of Louisiana. In this dead zone, dissolved oxygen is so low (less than 2 milligrams per liter, or 2 parts per million), that little can survive there. This dead zone was formerly as large as New Jersey (≈22,000 square kilometers, or ≈9,000 square miles). But since the 2010 BP oil spill, it has grown to the size of Massachusetts (≈27,000 square kilometers, or ≈10,000 square miles). A decade ago, fixing it would've cost $14 billion. The cost now? Who knows.

But that's a breath of fresh air compared to the hypoxic zones in the Baltic Sea (84,000 square kilometers) and the Black Sea (40,000 square kilometers, until recently) (Rabalais, Turner, and Wiseman 2002). Even a hurricane blowing through and re-oxygenating the water helps only briefly. Hypoxia in bottom-waters can reform in a matter of weeks, and once formed, seems to have a "memory" toward reforming again and again.

Professor Nancy Rabalais, executive director of the Louisiana Universities Marine Consortium, has published over 50 scientific papers on eutrophication and dead zones since 1990. In one of these, she and her colleagues provide a thorough account of the history, formation, and implications of dead zones, with particular focus on the Gulf of Mexico:

> Sediment core indicators clearly document recent eutrophication and increased organic sedimentation in bottom-waters, with the changes more apparent in areas of chronic hypoxia and coincident with the increasing nitrogen loads from the Mississippi River system in the 1950s. (Rabalais, Turner, and Wiseman 2002, 246)

In particular, nitrogen input has tripled since the 1950s, and the dead zone has steadily grown.

Intriguingly, another heavily eutrophic body of water, Chesapeake Bay, shows a strikingly similar pattern. "Results indicate that hypoxia and anoxia may have been more severe and of longer duration in the last 50 years, particularly since the 1970s. The sediment core findings corroborate long-term changes in Chesapeake Bay [phytoplankton blooms] since the 1950s"

(Rabalais, Turner, and Wiseman 2002, 249). Indeed, in 1949–1950, the low dissolved oxygen levels in the deep portion of the mid-bay channel clearly reflect hypoxia: by 1970, these regions were anoxic in the summer months; by the mid-1980s, anoxic conditions in the mid-bay channel began in May and extended into September, with hypoxic conditions covering much of the remainder of the bay bottom. The effect on the once-prolific fishing industry has been devastating: before 1965, crabs were abundant in deep water in both spring and autumn; today, crabs are restricted to shallow water, while those caught are so stressed that they often die before transport to market. In the adjacent Potomac the situation is even worse: "all crabs below 6 m are reported to have died in 1973, and crabs were driven ashore in large numbers in many late summer periods. Similarly, in the mid-bay region there have been reports of 'crab wars,' in which tens of thousands of crabs crowd into shoal waters and may actually leave the water" (Officer et al. 1984, 26).

So too, long-term datasets indicate that the northern Adriatic, the Baltic, and the Black seas have all followed the same pattern of an excess of nutrients, increased phytoplankton blooms, and worsening hypoxia. There is no easy answer to the eutrophication problem, but there is no escaping the fact that dead zones and algal blooms are a growing problem.

The mighty Mississippi begins as a tiny stream in northern Minnesota and travels 2,300 miles, draining 31 states and more than 40 percent of the continental United States, including more than 50 percent of the farmland. At its 20-mile-wide delta, it pours 4.5 million gallons of water per second, containing some 1.8 million tons of nitrogen per year, into the gulf. According to the US Geological Survey, this is a threefold increase since the 1950s. Excess fertilizer from fields, lawns, golf courses, piggeries, cattle ranches, and burning of fossil fuels, plus most of the wastewater from America's heartland, are transported to the gulf, year after year after year.

These nutrients trigger phytoplankton blooms, which can be a good thing, as they provide food for grazers and the carnivores that prey on them. But these blooms are not always a good thing. . . .

How Excess Nutrients Cause so Many Problems

Overenrichment of coastal waters has two primary deleterious effects: hypoxia and algal blooms. Algal blooms occur in response to the plentiful nutrients, just as adding fertilizer to a garden gives the plants a boost. The waters off Florida are a good example: recall that they have turned from turquoise to emerald because of the dense blooms of phytoplankton. Hypoxia leads to anoxia, which leads to toxic hydrogen sulfide. Organisms die. Algal blooms also lead to hypoxia, as well as toxicity complications up the food chain. Organisms die. Pick your poison.

Hypoxia

Hypoxic zones in the marine environment are normal in limited parts of deep oceans and fjords. They have been around throughout geological time. Even in the open ocean, there is a layer in the water column between about 200 and 1,000 meters (600–3,000 feet) called the "oxygen minimum zone," which functionally divides the upper, well-oxygenated ocean from the vast, dark depths beneath; many creatures won't leave it or cross it so it acts as a natural boundary. A similar naturally occurring deep benthic hypoxic region covers more than 1 million square kilometers (almost 400,000 square miles) of permanently hypoxic seafloor (Helly and Levin 2004). But the recent trend toward coastal and estuarine hypoxia and anoxia is not "normal," and it is increasing. Coastal and estuarine areas are by far the sea's most productive regions. The coral reefs of eastern continental shelves, the fisheries driven by upwelling off the western continental coasts, and estuaries where many of our fisheries' resources and food breed are now rapidly turning into wastelands.

The species composition, volume, and biomass of animals living on the seafloor are highly susceptible to a lack of oxygen—that is to say, the whole system reshuffles. The effects of hypoxia are essentially a one-to-one equation expressed as who lives and who dies. The answer is simple and straightforward: those species that require more oxygen suffocate, while species requiring less sometimes survive. You breathe, you die; you hold your breath, you live. Like nerve gas.

Polychaetes, segmented marine worms, are among the organisms least likely to succumb to spreading dead zones. Whereas the oxygen content of seawater in its normal state is about 5–7 parts per million (ppm), in dead zones it falls to 1–2 ppm or even to 0. Polychaetes are capable of surviving over five days in water with oxygen concentrations below 0.5 ppm. So are jellyfish. In contrast, whereas an oxygen concentration of 1.5–2 ppm was long thought to be lethal for shrimps, crabs, fishes, and mollusks, it is now clear that many perish even at 2–3 times that concentration (4.5 ppm) (Vaquer-

Sunyer and Duarte 2008). This is not merely a "sad loss" of living things; that loss also often leads to a cascade of change. Recall the godforsaken Black Sea. On the northwestern shelf, the mussel community in the 1960s covered an area of about 10,000 square kilometers (almost 4000 square miles), but by the 1980s the mussel biomass had declined by an order of magnitude and the community in which they used to live had shrunk to a mere fraction of its earlier size. With them went the filtration of literally billions of liters of seawater per day.

Sometimes heavy phytoplankton blooms and the early stages of a developing dead zone can mislead scientists. Daytime water testing may still indicate normal levels of oxygen, because the phytoplankton are busily making it through photosynthesis. But at night, when phytoplankton shift into their respiring mode, they use oxygen rather than manufacture it, so fish and other animals can suffocate.

Dr. Daniel Conley of Lund University in Sweden and several European colleagues have studied the biogeochemical causes and consequences of hypoxic events (Conley et al. 2009). They show that dead zones not only kill bottom-dwelling communities but also change water chemistry and the way in which ecosystems process nutrients. Hypoxic bottom-water conditions cause seafloor sediments to release dissolved inorganic phosphorus. In the Baltic, the volume of phosphorus released from sediments is an order of magnitude larger than the amount flowing in from rivers. This stimulates a positive feedback cycle of phytoplankton blooms that drive hypoxia. Hypoxia also lowers the rate of denitrification, that is, the process of gassing off excess nitrogen through anaerobic respiration by microbes in the sediments. By slowing the rate of nitrogen loss, the excess of nutrients in the water acts as a positive feedback for continued blooms of phytoplankton.

So dead zones actually enhance the volume of available nutrients that stimulate phytoplankton growth, which in turn enhances the dead zone. Furthermore, it appears that once an ecosystem has experienced a hypoxic event, it is more prone to repeatedly experiencing conditions of oxygen deficiency. Seafloor communities are made up of large, slow-growing species, such as sponges, corals, and ascidians, as well as echinoderms, worms, and other species that constantly bioturbate and bioirrigate the sediments, speeding up re-oxygenation and remineralization of nutrients. When these animals suffocate, they are replaced by smaller, fast-growing, rapidly colonizing species that don't mix or dig into the sediments. The sediments therefore remain anoxic and lack a buffer against episodic or persistent occurrence of hypoxia.

It is also believed that the interplay between nutrient/sediment processes and ecosystem hypoxia may act in a threshold-like manner. In particular, after reaching a shifting point, these processes act to stabilize the ecosystem at its "new normal," that is, resisting a return to its original, well-oxygenated status. So reversing the problem of hypoxia is no easy fix.

Algal Blooms: "And All the Waters . . . Turned to Blood"

Since the beginning of recorded time, man has been fascinated with red tides. Well, fascinated or bewildered. The first plague in the Book of Exodus gives a vivid description of a red tide in the Nile, "And all the waters that were in the river were turned to blood. And all the fish that was [*sic*] in the river died; and the river stank, and the Egyptians could not drink of the water of the river; and there was blood throughout all the land of Egypt" (7: 20–21). Homer's *Iliad* makes repeated references to red water, such as "wherewith to cross the wine-dark sea" (2.2.613). The classical Greek geographer Strabo attributed the name of the Red Sea to the color of the water, which is often thought to be due to red tides (Hoyt 1912). In AD 77, the Roman naturalist Pliny the Elder wrote, "There are sudden fires in the waters" (*Natural History* 2.111)— this most certainly must have referred to either a red tide during the daytime or its bioluminescence after dark. In 1835, Charles Darwin encountered red tides off the coast of Chile during his famous voyage on the *Beagle*, inspiring him to make detailed observations and descriptions of the microscopic "animalcula."

One of the most disastrous algal bloom events took place near Sitka, Alaska, in 1799: a large group of Aleut hunters on a fur sealing expedition stopped in a strait to dine on mussels; within hours, over 100 men were dead from paralytic shellfish poisoning. The passage was named Peril Strait to commemorate the event, as were two nearby regions: Poison Cove and Deadman's Reach. These sorts of events were once considered rare and shocking but are becoming more common. Persistent paralytic shellfish poisoning has resulted in the complete loss of the Alaskan wild shellfish fishery which once produced annual landings of 5 million pounds (Solow 2004).

This chapter opened with a series of dot-point examples of mass mortality events of marine animals from algal toxicity. Shocking as those examples are, they are far from the worst effects of eutrophication. To juxtapose two emotive examples for the sake of analogy, those mass mortalities are comparable only to the Columbine massacre—or 9/11, for that matter—as against, say, the planetary mortalities from smoking-related illnesses, year after year after year.

Like smoking, algae also cause many direct and indirect problems. Direct toxicity. Indirect toxicity through the food chain. Reduced water visibility and light penetration. Changes in food web structure. Selective starvation. Dead zones. Alas, those pesky dead zones. . . . Those bright, beautiful shades of emerald green waters off the Florida Keys may look harmless. But so do piranhas. . . .

Direct and Indirect Toxicity

Some types of algae release toxins that produce a poisonous cloud in the water. The common red tide dinoflagellate *Gonyaulax* produces skin irritation, respiratory stress, and burning to the mucous membranes, such as the eyes, simply by exposure. Similarly, the dinoflagellate *Pfiesteria*—often called "the cell from hell"—is associated with skin lesions and mass mortality in fish and marine mammals from its secreted toxins and has been associated with skin problems and Alzheimer's-like symptoms in humans (Barker 1998). Many types of algae are directly poisonous when ingested by copepods and other small plankton. Other common types concentrate up the food chain to wreak wider havoc. These are the ones that cause ciguatera (concentrated in fish), paralytic shellfish poisoning (concentrated in oysters, mussels, and clams), and domoic acid poisoning (which affects fish, mammals, and other vertebrates).

Another feature of algal blooms not often talked about is cholera. Cholera is generally associated with poor sanitation—but that's what eutrophication is all about, excessive "fertilizers" in coastal waters. The pathogen that causes cholera, *Vibrio cholerae*, lives in brackish and coastal marine waters, where it attaches to zooplankton with chitin shells—for example, copepods (Huq et al. 1983; Colwell 1996). A single copepod may carry up to 10,000 *V. cholerae* cells—more than 10 times the infectious dose. During times when copepods are abundant, such as plankton blooms, it becomes quite easy to ingest several copepods in a mouthful of seawater or untreated well water. In October 1992, a form of cholera broke out in the southern Indian port city of Madras (Colwell 1996). It spread rapidly, reaching north to Calcutta within months. By December of that same year, it reached Bangladesh; the disease rapidly spread through the entire country. Almost 47,000 cases were reported, including 846 deaths in Bangladesh alone.

Besides their toxins that affect us, some types of phytoplankton inhibit their copepod predators. For example, brown tides and cyanobacteria can reduce egg-production rates, and some diatoms can interfere with cellular growth (Cloern 2001). These processes restrict the flow of energy up the food chain and result in uneaten phytoplankton that die and exacerbate the hypoxic condition.

> Unrestrained runoff of nutrients and toxins, coupled with rising temperatures, will increase the size and abundance of dead zones and toxic blooms that may merge all along the continents. Even farmed seafood will be increasingly toxic and unfit for human consumption unless grown in isolation from the ocean. Outbreaks of disease will increase. (Jackson 2008, 11463)

Water Clarity

For several reasons, eutrophic conditions often result in reduced water clarity and limited light penetration. First, rich blooms of phytoplankton and zooplankton make for more individuals in the water acting as particles. Second, the decay of organic matter often results in colored, dissolved particles. Finally, because bottom-dwelling filter-feeders, such as clams, oysters, mussels, bryozoans, and tunicates, cannot survive in hypoxic conditions, their filtration of waterborne particles is suppressed. Reduced light penetration also leads to loss of seagrass meadows and macroalgal beds and, along with them, the organisms that shelter in them.

More particles in the water also means reduced visibility. Most fish, mammals, and crustaceans are visual predators, whereas most jellyfish species are tactile predators. This gives jellyfish a distinct competitive advantage in conditions of reduced visibility, such as was highlighted earlier for *Periphylla* in the Norwegian fjords. The correlation between reduced visibility and jellyfish population increases and fish population declines has been examined in Norway (Eiane et al. 1999), but has yet to be examined as a potential factor in other jellyfish domination scenarios like the Black Sea, the Gulf of Mexico, the Benguela Current off Namibia, and in Southeast Asian waters.

There are, however, several studies demonstrating that water turbidity interferes with mate choice in pipefish, sticklebacks, and gobies. So the possibility cannot be discounted that turbidity may have multiple negative effects on fish, for example, predation effects, reproduction effects, and others. Furthermore, the extent to which the reproductive effects of turbidity play a role in diminishing fish stocks in eutrophic areas has not been examined.

While phytoplankton blooms are increasingly associated with toxic species, even nontoxic phytoplankton can cause serious problems. Consider the North Atlantic right whale, which is in grave danger of extinction. The few remaining whales in the population feed seasonally off Cape Cod, gulping down gigantic quantities of concentrated patches of copepods. However, in some years a nontoxic species of phytoplankton called *Phaeocystis* blooms in such massive numbers as to obscure the patches of copepods, making it difficult for the whales to find enough to eat (Solow 2004).

Zooplankton Size and Visibility

A study by Dr. Shin-ichi Uye, a plankton researcher, as well as executive and vice president of Hiroshima University, focused on changes in copepod demographics with eutrophication. Uye (1994) compared the copepod communities of Tokyo Bay with those of Osaka Bay, both being eutrophic, but Tokyo more so than Osaka. Tokyo Bay is considered the most heavily polluted embayment in Japan. Historical data demonstrated that as Tokyo Bay has become more eutrophic over the last 40 years, smaller-size copepod species have become dominant. In Osaka Bay, overall copepod body size is larger, but smaller specimens dominate in more severely eutrophic areas.

Through the 1960s and 1970s, Uye chronicled Tokyo Bay's changes in the phytoplankton community, believed to be at the heart of the shift in the copepod demographics. As "weedier" groups of phytoplankton called dinoflagellates took over from the more "normal" diatoms, their grazers changed too. The single most dominant species of copepod now in Tokyo Bay, the small-bodied *Oithona davisae*, feeds only on dinoflagellates. The trade-offs and changes associated with dinoflagellates and diatoms are discussed in more detail below in chapter 13.

The preferences of *Oithona* for dinoflagellates are not the only thing that make it well suited to dinoflagellate-infested Tokyo Bay, for so too does its reproductive strategy. Many types of copepods spawn their eggs freely into the water, where they sink toward the bottom to hatch. In the anoxic bottom-water of the highly eutrophic Tokyo Bay, this is likely to result in the eggs dying. But *Oithona* carries its eggs in a sac until they hatch so that they are protected from the toxic bottom-water. It appears that the feeding and reproductive strategies of *Oithona* enable it to exploit the areas where other copepods are unable to thrive. However, *Oithona*'s success comes at a great cost to other species. Because *Oithona* has a far smaller body size than other copepod species, it is harder for visual predators to see. Good for the copepod, bad for the fish. In fact, the anchovy fishery has dramatically reduced in Tokyo Bay to only 10 percent of the catch in Osaka Bay. It appears that visual predators are more successful in Osaka Bay, where they can see their prey.

Enter jellyfish. In the period during which these changes—increasing eutrophication, demographic changes in the copepod community, and decreasing fish stocks—were taking place, two types of jellyfish were becoming more abundant: *Aurelia* and *Bolinopsis*. *Aurelia*, the moon jellyfish, is a highly invasive and opportunistic weedy species, while *Bolinopsis* is a close relative of *Mnemiopsis*, the species that invaded and wreaked havoc in the Black Sea.

Aurelia and *Bolinopsis* are nonvisual predators, and while they can take both large and small copepods, smaller ones are more easily entrained in their feeding currents. Thus, high production rates of smaller copepods suit them just fine. In fact, according to Uye, "in recent years the frequent occurrence of enormous numbers of jellyfish such as *Aurelia aurita* and *Bolinopsis mikado* has become a common phenomenon in Tokyo Bay" (Uye 1994, 517). Increasing eutrophication has therefore led to a shift in phytoplankton demographics, which in turn led to a shift in the copepods, which in turn created a niche opportunity for jellyfish over the fish.

A similar pattern has been documented in the Black Sea (Zaitsev 1992). Over the past 40 years of increasing eutrophication, the northwestern shelf waters have been characterized by a reduction in zooplankton body size and increasing jellyfish abundance.

The Effect of Eutrophication on Ecosystems

Eutrophication follows a predictable pattern through three consecutive phases (Wassmann 2005). First, there is the enrichment phase, which takes place in previously low-nutrient ecosystems. It is characterized by an abundance of life with increases in fish and shellfish yields, as well as overall benthic (seafloor) and pelagic (water column) biomass. Moderate eutrophication can be considered beneficial from a fishery point of view.

The second phase of eutrophication is when the initial and secondary effects become visible. Beyond a certain level of enrichment, no additional increase in harvestable resources occurs. Instead, dramatic changes to the ecosystem begin to appear. Phytoplankton bloom in incredible densities, and the incidence of harmful algal blooms increases. As phytoplankton flourish, declining light penetration depth decreases the area where macroalgae can thrive. Increasing frequency of hypoxic episodes leads to an increase in tolerant species, such as segmented worms.

The third phase of eutrophication is beyond the "point of no return," when the ecosystem can no longer recover to its original state after nutrient supply is reduced. This phase is characterized by anoxia and mass mortality. Sensitive species disappear. Opportunists take over. The sediments, devoid of living organisms, repeatedly slip back into an anoxic state.

Changes in Food Web Structure

Eutrophication promotes macro-scale changes in food web structure through small shifts in phytoplankton size. Large phytoplankton are eaten by large copepods and krill, which in turn are eaten by fish, seabirds, and whales. Small phytoplankton are eaten by small copepods, which are too small for visual predators but perfect for jellyfish. This dynamic is explored more fully in chapter 13.

Furthermore, many types of microbes are not photosynthetic, but instead obtain their nutrients by eating other organisms or their remains. They are called "heterotrophic" organisms because they get their nutrients from other sources rather than generating it themselves like plants do (*hetero*, meaning "other" + *trophic*, meaning "nutrition"). Experimental manipulation of eutrophic conditions has demonstrated that blooms of heterotrophic plankton lead to jellyfish blooms (Parsons and Lalli 2002).

Selective Starvation

Phytoplankton blooms often consist of dinoflagellates and other smaller microbes collectively referred to as "flagellates," rather than the larger diatoms. Herbivorous copepods, krill, and some pelagic fish will only eat the largest phytoplankton, that is, diatoms. When flagellates dominate, the typical food web that we are familiar with cannot thrive, but jellyfish do. The trade-off between large diatoms and small flagellates is also examined in detail in chapter 13.

The Effect of Eutrophication on Jellyfish

In general, eutrophication tends to reduce the diversity of jellyfish species but dramatically increase the biomass of one or a few. In other words, many jellyfish species are not weedy enough to exploit it, but those pesky few who are do quite well. Curiously, there are no cases identified where jellyfish blooms can be blamed on eutrophication alone (Arai 2001), but eutrophication is one of the contributing factors in many jellyfish blooms—particularly the worst ones. It therefore seems clear that eutrophication plays a strong role in degrading ecosystems to the point where other perturbations can have an unexpectedly severe effect.

Furthermore, eutrophication and overfishing have similar and synergistic effects, so teasing them apart can be difficult. In particular, they both cause a decline in species diversity and an initial increase in productivity of benthic and pelagic food webs, followed by progressive dominance of short-lived, pelagic species like jellyfish (Caddy 1993, 2000).

> In coastal areas, changes caused by eutrophication . . . appear to enhance jellyfish production and not fish production. This indicates that a food chain favoring jellies occurs in these areas, rather than there simply being a vacancy for an opportunistic feeder at a higher trophic level. Eutrophication of coastal waters now appears to be the most severe form of coastal pollution. (Parsons and Lalli 2002, 114–115)

Jellyfish polyps and medusae tolerate low oxygen conditions better than most fish and other higher organisms with high respiratory demands. In fact, experiments show that hypoxic conditions actually enhance jellyfish predation: where other organisms struggle to survive, jellyfish benefit by feeding on sluggish prey (Shoji et al. 2005).

An ecosystem shift toward algal blooms is not the only result of eutrophic conditions. Jellyfish also become more of a problem. In fact, several aspects of eutrophic conditions may actually promote jellyfish blooms.

Turbidity and Hypoxia Feedbacks

Recall the *Periphylla* example from the Norwegian fjords, where reduced water clarity favored tactile predators like jellyfish over visual predators like fish, and that reduction in fish in turn promotes turbidity. So too, it appears that jellyfish may provide a positive feedback to hypoxia in at least a couple of different ways (Purcell 2012). One is based on predation. Studies on the heavily eutrophicated Skive Fjord in Denmark found that especially severe cases of hypoxia occur in years when mass occurrences of jellyfish are also taking place (Møller and Riisgård 2007). It seems that as the jellyfish eat the zooplankton, so the blooming phytoplankton are not efficiently grazed and settle to the bottom to decompose, thus leading to more severe hypoxia, which kills the filter-feeding mussels, exacerbating the decline in water quality. Another feedback is based on decomposition of jellyfish blooms. As large numbers of jellyfish corpses sink to the seafloor, a massive pulse of organic and inorganic nutrients is released into the water, and the oxygen demand required to decompose their tissues may lead to localized hypoxia or anoxia (Pitt, Welsh, and Condon 2009).

Evidence indicates that many species of jellyfish can live in hypoxic waters, allowing them to outcompete and outsurvive fish and other invertebrates in eutrophic conditions, both in their medusa and polyp stages (Rutherford and Thuesen 2005; Condon, Decker, and Purcell 2001). Some species congregate in naturally hypoxic zones, such as fjords or the deep sea, while other species appear to capitalize on seasonally hypoxic conditions resulting from eutrophication. Physiological and behavioral adaptations make many jellyfish species well suited to the low-oxygen environments that our coastal seas are becoming.

Adaptations to Living in Hypoxic Environments

A particularly intriguing study by Dr. Erik Thuesen of Evergreen State College in Washington State and his colleagues found that the jellyfish *Aurelia labiata* is able to store oxygen in its "jelly" then use this intragel oxygen while in hypoxic or anoxic conditions (Thuesen et al. 2005). The oxygen reservoir can last about 2 hours in zero-oxygen conditions and can be about 70 percent recharged in another couple of hours. Similarly, three other species tested were able to regulate their oxygen levels in hypoxic conditions. These results suggest that jellyfish are well suited to eutrophic conditions, which are similar to those in which they evolved hundreds of millions of years ago.

Furthermore, the "skin" of jellyfish is only one cell-layer thick, enabling them to respire across their entire body surface. Such a large surface area for oxygen exchange may predispose them to tolerate low oxygen environments (Purcell et al. 2001a).

Storing oxygen and "breathing through the skin" are definite attributes for low-oxygen environments; so is the basic jellyfish lifestyle. Unlike many other types of organisms, jellyfish stay up in the water column. Even in the most heavily eutrophicated areas, the surface layers are liveable for those species that can avoid the dead zone below. A good example of this is the jellyfish blooms in Namibia, discussed earlier, where the jellyfish have created a sting-and-slime curtain that acts as a fish-exclusion zone.

Furthermore, it seems that some jellyfish use low-oxygen layers almost as a tool for their own benefit. Many species of jellyfish hunt along the interface of anoxic zones that prevent escape of fish and zooplankton (Purcell et al. 2001a). *Mnemiopsis* is known to concentrate in areas of very low oxygen in the Chesapeake as a refuge from its sea nettle predators. Similarly, along the edge of the anoxic zone in the Black Sea, the quasispherical comb jelly *Pleurobrachia* aggregates up to 70 per cubic meter (2 per cubic foot), forming a layer that looks "like beds of mushrooms" where it gorges on copepods (Vinogradov, Flint, and Shushkina 1985).

Below we examine several specific instances of jellyfish living in different types of natural and unnatural hypoxic environments. In all cases, jellyfish demonstrate a particular ease for not only coping with—but actually flourishing in—these types of habitats.

Living in Naturally Hypoxic Environments

Many studies have recorded various species of hydromedusae concentrated in hypoxic waters in the fjords of Norway and British Columbia (Purcell et al. 2001a). Elsewhere, two species, the Antarctic hydromedusa *Rathkea lizzoides* and the Palauan scyphozoan *Mastigias*, concentrate just above the anoxic zones in their habitats and make periodic excursions into them, apparently for feeding (Purcell et al. 2001a). These areas are naturally low in oxygen, and it appears that these jellyfish are naturally exploiting the unusual environmental conditions.

Living in the Oxygen Minimum Zone of the Deep Sea

Some species of jellyfish, such as the 20-meter-long (60 feet) siphonophore *Apolemia* and the deep-sea *Periphylla* are normally found in the oxygen minimum zone of the deep sea (Fosså 1992; Båmstedt et al. 1998). In fact, researchers using submersibles from the Monterey Bay Aquarium Research Institute have observed that many of the commonest species of jellyfish across all groups are most often found in the oxygen minimum zone (Purcell et al. 2001a).

Similarly, the siphonophore *Nanomia* occurs in such dense shoals—up to 1,000 colonies per cubic meter (28 per cubic foot)—that it has been reported to fool sonars and echo sounders (Barham 1963). *Nanomia* is related to the Portuguese man-o'-war, and has a similar construction: each colony has a gas-filled float, no larger than a grain of rice, above its many groups of tentacles and feeding persons. The discontinuity between the seawater and the massive number of gas-filled floats is picked up by the sonar, which senses it as a false bottom where *Nanomia* blooms occur.

While our own experience with *Nanomia* is often in the form of frustrated sonar readings, the way it clusters in the oxygen minimum zone makes it a formidable predator of other species, particularly those that become sluggish when they venture into the low oxygen environment.

Living in Anthropogenically Hypoxic Environments

While it seems clear that some species have evolved to live in naturally hypoxic conditions, what about species that evolved in so-called normoxic conditions? The Chesapeake Bay is a good laboratory for investigating this question, because it suffers from seasonal catastrophic eutrophication as a result of anthropogenic disturbances. While jellyfish species diversity has declined as eutrophication has worsened, those species that have survived it seem well suited to their oxygen-depleted environment.

Mnemiopsis can live for more than 3 days in water with less than 0.5 parts per million of oxygen, while polyps of *Chrysaora quinquecirrha* from the Chesapeake thrive and reproduce at the same low oxygen concentration. Even the more metabolically active medusae of *Chrysaora* can survive for over 4 days with oxygen at only 1 part per million (Purcell et al. 2001b).

Consider too, the hydromedusa *Nemopsis bachei*, which blooms in incredible abundance in the highly eutrophic Chesapeake Bay (Purcell and Nemazie 1992). Reaching densities up to 132 per cubic meter (4 per cubic foot), its predation effects make it the most important gelatinous predator in the springtime in the Chesapeake. It's not the biggest, or the fastest, or the smartest; it simply outnumbers everything else.

Medusae use their lower energetic requirements and behavioral adaptations to survive where most other species cannot. Similarly, polyps can outcompete other species for space and food in low oxygen conditions, and many live in shallow water above the oxygen-depleted zone. Furthermore, if conditions become intolerable, polyps can encyst until conditions improve.

However, not all medusae and polyps survive hypoxia and anoxia equally well, as differences in polyp survival in periodic anoxic bottom conditions are thought to explain the disappearance of many types of medusae from the Adriatic Sea in recent decades (Rutherford and Thuesen 2005), and may account for the disappearance of iconic species, such as the enchanting *Scrippsia pacifica* and the smaller but equally captivating *Polyorchis penicillatus* in California. Formerly so common as to be featured on textbook covers, stamps, and even university logos, sightings of these species are now few and far between.

More Food

One might intuitively think that more food due to eutrophication is a good thing. Fiesta time! But not necessarily—it depends on who it attracts to dinner. One of the main complications of an increase in primary productivity (i.e., phytoplankton) is that it inevitably leads to an increase in zooplankton, mainly copepods. Recall that phytoplankton and zooplankton blooms often lead to dead zones—as well as to jellyfish. The original burst in primary productivity is therefore wasted, as much of the energy potential never makes it up the food chain to fish, mammals, and birds. Worse still, it's not only wasted, but the decomposing phytoplankton, as well as the dying and decaying jellyfish, further contribute to the dead zone.

The high nutrient levels that trigger the eutrophication process result in the production of large numbers of zooplankton. While in theory this is a boon for all species that prey on zooplankton, and so on up the food chain, in practice that is not necessarily so. Jellyfish can reproduce and grow much faster than most other species to take advantage of the extra food, while other species still have to find their food, despite the reduced visibility. As jellyfish rapidly multiply into a bloom, larvae of other species trying to reproduce simply become more food for the jellyfish.

Intriguingly, it also appears that jellyfish are able to capitalize on effluent-enriched water by absorbing nutrients directly through their "skin" (Muscatine and Marian 1982; Wilkerson and Dugdale 1983). Specifically, dissolved amino acids are absorbed through epidermal cells that are specialized in the extraction of organic matter from seawater. Experiments have shown that 10–40 percent of the total metabolic requirement of jellyfish can be met through this route. Furthermore, crustaceans and fish (i.e., competitors of jellyfish) have limited ability to absorb nutrients directly and are therefore at a competitive disadvantage in sewage-polluted conditions.

Shift in Food Size

Eutrophication often leads to a shift from large to small phytoplankton. Recall that herbivores like copepods and krill will preferentially take the largest particle size they can handle, and so the larger species starve for lack of suitable phytoplankton. This leads to a community dominated by smaller copepods, which are preferred by jellyfish and not preyed on by visual predators, such as fish, which cannot see the smaller prey items. This mechanism of shifting from a fish-dominated ecosystem to one controlled by jellyfish is explored in greater depth in chapter 13.

Cassiopea: A Good Example

The so-called upside-down jellyfish *Cassiopea* appears to be capitalizing on eutrophication in the Bahamas (Stoner et al. 2011). A team of scientists from Florida International University examined the population density and medusa body size at 5 anthropogenically disturbed sites adjacent to relatively high-density human population centers, and 5 undisturbed sites adjacent to uninhabited watersheds. The researchers found that jellyfish were larger and more abundant in areas adjacent to human population centers.

The anthropogenic nutrient loading in these disturbed areas seems to create a perfect setting for the symbiotic algae in *Cassiopea*'s tissues that in turn supply a rich feast for their host.

In general, eutrophication works synergistically with other perturbations to lead to environments favorable to jellyfish. However, *Cassiopea* may be a case where eutrophication alone is the cause.

Oligotrophic Conditions: Not Enough Nutrients

Visualize for a moment snorkeling in the warm, shallow water off a beautiful tropical beach . . . the water is so clear that it seems like you can see forever.

Probably not part of your daydream, but the reason the water is so clear is because there are so few phytoplankton and other living things in it. That is to say, the water is devoid of life. Compare this with the emerald green water in the Florida Keys, the color of which is the result of too much phytoplankton. Tropical waters tend to be very stable and low in nutrients due to lack of upwelling—the scarcity of nutrients is called an "oligotrophic" condition—quite the opposite of the eutrophic conditions caused by excess nutrients. Another example of a typically oligotrophic ecosystem is the open ocean.

Flagellates dominate in oligotrophic conditions, because the nutrient levels are too low to support diatoms. Because flagellates are so small, they have a greater surface-to-volume ratio, which enables them to obtain the nutrients they require even when in short supply. Paradoxically, then, while conditions of excess nutrients favor jellyfish blooms, so do conditions of limited nutrients.

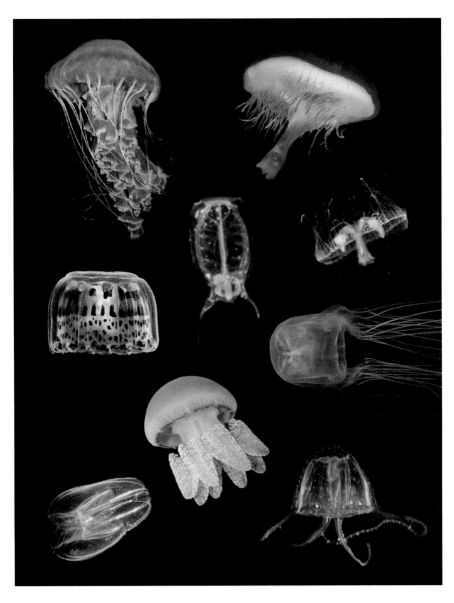

PLATE 1. Types of jellyfish, clockwise from upper left: *Chrysaora achlyos*, a scyphozoan 3 feet in diameter (photo by Gary Florin, Cabrillo Marine Aquarium); *Aequorea*
sp., a hydrozoan 2 inches in diameter (photo by Karen Gowlett-Holmes); *Craspedacusta sowerbyi*, a freshwater jellyfish 1 inch in diameter; *Chironex fleckeri*, the deadly
Australian box jellyfish 1 foot in diameter; *Carukia barnesi*, an Irukandji jellyfish
½ inch in diameter; *Catostylus mosaicus*, a rhizostome blubber 1 foot in diameter;
Bolinopsis sp., a ctenophore comb jellyfish the size of an egg; *Linuche unguiculata*, a
coronate ½ inch in diameter (photo by Ron Larson); *Thalia democratica*, a salp ½ inch
long (center; photo by David Wrobel).

PLATE 2. Problem jellyfish species. Top row: *Chrysaora quinquecirrha*, Chesapeake Bay, USA, 6 inches in diameter; *Carybdea rastonii*, Sydney, Australia, 1 inch in diameter (photo by Karen Gowlett-Holmes). Second row: *Periphylla periphylla*, Norway, 1 foot tall (photo by David Wrobel); *Mnemiopsis leidyi*, Black Sea, the size of an egg; *Phyllorhiza punctata*, Gulf of Mexico, 1 foot in diameter (photo by Sue Morrison); Third row: *Aurelia labiata*, Diablo Canyon Nuclear Power Plant, California, USA, 1 foot in diameter); *Crambione mastigophora*, Western Australia, 4 inches in diameter (photo by Caroline Williams). Bottom row: *Pelagia noctiluca*, Mediterranean and United Kingdom, 2 inches in diameter); *Nanomia cara*, Gulf of Maine, up to 12 feet long (photo by Marsh Youngbluth, Harbor Branch Oceanographic Institute).

PLATE 3. Jellyfish blooms. *A*, Thousands of Santa's hat jellyfish, *Periphylla periphylla*, caught in a trawl net in Lurefjorden, Norway (photo by Jennifer E. Purcell). *B*, Bloom of zillions of sea nettles, *Chrysaora fuscescens*, up to the size of basketballs, in Monterey Bay, California, November 2010 (photo by Arlo Hemphill).

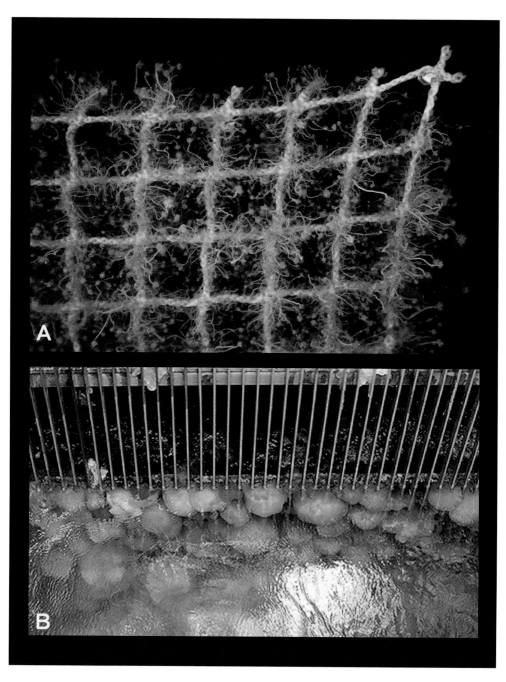

PLATE 4. Jellyfish clogging problems. *A*, Thousands of jellyfish polyps (hydroids) growing on a salmon aquaculture cage, Norway (photo by Thor Nielsen / SINTEF). *B*, Hundreds of jellyfish blocking the grills at the Hadera Nuclear Power Station near Tel Aviv, Israel, 5 July 2011 (photo ©AFP).

PLATE 5. *Nemopilema nomurai*, the sumo wrestler–sized jellyfish plaguing the fishing grounds in the Sea of Japan. *A*, A scuba diver tagging a large medusa (photo ©AFP). *B*, Hundreds of large *Nemopilema* in a fishing net (photo ©Asahi Shimbun).

PLATE 6. Overfishing: evidence of big fish and large numbers. *A*, Abalone shell mound, Santa Barbara, California, ca. 1920 (Census of Marine Life). *B*, Atlantic halibut, Provincetown, Massachusetts, ca. 1910 (Census of Marine Life). *C*, Black marlin, New Zealand, ca. 1933 (photographer unknown, provided by Callum Roberts).

PLATE 7. Heavy take of two species now in peril. *A*, Tuna in the Tsukiji Fish Market, Tokyo, 2007 (photo by Amanda Hamilton). *B*, Netful (60–80 tons) of orange roughy, Great Australian Bight, 1989 (photo by Karen Gowlett-Holmes).

PLATE 8. Eutrophication. *A*, Mississippi River sediment-laden, nutrient-rich fresh-water plume entering the Gulf of Mexico. The sediments will settle out but the dissolved nutrients will be carried far along the shelf, leading to the formation of hypoxia (photo by Nancy N. Rabalais, Louisiana Universities Marine Consortium). *B*, This is not a gravel road; the "gravel" is millions of dead fish and the "cracks in the road" are ribbons of oil in a shipping channel in Venice, Louisiana, September 2010 (photo by P. J. Hahn).

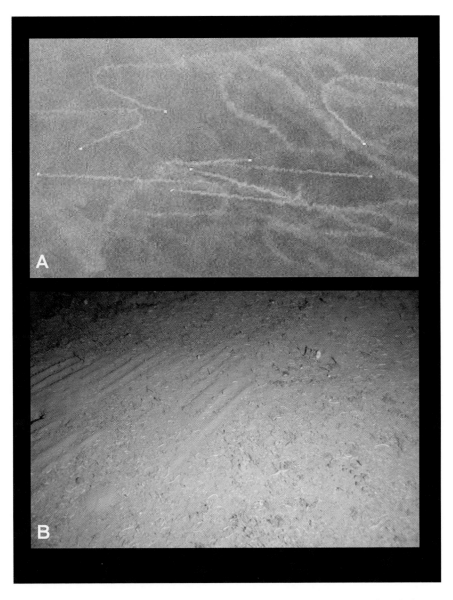

PLATE 9. Trawling damage. *A*, Trawling mud trails stirring up sediments that choke filter feeding organisms, Gulf of Mexico off the Louisiana coast, 24 October 1999 (photo by NOAA Landsat). *B*, Typical upper continental slope soft-bottom habitat showing furrows made by recent trawling (left) adjacent to less disturbed habitat from past trawling (right); fragments of benthic fauna (brittlestars) and overturned coarse sediments open up new habitat to be colonized by opportunistic and weedy species such as jellyfish polyps. Southern Australia 28 November 2004, depth of 318 meters (photo ©CSIRO Marine and Atmospheric Research).

PLATE 10. Pollution. *A*, Sign warning of unsafe fish at Los Angeles, California (photo by Gary Florin / Cabrillo Marine Aquarium). *B*, Flesh-footed shearwaters (mutton-birds) with normal plumage (above) and with abnormal white spotting (below), indicating radiation damage (photos by Jennifer Lavers). *C*, Permanent signage installed to warn of unsafe water flowing into the Tamar Estuary, Launceston, Tasmania, Australia (photo by Robin Smith).

PLATE 11. Introduced species. *A*, An urchin barren where kelp and abalone used to be plentiful, Santa Cruz Island, California (photo by Dan Richards). *B*, An aggregation of *Asterias amurensis*, the Japanese seastar, that has established itself as a dominant pest in southern Australian waters (photo by Gary Bell / CSIRO).

PLATE 12. Climate change. *A*, Polar bears are turning to cannibalism as their habitat and food sources diminish with climate change (photo © Picture Media). *B*, Permafrost bluffs eroding at near the Inupiuq village of Kaktovik on the Beaufort Sea, Alaska (photo © Hugh Rose/Accent Alaska).

PLATE 13. Variety of phytoplankton and microzooplankton found in coastal waters, mostly about 50–100 micrometers (about 2/1,000–4/1,000 of an inch) in diameter. Top row: ciliate, *Proboscia, Chaetoceros*. Middle row: tintinnid, *Mastogloia, Hemiaulus, Asteromphalus*. Bottom row: *Ditylum, Planktoniella*. (All photos by Pru Bonham, used with cooperation of CSIRO Marine and Atmospheric Research.)

PLATE 14. Effects of ocean acidification: Rows 1 and 2, left: the sea butterfly (pteropod) *Limacina helicina* in healthy condition from above the aragonite saturation horizon (1,000-meter depth) and in a degraded condition from unsaturated waters (2,000-meter depth), $\frac{1}{20}$ of an inch (photos by Dr. Donna Roberts, Antarctic Climate and Ecosystems Cooperative Research Centre). Rows 1 and 2, right: *Limacina helicina* alive, to $\frac{1}{3}$ of an inch (photo by Russ Hopcroft) and its predator the sea angel *Clione limacina*, 1 inch long (photo by Alexander Semenov / White Sea Biological Station). Row 3: the coccolithophore *Emiliania huxleyi* in normal (left) and with evidence of malformation from acidified conditions (right). Images taken with a scanning electron microscope, about 1/10,000 of an inch (images by Joana C. Cubillos). Row 4: the calcareous nannofossil *Discoaster* before (left) and during (right) the Paleocene-Eocene thermal maximum (PETM) ocean acidification event about 56 million years ago. Note the corrosion in the PETM specimen due to acidification. Images taken with a scanning electron microscope, about 1/10,000 of an inch (images by Patrizia Ziveri / VU Amsterdam).

PLATE 15. Fishing down the food web. Trophy fish taken aboard the charter boat Gulf Stream in the Florida Keys. *A*, 1950s, aboard Gulf Stream I (photo from Wil-Art Studio / Monroe County Library). *B*, ca. 1970s, aboard Gulf Stream II (photo from Dale McDonald Collection / Monroe County Library). *C*, ca. 1990s, aboard Gulf Stream III (photo from Dale McDonald Collection / Monroe County Library).

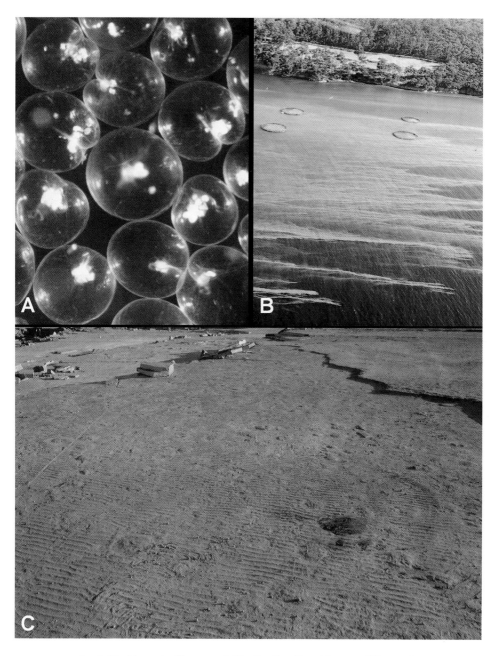

PLATE 16. *A*, *Noctiluca scintillans*, a red tide dinoflagellate (about 1 millimeter in diameter). *B*, A red tide of *Noctiluca* in a bay in southern Tasmania, Australia; the rings are salmon aquaculture pens (photo by Dr. Judi Marshall and Prof. Gustaaf Hallegraeff, University of Tasmania). *C*, Each of the dots and circles is a fossilized jellyfish, up to 3 feet in diameter, in a mass stranding event 500 million years ago, Late Cambrian, Mosinee, Wisconsin.

CHAPTER 8

Pollution Destabilizes Ecosystems

What if ocean pollution were to result in death of the tiny planktonic plants? These phytoplankton are the producers of the ocean. They are the basis of the marine trophic pyramid! Elimination of these plants would result in extinction of life in the oceans. Such a catastrophe is unthinkable.

—WILLIAM MASON AND GEORGE FOLKERTS, *Environmental Problems*

"Very nice country . . . Walked from town & fell asleep beneath a tree." These remarks were penned by Charles Darwin in his field notebook on Tuesday, 16 February, 1836, about his final "pleasant little excursion" along the banks of the Derwent River before leaving Tasmania (Banks and Leaman 1999).

The Derwent Estuary was one of Australia's finest deepwater ports and was also the center of the Southern Ocean whaling industry. But alas, like so many places in the far corners of the map, most people have never heard of it. It's at the southern end of the southern corner of the southern island of the southern continent in the Southern Hemisphere. Rotate your globe to Australia . . . now go down . . . no, farther down . . . if you get to Antarctica, go back up a smidgen. Or go to New Zealand and turn left.

Tasmania is a wild and magical place. The mystery of the striped and toothy thylacine: forever gone or master of stealth. The Tassie devil's contagious disease, passed during face-to-face romance or combat. Spotted quolls

Based on EPA estimates, in one week a 3,000-passenger cruise ship generates about 210,000 gallons of sewage; 1,000,000 gallons of gray water (shower, sink, and dishwashing water); 37,000 gallons of oily bilge water; more than 8 tons of solid waste; millions of gallons of ballast water containing potential invasive species; and toxic wastes from dry cleaning and photo-processing laboratories (PEW 2003).

A recent US National Academy of Sciences study estimates that the oil running off our streets and driveways and ultimately flowing into the oceans is equal to an *Exxon Valdez* oil spill—10.9 million gallons—every 8 months (PEW 2003).

In the United States, animal feedlots produce about 500,000,000 tons of manure each year, more than three times the amount of sanitary waste produced by the human population. More than 13,000 US beaches were closed or under pollution advisories in 2001, an increase of 20 percent from the previous year (PEW 2003).

and naked-nosed wombats roaming without fear of predators. Tasmania is a land of orchids and carnivorous plants and rainbows. Lots of rainbows.

Tasmanians are fervent "greenies." And that's a good thing, because somebody has to preserve this little corner of wonder. But underwater is a different story. The ravages of industrialization have taken their toll—perhaps it's a case of out of sight, out of mind.

In 1974, the Derwent was declared the most polluted river in the world (Bennett 1999). Zinc- and cadmium-contaminated oysters. Mercury-contaminated fish. Acids from a paper mill. Untreated or poorly treated effluent from thirteen sewage outfalls. Half a state's worth of agricultural waste funneling into the catchment. Too many decades of industrial dumping.

The major issues identified in the 1997 State of the Derwent Report include heavy metals, introduced species, high sedimentation rates, pathogens, loss of seagrass and wetlands, elevated nutrients, resin acids and low dissolved oxygen, and accumulated sludge deposits (Bennett 1999). Once inviting enough for Darwin himself to take a snooze along her shores, the Derwent is now a shocking story of power struggles, secrecy, fat-cat profits, and wanton disregard for the environment and the services it provides.

The Derwent is a microcosm for just about every estuary, bay, harbor, and coastal waterway in the world. We have been using rivers as giant colons to rid our urban waste (see plate 10). Beaches have become repositories for our soda bottles, beer cans, plastic bags, and used syringes. Ships' ballast tanks have become essentially comfy aquariums for travel-minded flora and fauna. This chapter and the next focus on these issues and what happens next, after we have abused and insulted our seas.

The morning of 19 September 2009, surely must have seemed like any other. Except on that day, 498,818 volunteers from 108 countries and locations picked up 10,239,538 items of marine debris, totaling some 7.4 million pounds (Ocean Conservancy 2010). Over 2 million cigarette butts and filters—just 15 seconds' worth of worldwide consumption. Over 1 million plastic bags. Nearly 1 million food wrappers and containers. Over 900,000 caps and lids. Beverage bottles. Cutlery. Cans. Straws. Stirrers . . . In one day. Please reread this paragraph, then pause for a moment before continuing. Let it sink in. Let yourself be shocked.

Pollution is, in the broad sense, any substance or object introduced into the ecosystem that has or can have a harmful effect. Many types of pollution come from nonpoint sources, such as wind-blown debris, automobile exhaust, and agricultural or urban runoff. There are essentially two main types of pollution: nutrient and nonnutrient. Nutrient pollution includes substances that are natural in the marine environment but are in unnaturally excessive amounts (eutrophic conditions) or severely limiting (oligotrophic conditions). We examined the issues relating to nutrient pollution in the preceding chapter; we shall now turn our attention to nonnutrient pollution. Both types are highest in coastal areas, an outcome of increasing urbanization. And both types can cause phytoplankton blooms, leading to hypoxia or anoxia, in turn leading to jellyfish blooms.

Nonnutrient pollution includes the garbage and toxic substances that are not natural in the marine environment and have a deleterious effect on marine life, such as petroleum hydrocarbons, heavy metals, plastic debris, and toxic waste. Marine pollution can be obvious, such as an oil spill or plastic drink bottles and beer cans along a coastline, but often the most harmful pollutants are those we cannot see. Some chemicals react in such a way as to deplete oxygen from the water, causing anoxia. Many types of toxic chemicals diffuse into the seawater or adhere to tiny particles that are taken up by filter-feeders or deposit-feeders. Stored in flesh or fat, these toxins concentrate over time or magnify up the food chain, eventually ending up on our dinner table either directly through contaminated fish or indirectly through fish meal in animal feeds, transferring to us through meat and dairy products.

A gobsmacking array of pollutants makes its way into the marine ecosystem. Some dissipate quickly, while others accumulate on the surface of sediments until they are taken into the food chain or resuspended from the sediments by trawlers, bioturbating invertebrates, or storm action. Many types of heavy metal and organic pollutants concentrate in the tissues of higher predators (a process called "biomagnification"), where they become hazardous to the predator or to humans as *their* predator.

Pesticides. Herbicides. Fungicides. Hydrocarbons. Fluorocarbons. Oil spills. Leaching plastics. Radioactive waste. Chemical weapons dumping. Endocrine disruptors and various gender-benders. Antibiotics triggering microbial resistance. In fact, about 70,000 man-made chemicals currently surround us in building materials, personal care products, lawns and gardens, the water we drink, the air we breathe, and the food that we eat (McTaggart 2000).

These contaminants come from a wide variety of sources. Industrial factories discharge chemical waste directly into waterways or incinerate it into airborne particles that eventually settle. Crop-based agriculture is a rich source of pesticides, whereas stock-based agriculture often uses huge quantities of antibiotics and growth hormones. Urban roadways are a substantial source of oil and rubber residues. Medications, cosmetics, micro-scrubbers, disinfectants, and a stupefying array of household contaminants make their way into sewer systems, septic tanks, and rivulets and waterways through residential drains.

Nonnutrient pollution often follows the same routes into the oceans as excess nutrients, and as one might expect, those regions with the highest eutrophication often suffer from the highest levels of chemical contaminants. Oil spill residues in the Gulf of Mexico. Routine fish consumption advisories for mercury, PCBs, herbicides, insecticides, and endocrine disruptors in the Chesapeake. Complex cocktails of pesticide residues, petroleum hydrocarbons, synthetic organic compounds, and heavy metals in Boston Harbor and the Gulf of Maine, where tumors and diseases in fish and shellfish are above the national average. Nearly 50,000 containers of radioactive waste at the Farallones off San Francisco. Carcinogenic DDT and PCBs still lurking in fishes and sediments off Los Angeles. Similarly, concentrations of DDT and

PCBs are still readily found in Mediterranean mammals, birds, and inverte-brates, despite having been banned more than 30 years ago. The Canadian Inuit have been harvesting beluga whales for thousands of years, but today the whales are so heavily contaminated that they are no longer safe to eat . . . *or even to touch*—when they are found washed up dead on beaches, they have to be removed by HazMat teams (DFO 2012).

In Japan, mercury and Minimata disease, fallout from Hiroshima and Nagasaki, nuclear waste dumping by Russia in the Sea of Japan, toxic whale meat, toxic tuna, and the Fukushima nuclear meltdowns are just a few of the concerns. Coastal waters and sediments around Japan remain heavily pol-luted with contaminants including dioxins, PCBs, butyltin compounds, heavy metals, floating litter, bauxite residue, and radionuclides.

"Just as the speed and scale of China's rise as an economic power have no clear parallel in history, so its pollution problem has shattered all precedents" (Kahn and Yardley 2007). In the East China Sea, 81 percent of the sea area has been rated category 4 for pollution on a scale of 1–5; half of the red tides in China are from this region, and are blamed on petrochemical waste and heavy metal sediments. Currently, China is completing construction on new coal-fired power plants at the breakneck pace of 2–3 *per week* (Harrabin 2007). This adds even more pollution, not to mention carbon dioxide, on top of the nation's already severely polluted air and water.

In the Northern Indian Ocean, a persistent brownish haze covers about 10 million square kilometers (4 million square miles)—roughly the same area as the continental United States—and is as dense as the smog one might ex-pect to find in the most polluted cities (SIO 1999). Preliminary data indicated that the scatter effect of the aerosols reduced the amount of energy absorbed by the ocean surface by up to 10 percent, impacting photosynthesis and the amount of moisture evaporating from the ocean and therefore altering the entire rainfall cycle.

In Australia, herbicides and other pesticides are pouring into the Great Barrier Reef from sugar cane farming and other agriculture, heavy metals persist in the Derwent Estuary, the fish at Gladstone have become so toxic as to close fishing, and the world's highest levels of mercury were found in a dead dolphin in 1999 off Adelaide (Grady and Brook 2000). Despite input reductions, continued input and persistent concentrations of heavy metals, PCBs, DDT, and other harmful organic compounds, as well as radionuclides, are still high in the Baltic (HELCOM 2007). Contaminants that cause illness in fish, including lymphocystis, liver nodules, skeletal deformities, parasites, and skin ulcers, the latter of which were found in 43 percent of cod, continue to plague the North Sea (Lang 2008).

If your head is spinning from all the chemical names and medical and ecological effects, welcome to the club. It *is* head-spinning: that's the point. An astonishing number of chemicals are causing an astounding array of problems. So much so that for most of us, tuning out seems like our only option. And so the problem continues to grow.

Bioaccumulation and Biomagnification

Although they sound similar, these two concepts are fundamentally different. "Bioaccumulation" is the buildup over time of toxic substances in the body of an organism. "Biomagnification" is where these accumulated substances are passed up the food chain, and compounded through repeated ingestion of contaminated organisms.

The health of organisms—and indeed, the human organism at the top of the food chain—can be affected by pollution directly through acute toxicity poisoning or indirectly through these substances being concentrated in prey. But these processes can and do also affect community composition in marine ecosystems through their cumulative effects on higher levels of the food chain.

The Dirty Dozen

Twelve persistent organic pollutants have been identified by the United Nations Environmental Programme as being the worst of the worst: polychlorinated biphenyls (PCBs), dichlorodiphenyltrichloroethane (DDT), hexachlorobenzene, dioxin, furans, dieldrin, aldrin, endrin, chlordane, heptachlor, toxaphene, and mirex (Alessi et al. 2006). These chemicals share certain characteristics that make them particularly hazardous. They are toxic to both humans and wildlife. They resist breaking down, remaining intact for a long time. They are highly soluble in lipids (fats), and so are able to accumulate in the bodies of humans, marine mammals, and other wildlife. They concentrate up the food chain. They pass from mother to fetus through the womb and to the child through breast milk. They cause nervous system damage, immune system diseases, reproductive and developmental disorders, and cancers. Singly, they are toxic. In synergy, we have no idea.

Pesticide Residues

Warning signs are posted on fishing piers and fences along the coastlines of Southern California, alerting fishermen not to eat white croaker (see plate 10). White croakers are common from San Francisco to Mexico. In fact, they are so easy to catch that they are often considered a nuisance to fishermen trying to catch a more sporting fish. But that's the problem. Many of the fishermen aren't there for sport; they are fishing to feed their families. And because the white croakers are so plentiful and easy to catch, they make an easy meal. And they're toxic.

In 1987, Donald Malins and his colleagues from the National Marine Fisheries Service found high concentrations of DDT and PCBs, as well as liver carcinomas and other lesions, in tissue samples from white croakers in the Los Angeles area (Malins et al. 1987). By examining over 100 fish as well as sediment samples from 5 sites, they found that these chemicals were in higher concentration in prey items in the croakers' stomachs, including fish, squid, worms, crabs, shrimp, and clams, than in the sediments. It was clear that the prey were concentrating the toxins and passing them along to the fish.

DDT and PCBs are not natural in the marine environment. They got there through surface water, groundwater, and aerial emissions from industrial plants, such as Montrose Chemical Corporation near the Los Angeles Harbor. Montrose manufactured DDT, the new "wonder pesticide," at its Torrance plant from 1947 to 1982. Wastewater was disposed into the county sewer system, which emptied into the ocean, while storm water that ran off over the contaminated soils and sediments in and near the site and aerial emissions exposed drinking water aquifers that served more than 100,000 people. An estimated 1,700 tons of DDT were discharged just between the late 1950s and early 1970s.

In 1948, the Nobel Prize was awarded for the discovery of DDT as an insecticide. But the 1962 book *Silent Spring* by Rachel Carson brought attention to a more sinister side to the chemical as an environmental hazard. As a result of the environmental movement largely spawned by the book, agricultural use of DDT was banned by the United States in 1972, and eventually worldwide by the Stockholm Convention in 2004.

The Environmental Protection Agency issued an administrative order on 6 May 1983 requiring Montrose to cease all discharges of DDT, but sediments remain contaminated. DDT is an effective toxin against many kinds of arthropods, not only insects; crustaceans in the sea are also poisoned by the powerful neurotoxin. Worms and other small organisms living in the soils ingest and absorb these toxins too; fish and other organisms eat many worms, and bigger fish eat many small fish, and the toxins concentrate in fatty tissues up the food chain.

DDT also affects fish-eating birds like eagles by causing their eggshells to be very thin and brittle, leading to premature breakage and the death of developing chicks. In humans, DDT is linked with a range of unpleasant health effects, including:

- an increased occurrence of diabetes and cancers of the liver, pancreas, and breast;
- developmental and reproductive toxicity, such as premature birth and "disruption in semen quality, menstruation, gestational length, and duration of lactation" (Rogan and Chen 2005, 770);
- interference with thyroid function;
- neurological problems including Parkinson's; and
- asthma.

DDT stores in the body fat and has a half-life of 6–10 years. Testing by the Centers for Disease Control and Prevention in 2005 detected DDT and its derivatives in almost all human blood samples (Eskenazi et al. 2009), while Food and Drug Administration tests commonly detect DDT in our food (USDA 2009).

In 1990, the state and federal governments filed a lawsuit against 10 companies, including four DDT-manufacturing plants, for damages relating to release of DDT and PCBs into the marine environment near Los Angeles. The chemicals were released in contaminated sewage up until the 1970s.

In December 2000, the US Department of Justice and the California Attorney General announced a settlement with Montrose Chemical Corporation of California and other polluters totaling about $140 million to fund cleanup of the DDT contamination and restoration of the marine environment.

Ocean sediments near populated areas often contain persistent organic pollutants like DDT, PCBs, and PAHs (polycyclic aromatic hydrocarbons, a group of over 100 chemicals that are produced during the incomplete burning of coal, oil, gas, wood, garbage, or other organic substances). These residues of industrial burning and pesticide use from decades past are relatively safe while sequestered in the sediments. But when resuspended, they are introduced back into the food chain and into our food supply. In particular, storms and human activities like bottom-trawling, dredging in shipping channels, and coastal construction (for example, the building of marinas), stir up the sediments on the seafloor, causing the finer particles to resuspend in the water column as a plume that may drift with the current for great distances. As they settle out, these fine particles can be harmful to filter-feeding, bottom-dwelling organisms, such as clams, mussels, sponges, and tunicates. The nutrient phosphorus also often concentrates in shallow sediments and when resuspended, as we have seen, can contribute to phytoplankton blooms, thereby hastening dead zones.

The increase in turbidity caused by bottom-trawling also reduces light penetration, which can impact on photosynthetic organisms, such as kelps and corals. As we have seen earlier, turbidity can also reduce the ability of visual predators to see their prey, thus favoring tactile predators like jellyfish.

Petroleum Hydrocarbons

Of all the various types of pollutants in the sea, one of the most visible — and emotive — certainly must be oil spills. Crude oil is essentially a cocktail of substances that range from sticky to tarry, soluble to glob-forming, smelly to noxious to seriously poisonous. The blackened sand, slick water, and drowned birds are just the tip of the iceberg. The fractions, or chemical components, that often do the most biological damage are those that we can't see; they dissolve into the water or evaporate away, leaving only their toxic residues.

Hydrocarbons, the product of decomposed organic matter from eons past, occur naturally in crude oil. The purpose of refining oil is to purify these substances for our use. Asphalt, propellants for aerosol sprays, solvents, gasoline, and jet fuel are just a few of the uses of various hydrocarbons. Methane. Propane. Butane. Hexane. Octane. Toluene. Xylene. Naphthalene. We hear these terms in various contexts of daily life. Many of them cause cancer or nerve disorders or affect the blood, immune system, lungs, skin, eyes, liver, kidneys, or developing fetuses.

When these compounds enter the environment, undesirable outcomes often occur, particularly when they affect the base of the food chain. Experiments with numerous types of phytoplankton exposed to low concentrations of hydrocarbons have shown that some species are strongly stimulated to grow, while others are moderately stimulated, while still others are actually inhibited (Dunstan, Atkinson, and Natoli 1975).

Therefore, it seems clear that even low concentrations of oil in the marine environment can alter the growth or cause the demise of certain phytoplankton species. Even minor alterations low in the food chain can be anticipated to have major cascading effects on species further up. The "elephant in the room," so to speak, since the BP oil spill in the Gulf of Mexico, is what effect the disaster will have on the local ecosystem, and, more importantly, on the gulf fisheries, America's second largest next to Alaska, a business on which many people's livelihoods rest.

According to industry experts (Patton 2010), shrimpers' catches were down by 75 percent from normal in the first month of shrimping season after the spill. The shrimp harvest in the western Gulf of Mexico for that year was projected to fall some 20 percent below the historical average. If the *Exxon Valdez* damage is anything to go by, it can be a long road back: that spill destroyed billions of salmon and herring eggs, and those fisheries are still at reduced levels. Part of the problem is the stimulating effect that oil spills have on microbial activity, "spawning an explosion of bacteria that feed on crude." As the bacteria decompose the oil, their respiration processes suck the oxygen out of the water, creating or enhancing dead zones.

It will be many years before a clear picture emerges on the economic damage to the fishing industry from the spill, but what about the immediate ecosystem damage? What about the birds, the invertebrates, the algae?

Crude oil coats everything it touches with a shiny, tarry, smelly film. Every rock. Every feather. Every hair. Every gill. Every root. Every grain of sand. Images of oil-coated birds following spills are ubiquitous. Even thin films of oil will usually prove deadly to birds, because any spot of oil acts like a pinhole in their natural waterproofing and insulation barrier. Once that barrier is broken—it doesn't matter how broken—hypothermia is inevitable. Preening only makes things worse, hastening death through poisoning by ingestion.

When the *Exxon Valdez* ran aground in Alaska in March 1989, only 17,000 barrels of 257,000 were recovered. Thirteen hundred miles of shoreline were impacted, along with outright deaths of "250,000 seabirds, 2,800 sea otters, 300 harbor seals, 250 bald eagles, 22 killer whales, and billions of salmon and herring eggs" (NPS 2009, 1). The timing seemed like it could not have been worse: seabirds were gathered in prebreeding aggregations and fish were spawning ... The Deepwater Horizon spill in the Gulf of Mexico, however, is *worse*, having spanned both mating and nesting season.

Marine mammals don't fare well, either. Those with fur, for example, seals and otters, become coated in oil, losing their insulating properties and becoming at risk of hypothermia. Several hundred harbor seals are believed to have died from inhalation of toxic fumes (Peterson et al. 2003). Cetaceans can have trouble breathing if their blow holes become clogged with oil. Even those not directly affected may still eat contaminated food or face stiffer competition from losses through the food chain.

Once thought to break down through rapid dispersion and microbial degradation, it is now clear that spilled oil has both an acute mortality effect and a chronic sublethal effect. A study conducted nearly 20 years after the *Exxon Valdez* spill found that 98,000 liters (26,000 gallons) of oil were still trapped in the sediments along the Alaskan shoreline (Walsh 2009). These hidden remnants continue to leach toxins into the water year after year. Even low level chronic exposure, such as from persistent subsurface oil, is now known to have a range of deleterious effects, including:

- ongoing deformities and mortality of fish embryos in oiled estuaries and stream banks;
- concentrated contamination in filter-feeding clams and mussels that becomes toxic to their mammalian predators (including humans); and
- a lengthened recovery process for the entire affected ecosystem due to compromised health, depressed reproduction, and enhanced mortality overall, and the trophic cascades stemming from it (Peterson et al. 2003).

Some of the most profound effects are found in the "other" species—the less charismatic ones without the fur and feathers. Macroalgae and bottom-dwelling invertebrates suffer from a combination of chemical toxicity, smothering, and physical displacement and scalding by pressurized water used in the cleanup effort. These lead to cascades of indirect effects that can last for decades and slow the recovery process of the entire ecosystem (Peterson et al. 2003). Loss of the protective cover of the structural algal canopy reduces the survival of young algae trying to recolonize. This in turn reduces the habitat for grazing snails and their predators, thus opening up niche space for opportunistic and weedy species like green algae and barnacles.

Heavy Metals

Recall the highly polluted Derwent Estuary in Tasmania. In 1974, the heavy metal loads were so high that the sediments were saturated. A follow-up study in 1996 found that there was little reduction in heavy metal levels since the original investigation—the toxins were continuing to leach from the sediments, and the fish and shellfish were continuing to concentrate them to dangerous levels (Whitehead et al. 2010). Today, fishing bans are still in place and regular warnings are still issued about water safety.

Heavy metals, often called toxic metals, are a worrying source of pollution. These include elements, such as mercury, cadmium, lead, chromium, zinc, copper, and arsenic, among others. Some are required in minute amounts for human health, while others, even in small amounts, can cause severe health problems or death.

Unlike organic pollutants, heavy metals do not decay over time. They can accumulate in sediments and be stirred up later by trawling or dredging activities, storms, or bioturbating organisms. Of greatest risk to us is when heavy metals are taken into the food chain by deposit-feeding organisms and concentrated through bioaccumulation or biomagnification. Heavy metals can cause changes to tissue matter, reproductive and growth abnormalities, and behavioral alterations.

Heavy metal pollution arises from various sources. For example, chromium and cadmium are used in electroplating, while mercury pollution typically comes from metal smelting, as well as chlorine and cement plants and coal-fired power plants. Common sources of copper pollution include erosion of overhead cables by railway traffic, combustion of waste, and automobile brakes. Large amounts of lead were emitted by automobile exhaust until it was banned in fuels in 2000.

The plight of mercury poisoning was brought to the world's attention in 1972 with the publication of a black-and-white photograph, *Tomoko Uemura in Her Bath*, by American photojournalist W. Eugene Smith. The startling image depicts a mother cradling her severely deformed, naked daughter in a traditional Japanese bathtub. Tomoko suffered from a form of mercury poisoning known as Minamata disease, named for the region in Japan where it was discovered. Tragically, she was poisoned by methylmercury while still in the womb through her mother, who had eaten local fish and shellfish that had been contaminated by discharges from a local chemical plant. Unknown at the time, the congenital form of Minamata disease occurs when the placenta removes the mercury from the mother's bloodstream and concentrates it in the fetus.

From 1932 to 1968, the Chisso Corporation's chemical factory at Minamata released methylmercury into Minamata Bay and the Minamata River, causing the deaths of more than 600 people and giving rise to claims of illness by over 21,000 people. Thousands of residents of Minamata and nearby regions suffered the effects of mercury poisoning from consuming contaminated fish and shellfish from this region. Mercury poisoning causes a severe neurological syndrome, including walking and coordination difficulties, numbness in the hands and feet, muscle weakness, and damage to vision, hearing, and speech; many victims also experience rapid onset of insanity, paralysis, coma, and death.

Another source of mercury poisoning, ironically, can be found in Japanese school lunches. Whale meat is obtained as a byproduct of an annual harvest for "scientific purposes" of about 1,000 whales and 20,000 dolphins. The meat is fed to school children "as an effort to pass 'traditional food culture' down to children," the *Wall Street Journal* reports (Twaronite 2011). But the World Wildlife Fund calls the "scientific purposes" argument a sham to circumvent the international moratorium on whaling (ABC 2005). Apparently, so too is the "traditional food culture" argument. While a few coastal communities have been eating whale since the 1700s, for most it is a recently acquired taste, becoming popular as a cheap source of protein in the postwar years (Head 2005). About 20 percent of schools in Japan serve whale meat, which they buy at deeply discounted prices from the Institute of Cetacean Research, which carries out the government's whaling program (Parsons 2010; "Whale Meat Increasingly Back on Menu" 2010). However, samples of the short-finned pilot whale tested at 10–12 times the level of mercury that is considered safe (Reuters 2007), far higher than even tuna, which is on most "do not eat" lists.

Outside of Japan, mercury is a common heavy metal that we get from eating fish, particularly large, predatory fish like tuna, swordfish, king mackerel, and shark. Mercury in the oceans is primarily derived from the exhaust of coal-fired power plants then dispersed into the atmosphere, shunted toward the poles by circulation patterns, and settled into the oceans (Jackson 2010).

There is no "safe" level of mercury. The Food and Drug Administration suggests a limit of 1 ppm of mercury in fish and seafood. To put this into context, the highest category of contaminated fish, which includes, ahi tuna, swordfish, and shark, reveal concentrations of more than 0.5 ppm of mercury;

general guidelines suggest limiting the consumption of these fish to no more than 400 grams (14 ounces) per week, or 200 grams (7 ounces) if they have 1 ppm of mercury. These suggested limits do not take into consideration mercury buildup from occupational sources or contamination by drinking water, inhalation, other contaminated foods, or cosmetics.

In comparison with fish, whale meat is far more toxic, because the whales live longer, eat more contaminated fish and squid, and concentrate more toxins in their fatty tissues. It was reported in *New Scientist* in 2002 that tests on whale meat for sale in Japan revealed extremely high levels of mercury: 2 of the liver samples in 26 tested contained over 1,970 ppm of mercury, nearly 5,000 times the government's contamination limit of 0.4 ppm.

> At these concentrations, a 60-kilogram adult eating just 0.15 grams [less than ⅒ the weight of a US dime] of liver would exceed the weekly mercury intake considered safe by the World Health Organization . . . Rats suffered acute kidney poisoning after a single mouthful of the most highly contaminated liver . . . On average, concentrations of mercury in whale and dolphin livers were 370 [ppm], 900 times the government limit. Average levels in kidneys and lungs were also high, about 100 times the limit. None of the samples was below the limit . . . While levels were lower in muscle, on average it still contained 2.5 to 25 times the limit. (Coghlan 2002)

Plastic Debris

> Plastics made up 80–85 percent of the seabed debris in Tokyo Bay, an impressive figure considering that most plastic debris are buoyant.
> —JOSÉ DERRAIK, "The Pollution of the Marine Environment by Plastic Debris: A Review"

Not all that long ago, the question "Paper or plastic?" was about sensitivity toward forests—or about conveniently sized plastic liners for bathroom waste baskets. Then came reports about turtles mistaking plastic bags for jellyfish. But the true extent of the plastics problem remains off most people's radar. And each year, only 5 percent of the 1 *trillion* plastic bags produced in the United States alone get recycled (Sivan 2011).

We like plastics because they are lightweight, strong, durable, and cheap—but these very same properties are the reasons they pose a serious hazard to the marine environment (Derraik 2002). It's not just about turtles. Nobody really knows how long plastics will last in the marine environment. According to Dr. Tony Andrady of the Research Triangle Institute in North Carolina, "Except for a small amount that's been incinerated, every bit of plastic manufactured in the world for the last 50 years or so still remains. It's somewhere in the environment" (Weisman 2008, 126). The problem is that plastics don't "degrade" or "break down" in the sea, they only "break apart." As they become smaller and smaller, it is easy for us to think that they have gone away, but they don't. Powder-sized particles are taken up by tiny filter-feeding organisms and make their way into the food chain. On land, ultraviolet light and warm temperatures break down the plastics, but water reduces the UV penetration and buffers temperature change. These effects are particularly reduced in the deep sea, where many of the plastics are accumulating (Macfadyen, Huntington, and Cappell 2009).

Vast regions of accumulated floating and drifting plastic debris span thousands of miles in the mid-ocean gyres: drink bottles, beach sandals, food containers, rubber boots, laundry baskets, children's toys, toothbrushes, rubber duckies. Ship captains have dubbed an area midway between San Francisco and Hawaii, the "eastern garbage patch" (not to be confused with the western garbage patch, off Japan, or others in other gyres). This area, where plastic vastly outnumbers plankton, is a slowly rotating clockwise swirl of plastic flotsam about twice the size of Texas, possibly the world's largest dump (Weiss 2006).

It is impossible to accurately estimate the amount of plastic debris accumulating in the oceans and seas, but the limited studies that exist suggest that the problem is (as the saying goes) "bigger than *Ben Hur*." A review on pollution by plastics in the marine environment included the following:

> In 1975 the world's fishing fleet alone dumped into the sea approximately 135,400 tons of plastic fishing gear and 23,600 tons of synthetic packaging material. [Another study] estimated that merchant ships dump 639,000 plastic containers each day around the world, and ships are therefore, a major source of plastic debris. Recreational fishing and boats are also responsible for dumping a considerable amount of marine debris, and according to the US Coast Guard they dispose approximately 52% of all rubbish dumped in US waters. (Derraik 2002, 843)

But don't rush to blame just the fishermen and coast guard. In 1982, it was estimated that more than 8 million debris items per day were entering the sea, made up of about 8 percent plastics; however, as our love affair with plastic products has increased, so too has its accumulation in the sea, which is estimated to have risen severalfold (Thompson et al. 2004; Barnes 2005; Gregory 2009).

For many marine species, plastic debris has become essentially another food group. Turtles mistake plastic bags for jellyfish. Seabirds mistake disposable lighters for fish. Fish mistake Styrofoam particles for plankton. These are some of the effects of plastic waste in the oceans.

- Turtles often mistake plastic grocery carry bags and food wrappers for their jellyfish prey. Feeding experiments demonstrate that sea turtles actively select colorful plastic items over colorless, and that plastic items accumulate in the gut for many months (Lutz 1990), clogging their digestive tracts and blocking the passage of food and waste. A study of 371 leatherback necropsies since 1968 found over 37 percent contained plastic in the gut (Mrosovsky, Ryan, and James 2009); interestingly, the study also found a rapid increase in the incidence of plastic ingestion from the 1960s to the 1980s, with a leveling off after that.
- Penguins, pelicans, albatrosses, shearwaters, auklets, seagulls, and other seabirds, as well as seals and dolphins, ingest plastics or become entangled in plastics. Netting, fishing line, ropes, and other debris can become entangled around the neck, legs, wings, flippers, tails, and flukes, leading to strangulation, cutting and infection, amputation, drowning, and asphyxiation, as well as dragging during swimming. Seals in particular are curious and playful and will often poke their heads into loops and holes; however, while loops slip on easily, the lay of their hair keeps collars from slipping back off (Derraik 2002). Birds that prey on plankton are vulnerable to confusion of plastic pieces for copepods, krill, or squid, while fish-eating birds like albatrosses often mistake larger, colorful plastic debris for their prey; intestinal obstruction and diminished hunger signals are among the most harmful effects (Azzarello and Vleet 1987).
- Planktivores mistake tiny drifting plastic particles for their prey. Samples from the North Pacific central gyre in 1998 revealed that plastics outweighed plankton 6 to 1 (Moore et al. 2001); however, by 2008, this figure had risen to a whopping 46 to 1 (Algalita 2009). The majority of fragments found were thin films like those from sandwich bags and pieces of monofilament or polypropylene line—all those shining and glinting little bits must look like a ticker-tape parade or a Thanksgiving feast. Even jellylike salps

were found with plastic fragments embedded in their tissues, a startling find given their brief lifespan.

- Suspension-feeding or filter-feeding animals like mussels, clams, anchovies, and sardines have been shown to ingest and accumulate microscopic plastic fragments. Experiments with mussels demonstrated that a brief pulse of particles accumulated in the gut, then translocated to the circulatory system within 3 days and persisted for over 48 days (Browne et al. 2008). Given these results, it is not hard to imagine that longer-term exposure to microplastics could have deleterious effects.
- Deposit-feeding organisms have also been shown to be affected by microplastics. Experiments with sea cucumbers have demonstrated that up to twentyfold more PVC fragments and over a hundredfold more fragments of nylon line were ingested than were expected from the concentration in the sediments (Graham and Thompson 2009). It therefore appears that these organisms were preferentially selecting plastic particles over sediment grains.
- Toxic additives leaching from plastics are a big concern, particularly in the acidic environment of stomach acids and digestive fluids. Furthermore, plastics readily act as both a magnet and a sponge for attracting and absorbing PCBs, DDT, and other organic pollutants from the marine environment, then act like a toxic bullet as they enter into the food web and concentrate up the food chain (GESAMP 2010).

Consider the northern fur seals of the Pribiloff Islands in the Bering Sea. In 1969, 100 percent of the seals returning to rookeries were free of plastic entanglement. In 1973, 38 percent were entangled. In 1976, scientists estimated that up to 40,000 fur seals each year were being killed by entanglement in plastics (Derraik 2002).

Dr. Jennifer Lavers of the University of Tasmania has studied the flesh-footed shearwaters (also called mutton birds) at Lord Howe Island. She found that 96 percent of the birds breeding on the island have plastic in their digestive tracts. In April 2011, one bird was found with 276 pieces of plastic in its stomach—15 percent of the bird's body mass—the equivalent of an average human ingesting 11 kilograms (25 pounds) of plastic. According to Lavers, "Once ingested, plastic can block or rupture the digestive tract and leak contaminants into the bird's blood stream resulting in stomach ulcerations, liver damage, infertility, and in many cases, death" (Lavers 2011).

But there is an even more sinister side to the picture. The process of attraction of contaminants to plastics is called "adsorption." Dr. Hideshige Takada, a geochemist at Tokyo University, has studied adsorption of toxins onto microplastics, and found that PCBs and DDE (a breakdown product of DDT) build up on floating plastics and can eventually accumulate concentrations up to 1 million times those in the surrounding seawater (Raloff 2001; Mato et al. 2001).

Speaking at a conference on the problem of plastics in the Mediterranean in 2011, Maria Damanaki, European commissioner for Maritime Affairs and Fisheries, stated, "Last July a Franco-Belgian research team announced the results of their research; there were almost 250 billion small pieces of plastic in the Mediterranean and an additional 500 tones [*sic*] of dissolved plastic litter on the surface of our sea" (Damanaki 2011, 2). To combat the plastics problem, the European Union has unveiled a program in May 2011 to subsidize Mediterranean fishermen to catch plastic rather than fish. The aim of the plan is to provide fishermen with an alternative income while reducing pressure on dwindling fish stocks (Harvey 2011). Of course, there is still bycatch. . . .

For many years, startling photographs and stories have raised our awareness about turtles confusing plastic bags for their jellyfish prey, seabirds with plastic 6-pack holders twisted around their necks, otters and sea lions choked by tangles of fishing line, and once-beautiful beaches strewn with plastic drink bottles, food wrappers, flip-flops, cigarette lighters, and children's toys.

But what we haven't seen is what happens to those plastics as they break down. They fragment. They shred. They break into teeny little pieces. Sounds good? Nope: keep reading.

Larger pieces of plastic are often eaten by larger animals or become entangled with them, while smaller fragments are eaten by smaller animals, which are in turn eaten by larger animals. Plastics stay in the gut. They don't break down. Debris like bread wrappers, potato chip bags, and garbage can liners block the digestive tract. Or animals feel full from accumulated particles and accidentally starve. Seabirds regurgitate plastic debris into the begging mouths of their young, which cannot regurgitate and so choke and die. By the tens of thousands.

The problem is much worse with microplastics. Microplastics come primarily from two sources: first, the microscopic particles resulting from physical and chemical breakdown of plastic waste in the oceans, and second, manufactured microscopic resin pellets (called "nurdles") and cosmetic and industrial scrubbers. They even come from our cozy polar fleece jackets, which release thousands of tiny plastic threads every time they are laundered (Leschin-Hoar 2011). Microplastics are particularly abundant in subtidal sediments (Thompson et al. 2004). One particularly revealing study compared the density of plastics found in 2 groups of surface-water samples from 5 sites sequentially offshore from Southern California, the first sampling after 2 months without rain, and the follow-up sampling just after a storm with 9 centimeters (3 ½ inches) of rainfall (Moore et al. 2002). The average density of plastics was 8 pieces per cubic meter (1 piece per 4 cubic feet); after the storm it was 7 times higher. Furthermore, the average mass of plastic was 2 ½ times higher than the mass of plankton, and even higher after the storm.

Experiments with different types of feeding animals (e.g., filter-feeders, deposit-feeders, detritus-feeders), have demonstrated that these particles are readily taken up by animals who cannot differentiate them from food (Thompson et al. 2004). Visual predators are often attracted to the colorful specks. The precise effects of these plastics in the food chain are not yet well understood. However, it seems plausible that the toxins in and on plastics may be capable leaching out during the digestion process, and perhaps concentrating up the food chain. It also seems likely that as plastic particles accumulate in guts and tissues of animals, these particles may concentrate in the guts and tissues of the animals that eat them.

While the effect of large plastic items is quite obvious—we see trash on the shoreline and entanglement of marine animals—the effects of microplastics may be much more difficult to identify. The blockage of gut or circulatory systems by microscopic particles can happen slowly, and the types of animals most affected by microplastics are typically off most people's radar, for example, worms, plankton, sea cucumbers, bivalves. However, as toxins enter the food chain via these animals and concentrate in their predators, we may see effects that we previously could not have imagined.

Dirty Little Secrets: Nuclear Waste

Lawyers, Liars, and Liabilities

Nuclear radiation is a bit of a hot topic at the moment, with the Fukushima disaster exploding into the world news in early 2011. But long before Fukushima rattled our feeling of safety, hundreds of thousands of tons of dumped nuclear waste were a percolating problem for marine life. Authorities on the

Fukushima incident say that "release of [11 million liters of] radioactive water into the ocean . . . shouldn't pose a widespread danger to sea animals or people who might eat them" (Australian AP 2011), citing dilution as the reason. However, from the information below, these conclusions seem worth pondering. And questioning.

Between 1946 and 1982, when ocean dumping of radioactive waste was legal, some 1.7 million curies (Ci) were disposed of at sea in containers designed to last only 100 years. A curie is a nonmetric unit of measurement of radioactivity that is roughly the activity of 1 gram of the radium isotope ^{226}Ra; it is also used to measure the quantity of material containing the number of atoms that would produce 1 Ci of radiation. The metric unit of radioactivity is the becquerel (Bq), which equates to 1 decay per second. One curie is equal to approximately 37 gigabecquerels. Therefore, 1.7 million Ci is a lot.

Radioactivity is classified into three types relating to strength. So-called alpha emitters, such as plutonium and americium, are very strong in their ionizing ability but not very strong in terms of distance. Beta emitters, such as strontium, caesium, cobalt, and iodine, are weaker in terms of ionizing power but travel greater distances. Alpha radiation is the most dangerous because the ionizing process essentially rips cells to bits; beta radiation is safer, in that it only damages cells, but it can reach more cells that it only damages a bit, and can be very dangerous in this respect. A third category, gamma radiation, is a type of invisible, very high-energy light. The legal ocean dumping of radioactive waste contained about 99 percent beta emitters. Alpha emitters were also present in low quantities. What do we expect to happen as these containers degrade and bleed their contents into the food chain?

The London Dumping Convention came into force in 1975, making it illegal to dump high-level radioactive waste and requiring permits for low-level waste dumping. Low-level waste is defined as radioactive waste that does not require shielding during normal handling and transport, whereas high-level waste requires shielding at all times. A 1983 amendment to the convention suspended all radioactive dumping at sea.

According to Dominique Calmet of the Nuclear Fuel Cycle and Waste Management division of the International Atomic Energy Agency, the main objective of radioactive waste disposal in the deep sea is "to isolate it from man's surrounding environment for a period of time long enough so that any subsequent release of radionuclides from the dumping site will not result in unacceptable radiological risks . . . sea dumping is essentially a strategy of dispersion/dilution rather than one of containment" (Calmet 1989, 47, 48).

Back to Fukushima: the real question isn't "will the effect be widespread," because the answer is subjective (Atlantic species are unlikely to be severely affected, but the fallout plume has spanned the Pacific). The real questions should be, "What is the potential for toxic transfer to local organisms, how long will the danger in the ocean persist, and in what ways might radioactivity be passed up the food chain and affect the ecosystem dynamics as a whole?"

On 27 March 2011, the Environmental Protection Agency RadNet testing program found 422 pico-curies per liter of iodine-131 in rainwater from Boise, Idaho (EPA 2011); this is more than *14,000 percent* the limit for safe drinking water; dozens of other cities have also tested many times the safe limit for radiation (Higgins 2011). The following day, 28 March, CNN reported that radioactive iodine had been detected in milk in both California and Washington, prompting increased testing across the United States. Then curiously, it was considered a step forward for the crisis-plagued Fukushima nuclear plant when, on 5 April 2011, CNN reported that in the preceding days, the radiation readings had decreased from "7.5 million times the legal limit . . . to [only] 5 million times the norm" (CNN 2011). That same day, the Japanese chief cabinet secretary announced that radioactive iodine had been found in fresh fish, prompting the same regulation measures for seafood as had earlier been applied to vegetables.

In another example, two different experts quoted by *Australian Geographic* on the same day indicated a problem but dismissed it outright (Australian AP 2011). One said, "Very close to the nuclear plant—less than 800 m or so—sea creatures might be in danger of genetic mutations if the dumping goes on a long time," while the other said, "It's not even clear in what way marine life could be affected, because the level of radiation isn't yet well understood. Fish would probably escape such an effect anyway, because unlike static species such as oysters, they move around and would avoid continuous exposure."

It seems that these experts are thinking in terms of instant vaporization, like in the movies, or perhaps just the type of radiation that penetrates through like a microwave and causes instant damage. But neither seems to be considering the more likely effect of radioactivity making its way into the food chain through worms, clams, plankton, and other lower food sources . . . or the half-life of the elements being dumped . . . or the ongoing dumping . . . or the long-term effects. . . .

In fact, Anders Møller of the Université Pierre et Marie Curie in Paris has been studying the ongoing effects of the Chernobyl nuclear disaster over the past 25 years. Møller has shown consequences of radiation exposure for a range of bird species that are permanent residents around Chernobyl and also for migrants through the area (Møller et al. 2007; Møller and Mousseau 2009). As top predators, seabirds are often used as sentinels of the marine environment. Symptoms of exposure from Chernobyl have included sperm abnormalities, reduction in brain size, infertility, and, as an "early warning system," the development of albino feather patches within 3 months of exposure.

Within months of Fukushima, researchers in Australia began to notice similar white feather patches on flesh-footed shearwaters (see plate 10), a long-lived seabird species which breeds in Australia, but spends the winter months in the Sea of Japan (Jennifer Lavers, personal communication). The source of the radiation poisoning likely originates through the prey ingested by the birds. It is estimated that the cleanup of Fukushima will take more than a decade, but the problem will persist long after in the environment both in Japan and abroad as radioactive elements are transported around the globe for the next 20–30 years (the average half-life of cesium and strontium).

Amid Japan's struggles with whether to bring back online its nuclear power plants that were shut down in response to Fukushima, and facing decades to know the full impact of the meltdowns on the food chain, Russia announced in August 2012 that it had dumped in Arctic waters "some 17,000 containers of radioactive waste, 19 ships containing radioactive waste, 14 nuclear reactors, including five that still contain spent nuclear fuel, 735 other pieces of radioactively contaminated heavy machinery, and the K-27 nuclear submarine with its two reactors loaded with nuclear fuel" (Digges 2012). The containment status of these dumped items is unknown.

The Effect of Pollution on Ecosystems

> The reality is that if we want to have coral reefs in the future, we're going to have to behave that way and recognize the magnitude of the response that's necessary to achieve it. (Scripps News 2008)

No shortage of examples exists to demonstrate the effects of pollution on ecosystems; in fact, the hard part is deciding which examples to use. The *Exxon Valdez* oil spill in 1989 drowned, suffocated, chilled, and poisoned hundreds of thousands of birds, sea otters, whales, seals, and fish from acute toxicity, oil saturation, and hypothermia. Untold numbers of invertebrates and algae were killed from the oil or the cleanup effort, leading to further complications. In addition, 20 years on, tens of thousands of gallons of oil are still embedded in shallow sediments, continuing to release toxins in sublethal but chronic levels, causing depressed reproduction and recovery.

The Gulf of Mexico oil spill in 2010 certainly must have caused unprecedented losses to wildlife, but sadly, it's just too soon to tell. "Exact counts of killed or sickened animals are impossible, given that the majority of carcasses sink into the ocean, rot unseen in marsh grasses or are consumed after death by predators," according to a spokesperson for the Center for Biological Diversity, an environmental group based in Tucson, Arizona (Calkins 2011). Official government data on the acute effects included 1,146 endangered sea turtles, 8,209 birds, and 128 dolphins and whales. However, the center estimates the harm in far more confronting numbers based on studying animals the following season: about 6,165 sea turtles, 82,000 birds (comprising 102 species) and as many as 25,900 marine mammals (including 4 species of dolphins and whales). Even these numbers do not take into account the massive number of invertebrates on which the larger animals depend, nor are other factors considered, such as chronic exposure, nesting on oil-contaminated sediment, toxic effects of the chemical dispersant, the fate of the 35-kilometer-long plumes that lie below the surface, and the effects to the spring bloom of planktonic invertebrate larvae that coincided with the accident.

Vast tracks of cold-water corals living in the deep sea in the Gulf of Mexico were found dead and dying in December 2010, coated with a "black, fluffylike substance," believed to be from oil. Researchers at the University of Florida described the strong toxic response as more of a "smoking cannon" than a smoking gun. "It could be the tip of the iceberg of all kinds of weird things we're going to see in the Gulf of Mexico in the next three to five years [due to the oil spill]" (Jones 2010).

While different species tend to react differently to various pollutants, in general, pollution tends to favor weedy species over others. Studies have demonstrated that flagellates flourish in waters treated with low concentrations of DDT or copper, whereas diatoms do not (Menzel, Anderson, and Randke 1970; Thomas and Seibert 1977).

Eggs and larvae aside, it appears that planktonic ecosystems barely blink when oil spills occur. One study found that zooplankton reestablished within 5 days (Johansson, Larsson, and Boehm 1980), while others have found that microbes and flagellates are stimulated by oil (Davenport 1982). The planktonic response to oil spills and the preliminary response to the Deepwater Horizon spill were reviewed as part of a term paper assignment by students at Scripps Institution of Oceanography, who concluded that "early research shows that the planktonic community [of the Gulf of Mexico] exhibits an encouraging level of resilience" (Abbriano et al. 2011, 295).

Earlier the subject of habitat damage by bottom-trawling was raised. One of the more insidious effects of bottom-trawling is the stirring up of sediments (see plate 9). Each pass of the trawl resuspends miles of chemical residues, DDT, PCBs, hydrocarbons, mercury, radioactive particles, and a confetti of plastics. All the harmful substances that have finally settled and been buried beyond reach of the food chain are once again made active. And these toxic clouds spread for many miles, diffusing into the water like fine dust, back into the food chain.

An interesting example of the effects of chemical residues is found in the sea otters of Southern California, where high levels of DDT and PCBs are associated with a high frequency and variety of diseases (McKay and Mulvaney 2001). Similarly, a recent population decline in orcas (killer whales) of Puget Sound in Washington is thought to be linked to high levels of PCBs detected in tissue samples.

A mass mortality event of bottlenose dolphins occurred in 1987–1988 along the Atlantic coast of the United States, where approximately 2,000 dolphins died. The cause of death was believed to have been a virus, although environmental contamination and algal toxins likely contributed to the severity. Other similar disease-related mortalities and morbidities have occurred in the Indian River Lagoon in Florida, including about 50 bottlenose dolphins killed within a 3-month period in 1982; 5 bottlenose dolphins found with severe skin disturbances in 2001; more than 150 sick or dead loggerhead turtles in 2000–2001; and an ongoing high prevalence of disease in sea turtles (McKay and Mulvaney 2001).

Nuclear Pollution and Ecosystems

Nuclear power is often marketed to the public as a "clean energy." And it is, in many respects. None of that nasty carbon dioxide that we've been hearing so much about. But nuclear plants have those occasional pesky meltdowns, alas.

As we've seen above, accidents as well as decades of radioactive waste dumping have left several chances for potential catastrophe. There's the obvious pathway up the food chain to our dinner tables. But there's also the less obvious pathway of causing disease in the bigger animals as they eat contaminated smaller animals, and this upsetting the balance in a way similar to overfishing.

There's also the issue of thermal pollution caused by nuclear reactors and the effect of unnatural heat in confined ecosystems. For every unit of electrical energy generated by a nuclear power plant, 2 units of heat energy are generated and released into the environment. Apparently nobody has yet worked out a way to harness this extra energy. Many nuclear power plants use what's known as a "once-through cooling system" to remove the excess heat and keep the reactors operating properly. A typical plant of this type draws in more than 1 billion gallons of water a day, or 500,000 gallons a minute, *for each reactor*. The water is drawn in through massive pipes, cycled through the power generating station, then discharged back into the same body of water at temperatures up to 25°F hotter than it came in (Gunter et al. 2001).

In addition to the obvious and heart-tugging problem of sucking in manatees, turtles, and other charismatic estuarine megafauna, there are the not-so-small problems of sucking in microfauna and creating a tropical lagoon "microclimate" where it doesn't belong.

Entrainment of microfauna and microflora. If large animals can get sucked into the intakes—and they do—in addition to jellyfish ingress incidents, of which there are plenty, one can only conclude that smaller plankton must be entrained in tremendous quantities. There doesn't seem to be any data on this, no records, no estimates . . . But imagine a constant vacuum, sucking in a billion gallons of water a day along with its inhabitants—fish eggs and larvae, phytoplankton, invertebrate larvae—instantly cooking them, then discharging them into a warmer-than-normal body of water. Some get pulverized or liquefied in the process, creating more surface area to help speed up the bacterial decay process. Certainly this must be the pressure-cooker means to eutrophication.

A tropical lagoon microclimate. The warm-water effluent from power plants completely alters the local ecosystem. If it's along an open coastline, then the thermal plume may extend only a matter of hundreds of meters, often turning the affected area into a moonscape. But if it's in a semienclosed embayment or estuary, the entire ecosystem can be catastrophically changed. The dissolved oxygen concentration is too low for most indigenous species, and many physiological processes have narrow thermal windows in which the controlling enzymes will work properly. A (nonnuclear) natural gas-fired power station at Torrens Island, South Australia, has been supplying Adelaide's electricity since 1968 (Painter 2011). According to a statement to Australian Parliament:

> The plume from the Torrens Island power station discharges into the Angus Inlet, which discharges directly into the Barker Inlet, which is an aquatic reserve and probably South Australia's most important fish nursery area. The temperature of the Angus Inlet, especially in hot weather . . . will go over 40 degrees Celsius [104°F]. This hot water will extend up to three kilometres into the Barker Inlet aquatic reserve. (Parliament of Australia 1997)

Since at least 1972, perhaps not so surprisingly, the tropical jellyfish *Cassiopea ndrosia* has taken up residence and appears to be doing quite well there.

The Effect of Pollution on Jellyfish

The most obvious effect of pollution on jellyfish is that it typically causes acute death or chronic toxicity in other species, effectively leaving jellyfish the last man standing. However, jellyfish are not entirely immune to direct effects from pollution. Very few studies have been performed on jellyfish with regard to pollution responses, except a few on hydrocarbons and petroleum. We know from these experiments that oil kills jellyfish and that polyps and young medusae develop abnormally with even low concentrations. In fact, jellyfish polyps have been used as an environmental indicator for hydrocarbon pollution, by monitoring subtle changes in their development and behavior (Spangenberg 1984).

Drifting and resting pieces of plastic must certainly become easily fouled with jellyfish polyps, potentially acting as expansion room for colonies building a seed bank before a bloom event. Similarly, floating and drifting plastic objects are likely to aid in dispersal of polyp colonies over potentially extremely long distances, as long as the plastic stays afloat.

You may be rubbing your chin or scratching your head, trying to figure out what all this toxic and nuclear stuff has to do with jellyfish. That's the point, probably very little. It seems likely that jellyfish are one of the few predators not affected by radiation and many types of toxins—mind you, there's no evidence; it just seems reasonable. Other animals around them with muscle and bone and blood and fat will store persistent organic pollutants, radiation, and heavy metals, and presumably be affected by them. Jellyfish, however, because of their watery, clonal, and short-lived nature, probably won't. First, they don't have much in the way of tissues to store toxins in. Second, they don't usually live long enough to build up high concentrations. Third, even if the polyps are storing radioactivity or persistent toxins, so little of the polyp is passed along to each medusa larva that it is probably of little consequence. And fourth, because the medusa is the sexual stage and is ephemeral, the chance of illness or genetic mutation due to radiation or toxins is very slim. So while fish are growing two heads and cancer tumors, whales are dying of leukemia and mercury poisoning, and shearwaters are glowing in the dark and choking on plastics, jellyfish will still be gorging on their plankton banquets just like nothing had ever happened . . . assuming the plankton survive.

Thermal Pollution

In general, warming stimulates greater polyp reproduction and faster medusa growth. However, this is not always the case. Two extremes of the effects of unusually warm water on jellyfish are demonstrated by two examples from Australia. In the first, recall the case of the tropical species *Cassiopea* thriving in the warm effluent of the power station in the normally chilly waters of South Australia. In the second, a similar thermal plume appears to be driving a local extinction of a different jellyfish species.

Local Extinction Caused by Power Plant? Catostylus mosaicus

Reports of the large and conspicuous blubber jellyfish *Catostylus mosaicus* (see plate 1) swarming in the Tuggerah Lakes of New South Wales date back to at least 1892 (Scott 1999). It appears that *Catostylus* was consistently so abundant as to make fishing difficult.

The Munmorah Power Station on the shore of Lake Budgewoi commenced operation in 1967. A report in 1971 included the fact that the jellyfish were so numerous that they blocked the cooling water intake screens. However, by 1974, the jellyfish had virtually disappeared (J. Bell, in Scott, 1999). Several possible explanations have been put forth. First, there is local speculation among fishermen and others that the Power Station put something in the water that killed the jellyfish. The facility is known to release chlorine and heavy metals into the lake system (Kennedy 1997). However, given the lack of further available information, it is impossible to evaluate this hypothesis.

A second possibility is that warm water effluent from the power station altered the temperature of the lakes sufficiently to make them inhospitable to the polyps of *Catostylus*, thereby killing off the "seed bank" of the population. A third theory is that the warm water effluent altered the temperature enough to disrupt the algae/copepod dynamic of the food chain or the symbiotic algae in the jellyfish's tissues, essentially causing a "bleaching event" similar to corals, and leaving *Catostylus* to starve.

A considerable number of studies have been performed on the effect of warm water effluent from power plants on local ecosystems, including in Australia. In some cases, species richness in the discharge plume area has been found to remain unchanged or to increase, while at others, species have been considerably reduced or eliminated altogether. It is thought that these differences could be due to local conditions, such as topography of the discharge area or intensity and duration of the temperature changes (Robinson 1987). However, "species richness" only measures total number and does not take into account whether the species are the same as those present before the disturbance or even whether they are native or heartier, more tolerant introduced forms. Despite the obvious lessening of effect on the local environment if power station effluents were to be discharged directly to the ocean, rather than to enclosed lagoonal areas, many stations continue to be built in shallow estuarine areas.

If indeed, the warm effluent from the Munmorah Power Station was the cause of the disappearance of *Catostylus*, as seems likely, the implication is intriguing: if *Catostylus* needs cooler water to survive, projected climate change may well drive the population south into Tasmanian waters, or drive it extinct. But then again, climate change effects would be slower, allowing the species and its associated ecosystem time to move, whereas thermal pollution progresses very fast in comparison.

Our knowledge of the direct biological effects of pesticides, plastics, and radiation on jellyfish is lacking. However, it seems reasonable to conclude that any type of biological or ecological stress put on an ecosystem as a whole or on some of its inhabitants carries the potential to be favorable to jellyfish, in the same way that overfishing weakens one or more links in the food chain. Overall, pollution is effectively a "dream come true" for jellyfish in most cases, disturbing ecosystems sufficiently to give jellies the competitive edge.

> The window for action is narrowing. As the [UNEP] Year Book under-lines, persistent issues are in many cases becoming more acute, whilst new ones are emerging.
> —ACHIM STEINER, United Nations undersecretary-general and United Nations Environment Programme executive director

CHAPTER 9

Biopollution: The Twelfth Plague

Invasive marine species are one of the four greatest threats to the world's oceans! Unlike other forms of marine pollution, such as oil spills, where ameliorative action can be taken and from which the environment will eventually recover, the impacts of invasive marine species are most often irreversible! (Globallast.org, IMO 2012)

Jellyfish Threatening the Sydney Olympics
(Australia, 2000)

Imagine being stung by a type of box jellyfish, not enough to kill you, but enough to distract you and slow you down. This is what the Olympic Committee was facing when *Carybdea rastonii* (see plate 2), a type of box jellyfish, was found swarming just weeks before the start of the Sydney Olympics in 2000. *Carybdea* cannot kill or make people sick, but its sting is sharp enough to bring tears to the eyes, and it was swarming in the exact location scheduled for the swimming portion of the triathlon events.

Understandably, there was great anxiety about even fractions of seconds being at stake due to distractions from stings, and of course the possibility of more severe reactions. Various management options were deemed unfeasible, such as protective clothing (unacceptable because it would slow down swim times), exclusion nets (extremely costly to net large sections of coastline), sweeping the area free of jellyfish (unreliable), and moving to a different area (not possible).

Luckily, just 1–2 weeks before the Opening Ceremony, the jellyfish bloom dissipated, and the triathlon events went off without a hitch.

The word "plague" conjures a feeling of dread and despair. Devastation. Helplessness. Its sources seem mysterious or beyond our control, and its effects seem unstoppable. The Bible listed 10 plagues that God inflicted on the Egyptians: the Nile turned to blood, frogs, insects, flies, sickness among animals, boils, hail, locusts, darkness, and death of the firstborn sons. The Black Death of 1348–1350 swept through Europe like, well, "like the plague," and killed some 75–100 million people (about half of Europe's population at the time). These plagues were limited in their geographic reach; nuisance species, in contrast, are not.

The statistics are mind-numbing and confronting:

- It is estimated that several *thousand* species are transported around the world each day on the shipping industry's ballast water "floating zoos and botanical gardens" (Carlton and Geller 1993; NRC 1995; Carlton 2009, 39–40).
- Every hour an average of more than 2 million gallons of foreign plankton are released in US waters (Carlton 1995).
- In a 1993 report by the US Congress Office of Technology Assessment, it was found that a minimum of 4,542 species of foreign origin had established free-living populations in the United States; an analysis of the economic impact of just 79 of these species documented $97 billion in damages, in 1991 dollars (OTA 1993).
- The San Francisco Bay estuary has recorded 234 exotic species that have established populations, with at least an additional 125 species of obscure or uncertain origin (called "cryptogenic") (Cohen and Carlton 1998).
- For the last 50 years, a new species has been introduced into the San Francisco Bay ecosystem every 14 weeks (Cohen and Carlton 1998).

Introduced species are organisms that become established in nonnative habitats, often through human-mediated transport. The terms "introduced species," "invasive species," "exotic species," "alien species," "nuisance species," and "xenobiota" are often used interchangeably. They refer to species threatening human use of natural resources or harming the natural resources in ecosystems. Often they are nonindigenous, but sometimes they are native—either way, they have gotten wildly out of control. Such plague species are considered "biopollution."

We may think of introduced species as mere opportunists, hitching a ride in a comfortable ballast aquarium, disembarking in a foreign land without a passport. Some introduced species, however, conquer their new home and make it their own—not unlike the nomads first coming to the Americas across the Bering Land Bridge, or the early Aborigines populating ancient Australia, or later, European colonists shaping the New World. They are the weeds of the animal world. But more than everyday weeds just trying to eke out a living, plague species exert such force on the ecosystem that they alter it, often irreversibly. Species that are easy prey or less able to compete may go extinct.

Familiar terrestrial examples of plague species include foxes and rabbits in Australia, the brown tree snake in Guam, and fire ants and fruit flies in the United States. Aquatic examples include the zebra mussel in the Great Lakes, the alga *Caulerpa taxifolia* in the Mediterranean and United States, and the Japanese sea star *Asterias amurensis* in southern Australia. And of course, *Mnemiopsis*, the comb jelly that invaded the Black Sea.

When we think of Vikings, we don't usually think of clams. But the first known human-mediated introduction of an alien species from North America to northwestern Europe was the soft-shelled clam, believed to have been transported across the Atlantic by the Vikings (Leppäkoski et al. 2002). Since that time, the introduced species problem has continually worsened.

The primary mode of introduction is generally in or on ships and boats. Jellyfish lend themselves perfectly to ship travel, with the medusa passively sucked into ballast holding tanks or the hydroid expanding across hulls as an unnoticed fuzz. Increasingly, the role of drifting plastic debris as an agent of species dispersal is being recognized. Plastics are buoyant and they persist in the marine environment for a long time—perfect properties for rafting of willing travelers, such as jellyfish polyps, algae, barnacles, and tunicates.

One of the world's worst cases of multinational economic loss and large-scale ecosystem collapse from an exotic species was due to a jellyfish: *Mnemiopsis*. As detailed in chapter 3, *Mnemiopsis* is widely recognized as one of the world's most harmful invading species. It ranks as one of the "Ten Most *Unwanted*" on a promotional poster by GloBallast and the International Maritime Organization (GloBallast 2011). Similarly, the Invasive Species Specialist Group of the International Union for Conservation of Nature has ranked it among the "100 of the World's Worst Invasive Alien Species" (ISSG 2008).

The Effect of Plague Species on Jellyfish, and Vice Versa

Plague species affect jellyfish in two primary ways. First, jellyfish themselves are perhaps the ultimate invading organism: both the medusa and polyp phases are well suited to long transport and rapid colonization. For example, *Mnemiopsis* has independently established itself across the Atlantic in the Black and the Baltic seas, and heaven only knows in how many other localized places, and it appears that the Australian spotted jellyfish *Phyllorhiza* has been independently introduced into California, the Caribbean and Gulf of Mexico, and the Mediterranean.

Perhaps the most splendid example of invasive jellyfish is in the story of the freshwater jellyfish, *Craspedacusta sowerbyi* (see plate 1). When most people think of Walden Pond, they think of the historic retreat of the celebrated writer Henry David Thoreau. Thoreau lived in solitude at the shore of Walden Pond for two years, starting in 1845. While there, he wrote his famous book *Walden*, a social critique and commentary on nature.

In September 2010, *Craspedacusta* bloomed abundantly in Walden Pond. The species is common throughout the world's freshwater lakes and slow-flowing streams and, curiously, seems to have a particular fondness for man-made reservoirs and backyard ornamental lily ponds. So there was nothing particularly abnormal about *Craspedacusta* blooming in Walden Pond, except that this was the first report of it being found there. The mode of transport of *Craspedacusta* between bodies of water is still not well understood, but is believed to be facilitated by birds: polyps are carried in mud that sticks to birds' feet between their toes, and is washed off when they land at their next wading place—the original form of ballast transport, one might say.

The second main effect on jellyfish is that invasive species produce a "wobble" in the ecosystem, that is to say, they disturb the trophic relationships such that jellyfish can more easily exploit the instability. As one might expect, estuaries are hotspots for introduced species. Estuaries generally have heavy shipping traffic, long residence time for water, and lots of shallow places for organisms to live. San Francisco Bay in California, the Derwent Estuary in Tasmania, and Port Phillip Bay near Melbourne are all highly disturbed by nonindigenous species and highly infested with jellyfish. A few of the more well-studied estuaries are profiled here, but the story is the same all over the world.

San Francisco Bay. It was noted earlier that the San Francisco Bay estuary has a new species introduced into the ecosystem about every 14 weeks; it is considered "the most invaded aquatic ecosystem in North America" (Cohen and Carlton 1995, i). One species in particular, the Asian clam, *Potamocorbula amurensis*, has spread in San Francisco Bay to the extent that its collective filter-feeding abilities have impacted the entire food web by decreasing the available phytoplankton. Meanwhile, jellyfish blooms are extremely common in San Francisco Bay. Two species of moon jellyfish (*Aurelia labiata*, a native species, and *Aurelia aurita*, introduced from Japan) compete for food and presumably polyp attachment space. Two diminutive hydromedusa species, *Blackfordia virginica* and *Moerisia*, both introduced, also bloom in vast numbers, but being tiny and transparent, typically escape notice. And the beautiful and showy *Maeotias marginata*, introduced from the Black, Caspian, and Azov seas into the brackish upper estuary in the 1990s, blooms in huge numbers within view of the famous Jelly Belly factory. Whether these jellyfish are a consequence or a driver of the shifting ecosystem is unclear . . . but it is clear that *something big* is happening.

Derwent Estuary, Tasmania. Having been declared the most polluted river in the world in 1974 due to large-scale heavy metal dumping and other effluents (Bennett 1999), it may come as no surprise that in early 2011, a biodiversity survey in the Derwent found no native species (Kempton 2011). *No native species* . . . Far out.

Asterias amurensis, the infamous Japanese seastar, now dominates the seafloor (see plate 11). The water column, however, is dominated alternately by *Aurelia* blooms and blooms of spectacular and carnivorous phytoplankton species *Noctiluca*, both of which have become worse in recent decades (Naidoo 2009; McLeod et al. 2012) and have caused huge financial losses to the salmon farms in the neighboring Huon Estuary. A variety of small hydromedusae and comb jellies also explode into massive blooms in the Derwent— perhaps they feed on seastar larvae.

Port Phillip Bay and Gippsland Lakes, Melbourne. Coastal Victoria has recently received a lot of press for its new species of dolphin, *Tursiops australis* (Monash University 2011). But the dolphin, along with the region's

other natives, are threatened by invaders. Port Phillip has 99 introduced and
61 cryptogenic species, making it one of the most invaded marine ecosystems
in the Southern Hemisphere (Hewitt et al. 2004). Among them are *Asterias*,
the starfish dominating the Derwent Estuary, and *Undaria pinnatifida*, an
unwelcome Asian kelp that is competitively displacing native species. Three
large jellyfish species form frequent blooms in the bay and lakes and are an-
ecdotally reported to be increasing (no long-term data exist to confirm): the
Victorian lion's mane (*Cyanea annaskala*), the cauliflower blubber (*Catostylus
mosaicus*), and, of course, the nearly ubiquitous *Aurelia*. The extent to which
these jellyfish blooms may affect the dolphins and other resident species has
not yet been investigated.

So too, the Mediterranean seems particularly subject to plague species and
jellyfish problems. Over 500 nonindigenous species of fish, invertebrates, and
macroalgae have been reported so far, but the bias toward large and conspicuous species makes this likely to be just a partial inventory (Galil 2007). Two
of these plague species are jellyfish that have turned out to be new to science:
Rhopilema nomadica and *Marivagia stellata*, as discussed below.

Rhopilema nomadica in the Mediterranean
(Israel to Turkey, since 1976)

As one might imagine, the Mediterranean and its inhabitants are very well
studied. Mariners, scientists, and bathers have been observing Mediterranean
waters for hundreds of years. So it is considered remarkable when something
new is found.

A large unidentified blue medusa with a severe sting appeared off the
Mediterranean coast of Israel in the mid-1970s. Its spectacular size—up to
80 centimeters (2 ½ feet) and 40 kilograms (90 pounds)—icy blue coloration,
and bright red, swollen beaded stings demanded attention. Surprisingly, it is
believed by some to have migrated in from the Red Sea, where it had been
misidentified some 50 years earlier. However, this migration hypothesis is
disputed by others on the basis that the species is rare in the Red Sea. An
alternative theory is that it came in ballast water from the Indo-Pacific, where
many species still exist in obscurity. Nonetheless, wherever it came from, it is
now flourishing in the Mediterranean, where as a new species, it was finally

named and classified in 1990 (Galil, Spainer, and Ferguson 1990). *Rhopilema nomadica*, the blue nomad.

Its population exploded in the 1980s, forming massive summertime aggregations off the Israeli coast and causing an environmental health hazard. The problems it caused included clogged fishing nets, reduced fish catches, and injuries to beach-goers.

In 1989, *R. nomadica*'s summertime population density was estimated at 500,000 per square nautical mile. This massive concentration is attributed to the species' high reproductive potential: it starts to reproduce early in life, with a 10-centimeter bell diameter, and continues throughout its life. A single settled polyp can give rise to more than 100 medusae within two months; it bears an enormous number of larvae; and it rapidly forms resting stages (average of 14 per 2 months per polyp) and can form several at a time (Lotan, Ben-Hillel, and Loya 1992).

By 1988, *R. nomadica* was observed off the coast of Lebanon. By 1991, it was also found off Syria, while its population off Lebanon had reached concentrations as high as those off the coast of Israel. In just 14 years, unrestricted invasion of its new habitat spread along 500 kilometers (310 miles) of coastline. The population expanded in a south-to-north direction, as medusae, larvae, and polyps were carried by the counterclockwise currents. In August 1995, the species was confirmed in the waters off Turkey (Kideys and Gücü 1995), although it was thought to have arrived in small numbers as early as 1992. By 1995, it had also reached Egypt, swimming against the current.

Most recently, yet another species, *Marivagia stellata*, the starry sea-wanderer, has taken up residence in the eastern Mediterranean. This lovely, lacy species was only discovered in 2010 and immediately was identified as new to science (Galil et al. 2010). It, too, has the capacity to bloom out of control.

Two other species have also recently become established in the Mediterranean. *Phyllorhiza punctata* is a native to the southern bays and harbors of Australia, but in recent decades has spread to the Gulf of Mexico, California, and the Mediterranean, where its presence has been recorded regularly since 2005. Similarly, *Cassiopea andromeda* is native to the Indo-Pacific but has set up camp around the Mediterranean, being found in the waters of Cyprus, the Aegean, Malta, Israel, and along the Levantine coastline (Galil et al. 2010). Populations of both species have not yet proliferated in the same way as *Rho-*

pilema nomadica. It may be that *Cassiopea* and *Phyllorhiza* require slightly different conditions for a population explosion, or it may be that with *Rhopilema* already dominating the resources, the opportunity to take over is not yet available. Both are only mild stingers but are certainly capable of causing ecological havoc as well as problems for fisheries.

Only time will tell whether the local ecosystem can successfully resist these species . . . along with *Mnemiopsis*. We may find in a few short years that the long history of fishing and beach recreation that made the Mediterranean a playground for the rich and famous, has become a thing of the past.

> As a weakend man easily succumbs to disease, so damaged ecosystems readily fall victim to attacking forces.
>
> —COLIN WOODARD, *Ocean's End*

As bad as plague species are now, brace yourself for the problem only getting worse. Several studies have demonstrated that ocean warming facilitates the invasion of nonnative species. Professor John Stachowicz of the University of California, Davis, and his colleagues studied ascidians (sea squirts or tunicates) at Avery Point, Connecticut, in Long Island Sound (Stachowicz et al. 2002). Ascidians are sessile (stuck to the bottom) animals, many of which form colonies. The team found that the three most common introduced ascidians spawn and settle earlier than the native species in warmer years, thereby colonizing available space so that it is unavailable for later native spawners.

Similarly, another study looked at the potential for climate change to facilitate invasions by testing the effects of warming on weedy species at Bodega Bay north of San Francisco, California (Sorte, Williams, and Zerebecki 2010). Nonnative species represent 71 percent of cover on the docks at Bodega Harbor. These researchers used clean plastic plates to study which species settle first, grow the fastest, and survive the longest under different conditions. They found that introduced species responded well to warming whereas natives responded poorly, leading them to conclude that warming is likely to have a disproportionately negative impact on native species and to predict that "as ocean temperatures increase, native species will decrease in abundance, whereas introduced species are likely to increase in this system" (p. 2198).

This era of global transport and anthropogenically influenced mixing of species has been called the "Homogocene" period. We are homogenizing our harbors. Laugh if you must, but it's not all that wide of the mark. As thousands of species are comingled from different parts of the world, one might intuitively think that this would increase the local biodiversity. However, this is rarely the case. Too often, nonindigenous species cause dramatic losses of biodiversity. *Mnemiopsis* in the Black Sea. Foxes in Australia. Cane toads in Australia. *Asterias* the samurai starfish in Australia. Hedgehogs in New Zealand. Brown tree snakes on Guam. Zebra mussels in America's Great Lakes. The green alga *Caulerpa* in the Mediterranean. Rats introduced onto thousands of islands. *Homo sapiens.* . . .

Once introduced species are established, it can be difficult, if not impossible, to eradicate them or even quiet them to a low hush. Biocontrol, using introduced predators to control populations of introduced prey, is a Pandora's box. A few of the more notorious attempts at biocontrol have ended very, very badly, with the target species not brought under control, and the so-called controlling species itself becoming a feral pest.

To paraphrase words of wisdom, whose author I have long since forgotten:

To control plague species, executive orders may be far less effective than recipes.

CHAPTER 10

Climate Change Changes Everything

Climate is what you expect. Weather is what you get. Climate is the average weather. Climate change is getting weather we are not used to. Climate change is changes to the average, to the chances of something happening: changes in probability, increases in the frequencies of what the weather does. Climate change is when climate goes beyond what is considered natural variability. What was predictable becoming unpredictable. Uncertainty beginning to dominate.

—MEREDITH HOOPER, *The Ferocious Summer*

Can Jellyfish Predict Fine Wine?
(The Mediterranean, since 1785)

Imagine being able to predict the finest vintages of wine, knowing in advance which will gather the highest honors and fetch the highest prices. Jacqueline Goy at the Musée Nationale d'Histoire Naturelle in Paris has done just that . . . well, sort of. She has studied the bloom patterns of the havoc-wreaking jellyfish *Pelagia noctiluca* in the Mediterranean over the last 200 years. *Pelagia* is easy to study because it occurs in huge densities when it blooms, it stings badly and so tends to attract human attention, and it is big enough to see (up to about 9 centimeters, or 3 ½ inches, in diameter). It is also purple. Well, mauve, if you want to be pedantic (see plate 2). It is often called the mauve stinger, or the pink meanie, or the purple people eater.

Notes on *Pelagia* occurrences have been kept by French scientists since 1785. In analyzing the data, Goy and her colleagues found that there was a repeating pattern, with major blooms occurring about every 12 years (Goy, Morand, and Etienne 1989). The blooms typically lasted for several consecutive summers, followed by jellyfish-free summers until the next bloom cycle. The swarms persist for days to weeks at densities that routinely exceed 100 jellyfish per cubic meter (3 per cubic foot), with occasional temporary densities up to 6 times that amount (Zavodnik 1987; Hamner and Dawson 2009). Furthermore, the blooms described with the most accuracy appeared to have invaded the entire western basin of the Mediterranean at once, suggesting that the controlling factor is climatic.

These mass occurrences coincided with periods of low rainfall, high temperatures, and high atmospheric pressure during the late spring and summertime. These same conditions result in the best wine vintages (White et al. 2006).

Ideal conditions for wine include a long, warm growing season that allows the grapes to ripen fully and develop a balance between acids and sugars. Too-high temperatures mean earlier ripening with less acidity and color and a higher alcohol content. Rainfall should only occur in winter and spring, as summer or autumn harvest rain can cause fungal diseases and damage the grapes, while too much cloud cover means insufficient sunshine. Wine has been called "the canary in the climate change coal mine" (Brown 2007), because subtle changes in climate are reflected in its taste.

Almost every year since 2000 has been celebrated as another "vintage of the century." So too, almost every year has brought massive blooms of *Pelagia*. However, while jellyfish and wines enjoy the good life, a subtle migration is underway: the good wine–producing regions are shifting toward higher latitudes. Each 1° increase in France's temperature is equivalent to moving 200 kilometers (125 miles) north. English and Canadian wines, once a joke, are now drinkable, while warmer regions in Spain and Australia are feeling the heat and producing spicier, richer wines with higher alcohol content. Back in France, the annual harvest has been getting earlier and earlier for nearly 30 years.

French wine regions (Bordeaux, Burgundy, Rhone, Champagne, and the combined regions for Port production) with wines rated "outstanding" or better in a given year by the UK company Vintage Wine Gifts, can serve as a measure of vintage excellence over the last century. The following synopsis should suffice to persuade skeptics. In particular, for the years 1900 to 2005:

- There were 22 years with two or more outstanding wine regions in France; of these, 16 corresponded with *Pelagia* bloom years, 4 corresponded to years without *Pelagia*, and 2 lacked data about *Pelagia* occurrence.
- There were 6 years with 3 or more outstanding wine harvests in regions of France; all 6 corresponded with *Pelagia* bloom years.
- There were 4 years in which 4 or more French wine regions had outstanding vintages; all 4 corresponded with *Pelagia* bloom years.

Since Goy's 1989 study, *Pelagia noctiluca* has become increasingly pestilent—maybe good news for wine lovers but bad news for everyone else.

Dr. Priscilla Licandro of the Sir Alister Hardy Foundation for Ocean Science in Plymouth, England, and her European colleagues, analyzed data collected from 1958–2007 at 15 data-sampling stations throughout the northeast Atlantic and found increased volumes of jellyfish, primarily *Pelagia noctiluca*, since 2002 (Licandro et al. 2010). They also documented large outbreaks of *Pelagia* in the summer and autumn of 2007 and 2008. Most importantly, they found that the frequency of this typically tropical jellyfish in winter months had increased since 2002, and that the blooms appeared earlier in the year and persisted for longer. In other words, seasonal conditions favorable to jellyfish have been broadening in space and time.

Northern and western Mediterranean coasts have long been associated with the rich and famous, carefree beach holidays, romance, and pleasure. The French Riviera, the Greek Isles, the Côte d'Azur, the Costa del Sol, Cannes, Monaco. Even the names conjure up excitement. But an increase in jellyfish means an increase in stings, and an understandably decreased interest in these places as tourist meccas. Recent major sting events and mitigation efforts include the following:

- By the beginning of August 2006, over 30,000 people had been stung in the Mediterranean that summer; Malta, Spain, Sicily, and northern Africa were badly affected, according to MaltaMedia (2006).
- In the first two weeks of August 2006, more than 14,000 people were treated for stings on beaches in Catalonia, after billions of jellyfish drifted

toward crowded sands, according to *Time* magazine (Abend and Pingree 2006).

- In July 2008, in just a few hours, 300 people were treated for stings at Barcelona beaches, with 11 requiring hospital treatment, according to the *New York Times* (Rosenthal 2008).
- In 2008, the eighth consecutive year that *Pelagia* swarmed, beaches along the French Riviera had to be closed down; Cannes installed a floating barrier for the first time in an effort to keep its beaches safe for swimming, according to the *Telegraph* (Samuel 2008).
- A 2008 report by the US National Science Foundation stated that, "some popular beaches in Spain are now regularly swept by jellyfish-catching boats, similar to the way that hockey rinks are regularly swept by Zamboni machines" (NSF 2011).
- In August 2010, several beaches on Spain's Costa Blanca had to be closed following a rash of 50 stings at Denia, a resort area that attracts about 2 million foreign tourists each year (Govan 2010).
- In July 2011, 6 beaches in popular resort areas of Spain were closed following stings to more than 100 swimmers, following warnings that British seas could be "turned into a 'jellyfish soup' this summer" (NewsCore 2011; SkyNews 2011).
- In July and August 2012, more than a thousand people were treated for stings along the Malaga coast of Spain, leading authorities to close beaches to protect swimmers; the *Daily Mail* has dubbed the region the "Costa del Sting" (Milligan 2012).

These beach closures and other precautions were not sensationalistic or hypersensitive. There is legitimate cause for concern over stings from *Pelagia noctiluca*. A sting from this species caused a Greek woman to go into ana-phylactic shock from a severe allergic reaction, and she nearly died. Interest-ingly, this was one of very few cases of true anaphylactic reaction to jellyfish. Anaphylaxis is often reported, but rarely accurately so. The vast majority of jellyfish stings do not cause anaphylactic shock, but this case was credible. To date, only two cases of anaphylaxis from jellyfish stings are considered cred-ible by authorities, the other victim being a dog (Togias et al. 1985; Portier and Richet 1902).

Most *Pelagia* stings do not cause anaphylaxis, but many cause asthma, breathing difficulties, and high levels of pain and can result in long-term dermatological and other medical complications. *Pelagia noctiluca* is one of the most venomous species in Mediterranean and British waters, and any increases in its blooms in space or time could have devastating effects on recreation activities and the tourism industry.

Climate change. Global warming. Call it what you want, changes in the climate are a hot topic in science these days. Sometimes it can seem like nearly everyone is abuzz, trying to explain nearly everything in terms of global warming. Flowers bloomed earlier this year—it must be global warming. Floods in Tennessee or Queensland, or cyclones, hurricanes, typhoons, monsoons, or squalls, or drought for that matter—it must be global warming. The coldest Christmas day on record for Hobart—this must be proof that global warming is a hoax (it's not). . . .

To a nonscientist, this might seem overwhelming, and perhaps look like everyone is jumping on the bandwagon. But the truth is that living things are so closely attuned to their environment, and temperature plays such an important role in defining the niche of nearly every species on earth, that the real surprise is that alarm about climate change has taken so long to become a major political issue.

Heat speeds up the metabolic rate of living things, while cold slows it down. And it doesn't take much change to make a big difference in the metabolic demands to sustain life. If an animal is catching enough food to stay healthy, whereupon a warming environment causes its bodily functions to speed up slightly, it is likely that the animal will not be able to catch enough food to maintain its health and over time become malnourished.

For aquatic animals, it is even worse. Warmer water holds less dissolved oxygen than cooler water. Therefore, a marine animal adapted to respiring at a certain rate may suffocate in only slightly warmer water, simply because it cannot work hard enough to extract enough oxygen. Over time, its food supply and feeding rate cannot meet its increased energetic demands. So a minor temperature change can make a *huge* difference to the health and longevity of other living things, and particularly to the equilibrium of a given ecosystem. Some species are better able to cope than others, so the food web structure will inevitably be altered.

Sea temperatures have increased by about 1°C (1.8°F) in the past century and are currently increasing at a rate of about 1–2°C (1.8–3.6°F) per 100 years. Most of this warming occurred over just the past three decades. It may not sound like much, but it's a lot. Some species benefit while others suffer. Organisms have to expend more energy to extract oxygen from the water. Mismatches occur in the signals for biological events between predator and prey. Enzymes for biological processes work differently. Higher rates of larval mortality occur. These and other implications of warming water are discussed in the sections that follow.

Climate Change in a Nutshell

So what's the big deal about carbon dioxide? Okay, a quick review of how climate change works. Carbon dioxide is released into the atmosphere when we burn fossil fuels, for example, in coal-fired power plants and automobiles. We call it a "greenhouse gas," because the portion that stays in the atmosphere acts like an insulator, letting in the sun's heat but not letting it back out. As we add more carbon dioxide to the atmosphere, the world gets progressively warmer. The global warming process becomes a self-enhancing feedback, whereby the warmer it gets, the more insulating the carbon dioxide becomes, and makes it warmer still.

Generally, global processes take vast lengths of time, while heating and cooling are natural processes that have occurred many times before in the earth's history. But today we are heating up our air and water faster than they have ever been heated in the earth's entire 4.5-billion-year history. And we are creating a climate that will be different from anything human beings have ever experienced—a climate that may not be how we like it. Large numbers of animals and plants will struggle to survive, because they cannot adapt fast enough to such a rapid change; this will reshuffle the species composition of communities. Rapid increase in carbon dioxide has been called "a vast, uncontrolled geophysical experiment that is warming the planet and making the oceans more acidic, with increasingly dire implications for natural ecosystems and humanity" (Jackson 2010, 3771).

It is not uncommon to hear people say that they aren't convinced that the earth is warming. Perhaps they don't see the effects in their own lives. Consider the Arctic, where examples of warming climate abound. In 2011, Canada's Northwest Passage and Russia's Northern Sea Route opened simultaneously with the smallest area of sea-ice on record (Black 2011). Also in 2011, explorers rowed to the normally ice-covered North Pole for the first time ever—*rowed* (Hough 2011). And in July 2012 NASA announced unprecedented melting of the Greenland ice sheet: while about half of Greenland's surface naturally melts in the summer, "melt maps derived from the three satellites showed that on July 8, about 40 percent of the ice sheet's surface had melted. By July 12, 97 percent had melted" (Cole 2012).

Currently a high-stakes debate rages on in Alaska, where tundra travel (mostly associated with oil and gas exploration) is permitted by the Department of Natural Resources only during the winter when the ground is sufficiently frozen to protect the fragile habitat from environmental damage. In 1970, the tundra was frozen enough for travel for about 220 days per year, but the frozen season has gradually receded to about 100 days by 2002 (Bader and Guimond 2005). Elsewhere, such as the village of Kaktovik on Alaska's north coast, subsidence of the soil as permafrost melts and erodes is threatening the viability of structures built on it like the airport, school, and offices (see plate 12).

While our polar regions are thawing, the consequent sea level rise is threatening the very existence of other regions far away. Consider the island nation of Kiribati, comprising 32 atolls dotted over more than a million square miles of Pacific Ocean where the equator meets the international date line. None of the atolls is more than a few feet high, making it arguably the most vulnerable nation in the world to the effects of climate change. Faced with being flooded off the map, in 2012 the government began buying up land in Fiji to relocate its entire population (Chapman 2012). So too, the atolls of the Maldives are being swallowed up by rising sea levels: the nation's plight and its passionate pleadings to the world to "do something" about climate change are featured in the 2012 documentary movie *The Island President*.

The oceans have acted hitherto as a carbon sink, buffering the effects of the atmospheric carbon dioxide. In fact, phytoplankton in the seas scrub carbon dioxide and sequester as much carbon as rainforests and all other terrestrial plants combined. The sequestering process involves the calcifying organisms in the oceans dying and being buried, thereby acting as a massive carbon pump by sending their carbon to the ocean depths. But as the atmospheric carbon dioxide concentration increases, the seas will become more saturated with carbon and their scrubbing effect may become less efficient—so less carbon is fixed, and less is sequestered, meaning reduced ability of the seas to absorb more carbon dioxide, thus accelerating global warming. Bad news.

It used to be thought that as the polar ice melted, it would increase the surface area of ocean so that phytoplankton could sequester more carbon dioxide. However, this prediction has proven to be false. Instead, it appears that the opposite is occurring: as sea-ice melts, a progressively declining amount of carbon dioxide is sequestered, so the ocean's capacity for further uptake is likewise diminishing (Cai et al. 2010). More bad news. . . .

The Earth System Research Laboratory is a division of the National Oceanic and Atmospheric Administration. The laboratory collects atmospheric data from atop Mauna Loa in Hawaii, where "the undisturbed air, remote location, and minimal influences of vegetation and human activity at [Mauna Loa] are ideal for monitoring constituents in the atmosphere that can cause climate change." The organization CO2Now uses the laboratory data to track changes in carbon dioxide and temperature, and communicate them to the public in meaningful ways on their website (http://CO2now.org/). It's a sobering read.

So where does the carbon dioxide come from, and where does it go? According to CO2Now.org, 88 percent of humanity's carbon emissions, some 8.5 billion tons per year, comes from the burning of fossil fuels and the making of cement, while the other 12 percent, some 1.2 billion tons, comes from deforestation and land use changes. Furthermore, 47 percent of our carbon, 4.5 billion tons per year, goes into the atmosphere; 27 percent (2.6 billion tons) is taken up by terrestrial plants like trees; and 26 percent (2.5 billion tons) is taken up by the oceans.

The world's cement industry produces 5 percent of man-made carbon emissions, making it an important sector for mitigation. That leaves 83 percent from the burning of fossil fuels. *Eighty-three percent.* Oil. Gas. Wood. Coal. A lot of coal. Recall that China is building new coal-fired power plants at the mind-numbing rate of 2–3 *per week* (Harrabin 2007). According to Greenpeace, "Coal fired power plants are the biggest source of man made carbon dioxide emissions. This makes coal energy the single greatest threat facing our climate" (Greenpeace 2012a).

In 2009, fossil fuel CO_2 emissions decreased by 1.3 percent. These emissions were second highest in human history, just below 2008 emissions. They are 37 percent higher than in 1990.
—CO2Now.org

The Effect of Climate Change on Ecosystems

The humpbacks that traverse the world's oceans likewise will no longer be able to fill their capacious bellies, nor will the innumerable seals and penguins that cavort in southern seas. Instead, we'll have an ocean full of jellylike salps, the ultimate inheritors of a defrosting cryosphere.

—TIM FLANNERY, *The Weather Makers*

When we think of climate change effects on ecosystems, we may picture polar bears stranded on ice floes or resorting to cannibalism as their food sources dwindle (see plate 12), or we may recall stories of penguins nesting on bare rock or vast tracts of corals lost in mass bleaching events. But what about the little stuff? An interesting study on California zooplankton suggests a worrying trend where a moderate warming of surface waters led to a major decline in biota. Professors Dean Roemmich and John McGowan of Scripps Institution of Oceanography analyzed 43 years of temperature and zooplankton samples and found that zooplankton volumes decreased by about 80 percent from 1951 to 1993 (Roemmich and McGowan 1995). During this time, surface temperatures warmed by an average of more than 1°C (1.8°F). While zooplankton volume was consistently higher in northern samples relative to those from the south, the downward trend in abundance was similar in both regions.

Roemmich and McGowan suggested that the thermocline was the mechanism behind the observed pattern. The thermocline is an abrupt shift in temperature in the water—you can feel this when you wade into a lake in the summertime, when you suddenly step into water where your feet and knees are considerably colder than your waist. In particular, as the sea surface warms but the bottom-water does not, the temperature difference across the thermocline increases. This creates stratification, which limits the effects of upwelling by essentially insulating the nutrient-bearing layers from the sea surface. It's not that upwelling does not occur, it's just that the nutrient-rich waters do not mix with the upper, warmer, nutrient-poor waters, leading to reduced phytoplankton and, consequently, reduced zooplankton. The authors also noted a recent decline in the sooty shearwater, a zooplankton-feeding seabird that lives in the area.

As stated earlier, warm water holds less oxygen than cooler water. So it might seem like organisms can just swim deeper where it is colder to get the oxygen they need. Similarly, with coastal regions of the world becoming increasingly hypoxic through eutrophication, it might seem like going farther out would be the answer. But it doesn't really work like that. For one thing, farther and deeper waters have less light and less biodiversity, making the task of finding a meal harder than normal. Familiar corals and algae don't grow there, nor do the rich communities of fish and invertebrates that associate with them. It's teeming with life, but "the neighbors and the decorating" would be foreign to the local coastal bottom-dwelling species. As for pelagic species that spend their whole day swimming, most would starve if they ventured too far from shore where most of the world's marine productivity occurs. Fascinating ecosystems are found in these deep seafloor and pelagic communities, but the species that thrive there typically have special adaptations that enable them to do so.

There is another barrier much more daunting than food availability: the oxygen minimum zone, which is a naturally occurring layer of low oxygen in the midwater out beyond the continental shelves. Some species use it as a refuge from predators or energy expenditure, but to most species it represents an impenetrable barrier. So in practical terms, the "usable" part of the ocean as far as most species are concerned is the part of the water column above the oxygen minimum zone and not too far out away from shore. It's essentially a fairly narrow tube encircling the continents.

Dr. Lothar Stramma of the Christian Albrechts University of Kiel in Germany and his associates studied the oxygen minimum zone around the world. They found that climate warming is causing the tropical hypoxic zones to expand horizontally and vertically, and that subsurface oxygen has decreased adjacent to most continental shelves (Stramma et al. 2010). In particular, the team found that the low oxygen area where mobile macroorganisms are unable to abide has increased by 4.5 million square kilometers (about 1,740,000 square miles) since the 1960s—an area half the size of the United States, including Alaska and Hawaii. Thus, the most productive and liveable habitats of the upper nearshore waters are being compressed by the oxygen minimum zone expanding upward and eutrophication expanding out along the coasts.

And what about species that can't swim away? Consider coral reefs, which provide an incredibly rich and diverse habitat for other organisms. Reef-building corals in the tropics are already living close to their thermal tolerance limit and easily become stressed when exposed to small increases (1–2°C, or 1.8–3.6°F) in water temperature (Hughes et al. 2003). Coral bleaching is the term given to the process whereby stressed corals evict their resident algal symbionts, not only leaving the corals without color, but often resulting in their death.

Calculations using different climate change models all indicate that sea temperatures will exceed the thermal tolerances of tropical reef-building corals within the next few decades (Hoegh-Guldberg 1999). It is therefore likely that bleaching events will become more frequent and intense, occurring annually within the next 30–50 years. It is expected that Australia's Great Barrier Reef will face severe bleaching events every year by 2030. To put this into perspective, babies born today may never know healthy reefs as teenagers and adults.

One of the best ways to understand the potential effect of a perturbation on ecosystems is to look at the base of the food web. A recent comprehensive study on global phytoplankton patterns has done just that by assessing changes in ocean transparency from almost half a million measurements between 1899 and 2008 (Boyce, Lewis, and Worm 2010). Water transparency is affected by the density of phytoplankton and can be easily assessed by measuring chlorophyll, a pigment in plants including phytoplankton. The researchers divided the world's oceans into 10 logical regions and found that chlorophyll had declined in 8 of these regions. They estimated the global rate of decline at about 1 percent per year. Multiple lines of evidence suggest that these long-term global declines in phytoplankton are related to climate variability, particularly to increasing sea surface temperature. In general, cooler phases of climate cycles tend to have higher abundance of phytoplankton, whereas abundance is reduced during warmer phases.

Some may wonder whether huge increases in open-ocean fishing have contributed to the observed decline in global phytoplankton. In particular, as more fish are removed from the ecosystem, zooplankton increase, leading to higher predation on phytoplankton, which of course then reverberates back up the food chain as a decrease in zooplankton and then fish. This is possible, but less likely than the climate explanation. Open-ocean fishing has primarily targeted large species, whereas these large species do not directly consume phytoplankton. Rather, as these large species are fished out, their smaller fish and prawn food sources, often referred to as "mid-trophics," are able to flourish, which in turn puts more predation pressure on copepods and other small zooplankton, thus leaving phytoplankton with fewer grazers.

Regardless, diminishing phytoplankton is likely to spin off two different and inconvenient self-enhancing feedback systems. One has to do with ecosystem function. Essentially, decreasing phytoplankton means that there is less available food at the base of the food web, which means that there is increasing grazing pressure on dwindling resources, and so on. The second feedback system has to do with ecosystem service. Decreasing phytoplankton means decreasing ability to absorb atmospheric carbon dioxide, which means accelerated warming, which means decreasing phytoplankton, and so on (NRC 2003). As links in the food chain starve and die off, ecosystems restructure. Enter jellyfish.

Global climate change involves a variety of impacts that are expected to result from the greenhouse effect. It differs from ordinary changes in climate by being a shift in averages over time, a sort of "ratcheting of the norm" to a state that is not normal for us. Time is a crucial part of it. One storm is not climate change, but an increase in storm frequency is. A warmer than average summer is not climate change, but a shift in average summertime temperature is.

Although the projected intensity and ultimate human impacts of warming are often shrouded in uncertainty, many direct physical and biological effects of warming are fairly straightforward. The warmer the earth and oceans become, the more polar sea-ice melts in summer. Sea levels rise. Phytoplankton communities reshuffle. Species that depend directly or indirectly on sea-ice or phytoplankton will be variously affected. Antarctic krill that feed on phytoplankton under the sea-ice may vanish, and with them, their whale and penguin predators. Warming also causes corals to expel their algal symbionts, causing coral bleaching. If corals cannot quickly obtain new symbionts, they will perish. Even a mere 1–2°C (1.8–3.6°F) above the "normal" high temperature is enough to initiate coral bleaching.

Changes to climate affect an ecosystem in as many ways as there are species in it, perhaps more. Each species can be expected to respond in its own way—some benefit from change, while others fail to cope and go extinct. Some evolve behaviors or physical adaptations suited to their new world. In general, the best predictor of "who will live and who will die" is variability, or what scientists call "plasticity." Morphological plasticity is the ability of a species to take on different forms in different conditions as a way of adapting to different environments. Physiological plasticity is the ability to respond to different conditions in different ways, thereby increasing the probability of survival. Weedy species tend to be more adaptable to a wider variety of ecological conditions than, for example, orchids or penguins.

In general, the early life stages of species are more vulnerable to environmental changes than adult stages, both in terms of changes from day to day and of changes from the norm. Stages like fertilization of eggs, embryonic development, and larval growth can be particularly sensitive. Of all environmental changes, temperature is probably the most important. Many species won't spawn, breed, or even survive if the temperature is on either side of a narrow range. As stated above, warm water holds less oxygen than cold water, so difficulty in extracting oxygen can be fatal at particularly stressful times in development for some species. Unpredictable spring temperatures may stimulate spawning or larval release during a period of reduced phytoplankton supply, thereby reducing larval survival.

Many species are harmed by warming temperatures (McKay and Mulvaney 2001). Many Mediterranean populations of commercial sponges were decimated in an unprecedented die-off between 1986 and 1990; the event was thought to be bacteria-related, which would have been exacerbated by a thermal rise. A similar event that occurred in the West Indies in the 1950s was believed to be caused by increased temperature. Other mass mortality events linked to rising temperatures include millions of sea fans in the Ligurian Sea in 1999; mass echinoderm deaths in Japan and the Gulf of California in the 1980s; and a northward spread of a protozoan pathogen on commercial oysters in the southeastern United States in the 1990s.

One of the most compelling studies on the effects of climate change on ecosystems was performed by Doctors Richard Kirby and Gregory Beaugrand of The Sir Alister Hardy Foundation for Ocean Science in the UK. The researchers examined North Sea temperature and plankton data from 1958 to 2005, demonstrating that an ecosystem regime change took place during the 1980s (Kirby and Beaugrand 2009). Notwithstanding decades of heavy fishing, it appears that a 1°C (1.8°F) rise in sea surface temperature had a stronger effect on reshaping the ecosystem than any other pressure. It was so strong, in fact, that it sometimes caused the ecosystem to respond in unexpected ways. In particular, the effect of temperature on planktonic larval stages, as well as other types of plankton like jellyfish, is overwhelming, swamping other effects. The authors called this phenomenon "trophic amplification," where the effect of climate warming is intensified through indirect pathways in the food web. It seems that in the North Sea, fishing enabled jellyfish to increase so that the balance was already "tipped" when the temperature shift occurred. This synergistic effect can be compared to the case of the Black Sea, where overfishing and pollution are believed to have lowered the resistance of the ecosystem to the invading comb jellyfish *Mnemiopsis leidyi*, with disastrous effects. The authors concluded that the North Sea now favors jellyfish due to the 1°C (1.8°F) temperature change, which is expected to double by the end of the century.

The effects of climate change do not only involve adapting to shifts in temperature. Climate change alters the timing of key biological events and how predators and prey interact in an ecosystem. Any perturbation to an ecosystem will generally reshuffle the dynamic; big perturbations shuffle more than small ones; fast ones shuffle the dynamic more than slow ones. Remove one strand from the fabric, and the rest may unravel.

We are accustomed to thinking of "big things" when we think of ecosystems: lions, bison, polar bears, whales, sharks. But despite their size, these "charismatic megafauna" are only a small part of an ecosystem. The whole marine ecosystem depends, more than anything, on phytoplankton. Phytoplankton depend on physical conditions, such as sunlight, nutrients, and water temperature, and act as a bridge between the physical and the biological by being the very basis of the food chain.

Moreover, copepods, those super-abundant tiny crustaceans that comprise the majority of zooplankton, aren't just a primary food source for forage fish and jellyfish; they also play a fundamental role in helping to temper the rate and extent of climate change. As we have seen, the ocean absorbs about a third of the carbon dioxide in the atmosphere. Much of this carbon is "fixed" by phytoplankton, meaning that it is converted into energy in the form of body mass. When the ecosystem is healthy, copepods graze truly massive quantities of phytoplankton. Copepods transfer much of this carbon from the phytoplankton to the seabed, through sinking fecal pellets and through their own sinking bodies when they die. Furthermore, many copepod species migrate vertically every day into deeper waters to avoid predation; their time at depth facilitates the process of removing excess carbon from the carbon cycle by their waste becoming more readily locked in the sediments.

Phenology, the Timing of Important Biological Events

The timing of important biological events like spawning, hatching, and mating is typically synchronized with environmental cues, such as temperature, salinity, or day length. Generally, these cues serve species well, signaling when the conditions are likely to be safe for larvae or when food is likely to be most plentiful.

But not all species respond identically to the same sets of cues, so mis-matching of events can occur when the signals conflict. Many types of in-vertebrate larvae and phytoplankton develop faster at higher temperatures (Hoegh-Guldberg and Pearse 1995), but not necessarily at the same pace. Furthermore, species responding to other spawning stimuli like day length or chemical cues may be poorly synchronized with species responding to tem-perature alone as a cue for spawning and growth.

In general, cool-water copepods tend to be large species and to have a high biomass that peaks in the spring, while warm-water copepods tend to be smaller species that peak in autumn and have a lower biomass. However, this general rule can become problematical as waters warm: the cool-water copepods shift their reproductive cycle, tracking water at their preferred temperature, thereby leaving spring-spawned fish larvae without food. This is believed to have contributed to the decline of North Sea cod (Beaugrand et al. 2003).

In the North Sea, where the ocean has been warming, dinoflagellates are now peaking 23 days earlier than they did 45 years ago, while copepods have shifted forward by 10 days but diatoms collectively have shifted zero days in the spring and 5 days earlier in the fall (Edwards and Richardson 2004). There-fore, North Sea copepods are now peaking before their food becomes plenti-ful, which presents potential problems for energy transfer up the food chain. It seems likely that as environmental signals are further altered by climate change, such mismatches will become more frequent and more pronounced.

Many species in temperate and polar regions have tight windows of feeding and breeding seasons. Therefore, any mismatch between predator and prey reduces the already limited period for predator species to meet their needs. Phytoplankton appear to be more sensitive to changes in environmental cues than their zooplankton counterparts. Whether they are able to re-synchronize may be influenced by stress from other ecosystem perturbations.

The Effect of Climate Change on Jellyfish

Jellyfish repeatedly have been called sensitive beacons of climate change. This is a good metaphor. In general, small species low on the food chain, such as phytoplankton, copepods, and jellyfish, exhibit greater sensitivity to chang-ing climatic conditions than larger species higher on the food chain, such as vertebrates. Jellyfish are able to exploit the entire constellation of effects of even minor temperature changes, including fluctuating levels of dissolved oxygen, water column stratification, phytoplankton and copepod bloom-and-bust cycles, nutrient enrichment, and acidification. And even relatively minor changes in jellyfish abundance can have profound effects on the food chain and fisheries yields (Brodeur et al. 2008a). Therefore a jellyfish bloom can be an excellent early warning signal that something is wrong. Furthermore, because jellyfish have both pelagic and benthic stages and their blooms are conspicuous, they are among the easiest and most informative species to monitor.

For the most part, climate change is a dream come true for jellyfish, a perfect set of conditions. Warmer waters mean faster metabolism but less oxygen. Jellyfish grow faster. Other species struggle, so niches open up. More phytoplankton bloom, so more zooplankton bloom, which means there is more food available, but the biomass of other species as competitors is reduced. It's a perfect world.

The most obvious effects of climate change on jellyfish are physiological. In general, slight increases in temperature tend to increase the reproductive and growth rates of jellyfish, as well as lengthening their reproductive season. This response tends to be more pronounced for temperate species, whereas a rise in temperature may inhibit tropical jellyfish, which are probably already near their thermal maximum. Indeed, many temperate super-blooms have been associated with warm conditions (Purcell 2007).

Climate change is not just about increasing temperatures; it is also about climate variability and unpredictability. Because jellyfish can respond very quickly to environmental fluctuations, can bloom rapidly with explosive population growth, and can survive on minimal nourishment yet exploit excess food, they can be "first on scene and last to leave" when anomalous conditions arise. For example, jellyfish abundance has been linked with large-scale oscillations like El Niño, discussed below.

In general, jellyfish physiology is perfectly suited to climatic and environmental extremes as well as to rapid climate changes. Temperate *Aurelia* exploit the colder temperatures of winter to expand their polyp population, then exploit warmer seasons when more food is available by producing massive numbers of juvenile jellyfish. But even closely related species or varieties can differ dramatically in their response; for example, polar and subpolar *Aurelia* typically strobilate only in springtime, whereas temperate *Aurelia* typically strobilate in early spring and autumn, and tropical *Aurelia* strobilate all year.

The "million dollar question" is the bloom potential of dangerous species—box jellies and Irukandjis—under conditions of increasing warming. People want to know whether these species are likely to become more numerous and if they are likely to spread. Dr. Anthony Richardson of CSIRO and his colleagues wrote:

> Warming could expand the distribution of many of the most venomous tropical jellyfish species toward subtropical and temperate latitudes. The northeast Australian coast is home to two types of box jellyfish: the sea wasp *Chironex* and the species complex known as Irukandji; stings from both can be fatal to swimmers and force beach closures. With warming, there is the potential for box jellyfish to move southward toward more populated areas, which would have severe repercussions for the tourist industry. (Richardson, McKinnon, and Swadling 2009, 9)

This is a reasonable and valid concern. They *could*.

On the other hand, some have taken it further with more drama. One of the most implausible of claims is that dangerous jellyfish are already on the move south toward highly populated and popular Gold Coast and Sydney beaches. However, these claims are premature and lacking in evidence. As of mid-2012, no long-term sampling studies had taken place in the relevant latitudes in order to assess the question of species or population migration. It would certainly be noteworthy if a northern species were found farther south, but that still would not constitute demonstration of an established population, as implied by the migration discussion.

Furthermore, while a southerly migration of northern species is lacking in evidence, other species of Irukandjis already live in the south—and have for some time. A severe Irukandji-type sting in the 1950s was reported from Fraser Island (south of the Great Barrier Reef), so it would be inaccurate to suggest, as some have, that Irukandjis have been newly reported at Fraser Island. Similarly, Irukandji syndrome cases have been reported more or less yearly as far back as the 1960s in the Sydney and Brisbane regions, while a severe case of Irukandji syndrome was reported far south in the Melbourne region in 1999. While it is understandable and probably healthy to be concerned about the presence of dangerous animals, statements about "Irukandjis heading south" are nonsensical, because some species are already there. It gets good press, though!

Amid all that hullabaloo, some outstanding research is giving us some clues about the future. A recent study by Dr. Susan Jacups of Charles Darwin University in Australia's Northern Territory examined the relationship between sea surface temperature and box jellyfish stings (Jacups 2010). She found that box jellyfish "arrive" when the water is above 26°C (79°F), and that the sea surface temperature in tropical Australia has progressively warmed by almost 1°C each year since 1900. Not surprisingly, as the mean water temperature increases, so too is the period of the year when water temperature is above 26°C growing longer, that is, the box jellyfish season is lengthening.

Box Jellies and Irukandjis Going Extinct?

Another way of looking at how these jellyfish may react to warming waters is in failing to adapt. Extinction. Yes, box jellies and Irukandjis are scary in the danger that they present to us, but they are nonetheless highly specialized in their ecological needs, which puts them in a precarious position in rapidly changing environments.

We think of jellyfish as pelagic creatures, that is, they spend their whole life swimming or drifting on currents. But this is not strictly true. As explained in chapter 4, most jellyfish have a two-phase life cycle, with a pelagic swimming medusa phase and a tiny benthic (stuck to the bottom) polyp phase. While medusae may swim or drift to more favorable waters, the polyp has no such option. If conditions are unfavorable, the polyp simply dies. And without polyps, there is no next generation.

Furthermore, jellyfish don't just swim to a new latitude and begin a new life. Without suitable food and habitat, they will perish. In other words, it is not just a matter of an animal or species moving, but the whole habitat needs to move to support them—the mangroves, the small fish and prawns, the coral reefs—and this takes time.

Extinction of a robust species doesn't generally occur in one generation. Extinction also takes time. As waters warm, their capacity to hold dissolved oxygen is reduced and respiration becomes more difficult. This raises an organism's energetic output, and therefore its nutritional threshold for survival. The point is that it takes time to adapt or go extinct at the population or species level. This isn't a matter of a few jellyfish moving south and all of a sudden it's a migration. Box jellies and Irukandjis are like Formula One racing cars in comparison to their sedan and moped brethren. Box jellies and Irukandjis are not hardy weeds; they are more like fragile and vulnerable orchids. Well, okay, deadly orchids.

The Effect of Water "Spiciness" on Jellyfish

Often, temperature and salinity anomalies occur in tandem. Seawater masses are often either warm and salty or cooler and fresher. Therefore, the measure of the two variables combined is known as "spiciness," where the warmer, saltier water has a higher spiciness index (Raskoff 2001).

Many species of coastal and estuarine jellyfish are accustomed to large and frequent fluctuations in temperature and salinity. During and after heavy rains, the water becomes less saline; in summer, the water is warmer, whereas in winter, it is cooler. In some cases, fluctuations can be extreme. It is therefore common to find broad tolerance ranges for temperature, salinity, and even dissolved oxygen in species living in these coastally affected habitats.

In the Pacific, El Niño events are often associated with spicier water, but not all jellyfish react in the same way. The famous spotted jellyfish *Mastigias* at Jellyfish Lake in Palau underwent a mass die-off in the spicy conditions brought on by the 1997–1998 El Niño (Dawson, Martin, and Penland 2001). In contrast, other species are stimulated by spicy conditions to bloom, for example, *Pelagia noctiluca*, which blooms in 12-year cycles in the Mediterranean, and the lion's mane *Cyanea* in China's Yangtze Estuary (Goy, Morand, and Etienne 1989; Xian, Kang, and Liu 2005). Perhaps prophetically, 4 species often associated with large blooms were found to have a high tolerance to a broad range of salinities (Holst and Jarms 2010).

Effects of Temperature on Jellyfish

Few know jellyfish blooms as well as Dr. Jenny Purcell of Western Washington University. With more than 70 papers spanning more than 30 years of research, Jenny has built her career on identifying when and how jellyfish blooms occur. One of her landmark papers was an in-depth review on climate change and jellyfish blooms in which she concluded, "Global warming could result in expanded temporal and spatial distributions and larger populations of jellyfish and ctenophores" (Purcell 2005, 462).

Probably more than any other factor, temperature both stimulates and controls jellyfish populations. Extreme examples of this were given above in the section on thermal pollution. The actress Mae West is reputed to have said, "Too much of a good thing can be wonderful!" In the case of temperature to jellyfish, this can be quite true . . . but not always. . . .

In general, jellyfish polyps bud faster and medusae grow faster in warmer conditions, whereas low temperatures slow reproduction and metabolism. This is particularly true for most temperate and estuarine species, which tend to have a wide range of tolerance. All measures of medusa production of the pestilent moon jelly increase with rising temperatures: greater percentages of polyps produced medusae, the process occurred more quickly, more meta-morphic cycles occurred, each polyp released more larval jellyfish (ephyrae), and the proportions of ephyrae relative to buds increased (Purcell 2007). Fur-thermore, 13 of 16 species studied showed bloom increases during warmer conditions.

Intriguingly, the polyps of numerous species produce more juvenile medu-sae in warm temperatures but more buds when it is cooler. This tendency to allocate energy in different directions depending on temperature could have severe repercussions in view of climate change predictions. If average temper-atures in both summer and winter were to become 1°C (1.8°F) warmer than to-day, then in winter, jellyfish polyps would bud more clones than they currently do, while in summer, many more polyps would produce far more medusae. It is easy to see how even a slight rise in temperature may make an exponen-tial difference in the magnitude, frequency, and duration of jellyfish blooms.

Similarly, comb jellyfish in Narragansett Bay, Rhode Island, have doubled in the last 30 years; furthermore, thanks to a 2°C (3.6°F) increase in average winter temperatures, they are now surviving the winter (Whiteman 2002). Warm spring temperatures in the Black Sea and Chesapeake Bay correlated with a larger body size in *Mnemiopsis*, which favors egg production (Purcell et al. 2001b).

Whereas temperate species tend to be stimulated by warmer temperatures, the opposite is often true for tropical species, which are probably living near their thermal maximum of about 34–35°C (93–95°F). Recall that the tropical jellyfish *Mastigias* underwent a sharp population decrease during an El Niño warming period (Dawson et al. 2001); curiously, so too the population of the deep-sea jellyfish *Colobonema sericeum* declined considerably during a warm-ing period (Raskoff 2001). *Colobonema* is most well known to documentary enthusiasts as the jellyfish that casts off its bioluminescent tentacles, which then act as wriggling, glowing distractions while the jellyfish makes a beeline for somewhere else—it is the same concept as a detachable lizard's tail, but with the added glow-in-the-dark twist. It is hard to imagine such an entertain-ing, splendid species in decline.

While some jellyfish are limited by warm temperatures, other species are limited by cold. Native species are often killed or their reproduction is inhibited by cold temperatures, for example, the pink meanie *Pelagia* in the Adriatic and the tiny hydromedusa *Moerisia lyonsi* in the Chesapeake Bay (Purcell 2005). In the case of *Mnemiopsis* spreading from its Black Sea invasion, it was unable to survive the cold winters in the Sea of Azov ($< 4°C$, or $39°F$), and consequently failed to establish a breeding population there.

The positive effects that higher temperatures have on abundance of many types of jellyfish makes it likely that blooms will increase with global warming, particularly as populations expand poleward. Furthermore, because many species have a broad thermal tolerance range, they are as likely to do well in fluctuating conditions as in rising temperatures. However, it also seems likely that species favored by warmer conditions will edge out those that are not.

The Effect of Salinity on Jellyfish

For some species, such as the lion's mane, high salinity triggers population blooms. This species is flourishing in the saltier waters of the Yangtze Estuary created by the damming of the river in 2003.

It is widely believed, though unproven, that the polyps of the deadly box jellyfish, *Chironex fleckeri*, live primarily in the brackish portions of tropical rivers and are stimulated by the flushing of freshwater with the spring rains.

While medusae are often stimulated by moderate increases in salinity, polyps have varying responses to salinity. Polyps of the edible southeast Asian species *Rhopilema escuelenta* need medium to high salinities for survival and reproduction (Lu, Liu, and Guo 1989), whereas experiments on the weedy invader *Moerisia lyonsi* suggest that its highest rate of asexual reproduction occurs in water with high temperature and low salinity (Ma and Purcell 2005), but the polyps of the Chesapeake sea nettle, *Chrysaora quinquecirrha*, are most productive in waters of intermediate salinity (Purcell et al. 1999).

Many jellyfish species tolerate broad variations in salinity. *Mnemiopsis* can tolerate all the way from freshwater to hypersaline; this is one of the key factors in its weediness and highly invasive tendencies. And as we shall see in the next vignette, a close relative of *Mnemiopsis*, a creature called *Bolinopsis*, lives in the hypersaline waters of a reverse estuary in southern Australia, where it seems to be poised for explosive population growth.

South Australian Breeding Grounds in Peril
(Spencer Gulf, South Australia, since 1998)

"*Mnemiopsis*" has become a bad word all over the world since its decimation of the Black Sea ecosystem in the 1990s. It is on every watch list for nonindigenous species hazards. Curiously, its first cousin *Bolinopsis* has attracted very little of the same attention.

The differences between *Mnemiopsis* and *Bolinopsis* are entirely taxonomic rather than biological (see plates 1 and 2). Both are in the same grouping of lobed comb jellyfish; both prey on copepods, invertebrate larvae, and fish eggs and larvae; and both reproduce in the same way, apparently triggered by the same stimuli. But in *Mnemiopsis*, the lobes are attached near the base of the body, whereas in *Bolinopsis*, they are attached closer toward the mouth. Therefore, the difference between the two is about an inch of lobe length. No grand biological difference to keep *Bolinopsis* from becoming a diabolical pest. No ecological niche differences to prevent *Bolinopsis* from taking over like its cousin. That extra inch of lobe may go down in history as being the most politically inane reason for letting an ecosystem and its fisheries crash.

It's not that *Bolinopsis* isn't already proven to be a pest, it's just that it is not on the watch lists. In Japan, a species of *Bolinopsis* blooms in Tokyo Bay, reaching concentrations that rival those of *Mnemiopsis* in the Black Sea. Their very high consumption rate is thought to depress the mass of planktivorous fishes through competition, in turn leading to decreases in copepods and increased algal blooms (Hiromi, Kasuya, and Ishii 2005). *Bolinopsis* is thought to be increasing its populations in some Japanese coastal waters (Uye and Kasuya 1999).

In the northeast of the Great Australian Bight (the large concavity on the southern shores of the continent) lie two parallel gulfs. The westernmost one is Spencer Gulf, a rare and sensitive type of estuary known as a "reverse estuary." A typical estuary has freshwater input from an upstream river and marine inflow from the ocean; the sill moves up and down the estuary depending on rainfall. More rain means more freshwater, so the sill moves farther down toward the sea; less rain means less freshwater, so the sill moves closer upstream.

Spencer Gulf is unusual in that it operates in reverse. As it has no major river to bring in freshwater, evaporation results in the head of the estuary being hypersaline, saltier than the ocean. This also means that it has very little flushing, so that the residency time for creatures and chemicals is much higher the further up you go. To make matters worse, a massive desalination plant is under construction at the top of the estuary. Given the problems with desalination plants outlined in chapter 1, this seems likely to prove itself a mistake.

The Spencer Gulf is a nursery to many species of commercially important fish and shellfish—as well as a breeding ground for the giant cuttlefish—and the many other species that provide food for them or are simply part of the delicate balance of nature in the reverse estuary. Any threat to the region could not only severely cripple this rare ecological habitat but also decimate the profitable fisheries that rely on Spencer Gulf to supply the next generation of stock.

In 1998 and several times since, very large numbers of *Bolinopsis* were observed blanketing the surface of the upper gulf. They were packed in "cheek by jowl," so the top 6–12 inches of the water column was eerily rippled with their mucousy bodies. On a calm day, you could see their comb rows pressed against the under-surface of the water, giving the surface a strange caterpillar orgy–like appearance as far as the eye could see.

Locals have recounted similar observations, and some think it has gotten worse in recent years, more widespread, and is lasting longer through the season. But nobody has gone to gather any usable data. I tried. I applied for a grant to live and work in that remote region, but my proposal was rejected on the basis that the problem lacked data. Go figure.

I also tried to get fisheries organizations to look into it, but they lost interest when they heard that it wasn't *Mnemiopsis*. I now believe that in my lifetime, we are likely to see a severe crash of the Spencer Gulf ecosystem due to political insistence on the difference an inch can make. An inch.

Fluctuations in Rainfall

High coastal and inland precipitation leads to increased urban runoff and river flow, thus increasing the nutrients flowing into the sea, leading to phytoplankton blooms followed by zooplankton blooms. In general, higher food availability leads to expansion in jellyfish populations. However, increased rainfall also leads to flooding and reduces the salinity of estuaries and coastal waters. While some jellyfish and their polyps have broad salinity tolerance ranges, others cannot survive substantial freshwater input. Paradoxically, species like

Mnemiopsis with broad salinity tolerance can use fresher waters as a place of refuge from predators, such as *Chrysaora* and *Beroe*, that cannot thrive in such conditions.

In contrast, extended drought reduces freshwater input, raising salinity. Increased salinity endangers many species of fish and invertebrates that are unable to "osmoregulate," or rid their bodies of excess salt, whereas in general, jellyfish thrive in increased salinity. These direct effects of spicier water and indirect effects of reduced competition for resources as other species fail to thrive suggest that jellyfish are likely to flourish in droughts.

Effects of Light on Jellyfish

By contrast with studies on temperature and salinity, there have been relatively few experiments on the effect of light on jellyfish. Many organisms use changes from light to dark or vice versa, or changes in day length, as cues for spawning. While the effects of changing climatic conditions on temperature and salinity are obvious, the effect on light may be less direct, for example, cloud cover or water column turbidity.

Light has been found to be either a stimulus or an inhibitor of strobilation in polyps of several species, such as moon jellies and sea nettles; other stimuli include temperature, food, and various chemicals. Various sources of light-reflecting or light-absorbing gases and particles in the air are now changing the amount of sunlight reaching the earth's surface, including greenhouse gases, soot, dust, and clouds. The extent to which these changes may promote or inhibit jellyfish blooms is unclear.

As discussed in chapter 2, reduced light levels in certain Norwegian fjords are associated with increased jellyfish abundance—clearly demonstrating that tactile predators like jellyfish are able to exploit ecosystems in which visual predators like fish can no longer thrive. Moreover, many deep-sea organisms, including jellyfish, undergo long vertical migrations each day: up to the surface at night to eat, then back down below the lighted zone by day to digest and keep away from visual predators, demonstrating that jellyfish are well accustomed to using light to their advantage.

One can therefore extend these principles to any light limitation, whether from greenhouse gas–induced dimness, eutrophication, particles stirred by trawling, or land-based sedimentation. Furthermore, most jellyfish blooms are in some way attributable to changing weather patterns, whether over seasons or over decades; light is just one of the variables that they respond to.

Notes on El Niño/LaNiña and Other Oscillations

Most people have heard of El Niño ("It has to do with weather"). But what is it exactly, and what does it have to do with jellyfish? The name *El Niño* is Spanish for "little boy," in reference to the Christ child. In some years, Peruvian fishermen experience anomalous weather around Christmas time—warmer water, heavy rainfall, and reduced upwelling, leading to loss of fish.

El Niño, or more properly, the El Niño Southern Oscillation, and its "cousins" the Pacific Decadal Oscillation and North Atlantic Oscillation, may be thought of as short- to medium-term climate change phenomena that toggle back and forth, whereas "climate change" and "global warming" are unidirectional changes over a longer period.

As one might expect, jellyfish are so closely dependent on environmental conditions that their blooms are affected by climate oscillations. Examples include the following:

- Early studies in the North Sea suggested that *Cyanea* and *Aurelia* bloom patterns were correlated with the cold phase of the North Atlantic Oscillation, which differed from typical bloom patterns elsewhere (Lynam, Hay, and Brierley 2004, 2005). However, when a longer time sequence was studied, it quickly became clear that North Sea jellyfish behave like others: warming waters stimulate blooms and have been doing so strongly since the mid-1980s (Attrill, Wright, and Edwards 2007).
- In Jellyfish Lake in Palau, medusae and polyps lost their algal symbionts due to a substantial increase in temperature and saltiness, apparently initiated by the 1997–1998 El Niño. As a result of the loss of symbionts, the medusae died off and polyps failed to metamorphose, leading to a total disappearance of jellyfish from the lake (Dawson et al. 2001).
- In Monterey Bay, California, sea nettles were "several orders of magnitude less abundant during the 1991–1992 El Niño event than in the preceding non–El Niño years" and were thought to have shifted north to cooler waters (Raskoff 2001, 122).
- In California, during the 1991–1992 and 1997–1998 El Niño events, the shallow-water hydromedusa *Mitrocoma* suddenly increased in abundance and widened its depth range, compared to non–El Niño years. In contrast, the deepwater hydromedusa *Colobonema* narrowed its depth range and had a population decline, whereas it was found in very high numbers between El Niño events (Raskoff 2001).

- *Pelagia* in the Mediterranean has bloomed at 12-year intervals over the past 200 years when low rainfall, high temperature, and high atmospheric pressure have occurred (Goy, Morand, and Etienne 1989).
- In the Gulf of Alaska and Bering Sea, sea nettles increased during the 1980s and 1990s following a regime shift in 1972 with sea surface warming (Anderson and Piatt 1999; Brodeur et al. 1999).
- In Narragansett Bay, Rhode Island, the comb jelly *Mnemiopsis leidyi* responded to warming in the 1980s and 1990s by commencing their blooms earlier and reaching greater peak abundances, relative to the cooler 1950s and 1970s. This warming coincided with a positive phase of the North Atlantic Oscillation (Purcell and Decker 2005).
- In the Chesapeake Bay between 1960 and 1995, *Chrysaora quinquecirrha* abundance was highest in years when temperatures were warmer, and medusa numbers were lower in cooler years (Purcell and Decker 2005).
- Off Greenland, two forms of the hydromedusa species *Aglantha digitale* showed inverse responses to water temperature and salinity: a white form was linked with warmer temperatures, whereas a red form was linked with colder Arctic water of higher salinity (Purcell 2005).

While jellyfish abundance responses are not always consistent with oscillation patterns, for the most part they are tightly linked. In particular, jellyfish blooms often rise and fall with warming and cooling. This suggests that progressive global warming is likely to have an increasingly positive effect on the stimulation, expansion, and maintenance of jellyfish blooms.

Laboratory experiments on polyps demonstrate that they survive and reproduce even after severe and abrupt changes in temperature and salinity (Purcell 2005). Changes that are less abrupt (e.g., gradual warming), periodic (e.g., tidal or seasonal), or short term (e.g., rain) will be easier to tolerate, adapt to, and exploit.

It seems clear that both short-term oscillations and longer-term progressive warming patterns have a strong influence on jellyfish blooms. Most of the vignettes throughout this book attest to longer, larger, and more prolific blooms of temperate species, as well as increasing distributions into higher latitudes. It also seems likely that some, if not many, seasonal blooms will overwinter in active condition as waters become more favorable to their needs.

For tropical species, the outlook under intolerable warming conditions is less clear. Blooms are likely to become smaller in magnitude and shorter in duration. Whether they are able to shift their distributions into cooler waters will largely depend on whether enough habitat exists to support them.

Jellyfish Promote Climate Change

While the effect of climate change on jellyfish is intuitive and has long been recognized, the inverse has only recently come to light. In a fascinating study published in 2011, Dr. Rob Condon of the Virginia Institute of Marine Science and his colleagues found that jellyfish release extremely carbon-rich forms of "colloidal and dissolved organic matter" (i.e., mucus and fecal matter), which, curiously, is used by bacteria preferentially for respiration (i.e., carbon dioxide production) rather than for growth (i.e., building body mass) (Condon et al. 2011, 10225). Essentially, jellyfish blooms turn these bacteria into carbon dioxide factories. Furthermore, this special type of jellyfish "goo and poo" favors rapid growth and dominance of particular types of bacteria over others, thus altering the microbial community so that a higher percentage of the bacteria are producing more carbon dioxide. Thus, climate change promotes jellyfish blooms and jellyfish blooms promote climate change, and like all robust feedback loops, where it stops nobody knows.

And if all that's not bad enough, when voracious jellyfish blooms eat large quantities of larvae and copepods, this shunts food energy away from species whose sinking fecal pellets and carcasses help to sequester carbon in ocean sediments and instead toward dead-end gelatinous biomass. Both jellyfish biomass and their metabolic waste are far less efficient at sequestering carbon than those of diatoms, copepods, and fish, thus further enhancing an ecosystem shift that favors climate change.

And One More Thing

Just in case you lie awake at night wondering what you can do to help save the planet. . . .

Imagine my surprise on opening my morning paper on 17 February 2007 and finding an article entitled "Green Answer Is Blowing in the Wind" (Overington 2007).

I read on: "All forms of flatulence—from cats, dogs, even from Dad—contain methane, a greenhouse gas thought to contribute to climate change."

[*blink, blink, blush*] "For just $8, a Sydney-based company, Easy Being Green, can now make your cat carbon-neutral, so it can 'live guilt-free for a year.'" At this point, I was having a hard time reading the print through my tears of unrestrained laughter, which also caused my head to bounce up and down. Dogs cost a bloated $35 a year—and they can even make Granny carbon-neutral for just $10 a year. I could not have possibly made this up . . . If you want to shop around for a competitive long-term, sustainable, eco-friendly, carbon-neutral solution to all your cat problems, be sure to look into the exciting new technology for cat sequestration.

PART III

The Weeds Shall Inherit the Earth

Wildlife will consist of the pigeons and the coyotes and the white-tails, the black rats and the brown rats and a few other species of worldly rodent . . . and the houseflies and the barn cats and the skinny brown feral dogs and a short list of additional species that play by our rules . . . Earth will be a different sort of place soon, in just five or six human generations. My label for that place, that time, that apparently unavoidable prospect, is the Planet of Weeds. Its only redeeming feature, as far as I can imagine, is that there will be no shortage of crows.

—DAVID QUAMMEN, "The Weeds Shall Inherit the Earth"

CHAPTER 11

The Allee Effect, Trophic Cascades, and Shifting Baselines

It would be similar to studying the Serengeti after all the large grazers and carnivores were eliminated; one could still appreciate termites and other small grazers, but one's expectations of nature pale beside what it used to be. Here, we may understand the kelps; however, they are but a beautiful gossamer veil, undulating peacefully in the ocean, offering no hints of the marvelous species that should live there but for human greed. (Dayton et al. 1998, 320)

The Jellyfish Control Act
(Chesapeake Bay, 1966-1972)

On 2 November 1966, the Congress passed a piece of legislation called the Jellyfish Control Act (16 U.S.C. §§ 1201–1205; 1966, amended 1970 and 1972). The act authorized the secretary of commerce to "conduct studies, research and investigations to determine the abundance and distribution of jellyfish and other pests and their effects on fish, shellfish and water-based recreation."

One of the main focal points of the act was to gain knowledge and develop methods to control and eliminate sea nettles, *Chrysaora quinquecirrha*, in Chesapeake Bay (see plate 2). Up to $1 million annually through 1977 was spent for this purpose. Many antinettle methods were proposed or tested; most failed. Chemicals that killed polyps also killed oysters and crabs. Scientists tried to raise a known polyp predator, the striped sea slug (a benthic creature resembling a tiny shag rug with tentacles). That didn't work either. The navy accidentally stumbled on a remedy so promising it was patented: "When we hit this one band of frequencies [2 to 3.2 kilohertz], I noticed all these jellyfish on the surface trying to get out of the water," recalls codiscoverer Edward Thomas. "Every time the frequency sweep was repeated, the result was the same: Thomas would spot nearly a thousand nettles, alive but bobbing around the submarine doing a jellyfish impersonation of the painting *The Scream*" (LeGrand 2010). In the end, that remedy wasn't used. However, an enormous amount of high-quality research was conducted into the sea nettle's biology and ecology, much of which has formed the bases and primary reference points for many more recent studies on nuisance species around the world, including continued studies on *C. quinquecirrha*.

Long before sea nettles became a problem, the Chesapeake was the largest and most productive estuary in North America, characterized by extensive seagrass meadows and oyster beds, with abundant clams, blue crabs, scallops, and fish (Jackson 2001). Oysters were so large they had to be cut in two to be eaten, and their shells, still plentiful today in aboriginal middens, were the size of men's shoes. It has been calculated that before the 1870s, the truly enormous filtration power of these suspension-feeders could filter "the equivalent of all of the water in Chesapeake Bay in less than 1 week" (Jackson 2001, 5414). Today, it would take 46 weeks. The first report of hypoxia was in the 1930s, when ecological research was still in its infancy. As overfished oysters filtered less and less water, this likely amplified the effects of nutrient runoff from increasing farming and urbanization, triggering a eutrophication cascade that persists to this day.

Chrysaora quinquecirrha lives in Chesapeake Bay and its tributaries, where the summer and autumn blooms spell financial nightmare to the tourism and commercial fishing industries and often hamper recreation and leisure activities. At its peak in July and August, it reaches densities up to 16 medusae per cubic meter (1 per 2 cubic feet; Purcell 1992). Its common name, the sea nettle, is well deserved, considering its sharply painful sting. The sting is not life threatening, but it brings tears to the eyes of children and adults.

An unlikely key figure in Chesapeake jellyfish research was David Cargo. He had no PhD, but this Army Air Corps veteran made up for it with his nononsense approach to solving problems. Working at the Chesapeake Biological Laboratory, Cargo spent over three decades researching the bothersome sea nettle. From spring through fall, clicker in hand, his lunchtime routine was to "walk to the end of the 750-foot pier, counting only nettles on the south side; turn around, head shoreward and count the north side" (LeGrand 2010). Other data, such as air and water temperature, tides, salinity, and wind speed, were added later. Cargo's efforts toward one day being able to forecast "dry weather with a strong chance of midsummer nettles" earned him the label of "the Chesapeake version of Punxsutawney Phil."

One of the clear outcomes of the research spawned and sponsored by the Jellyfish Control Act is that jellyfish blooms cannot be controlled; however, they can be predicted. Cargo found that low spring precipitation resulting in high salinity, along with warm spring temperatures, correlate with robust blooms (Cargo and King 1990).

Another interesting outcome of Cargo's research, and of those who followed in his footsteps, is the recognition that the sea nettles are declining, but apparently for ominous reasons. Whereas in some places the blooms are so thick it looks like you could walk across them, overall their population has plummeted down to about 10 percent of their pre-1985 abundance. Good news for those interested in aquatic recreation, but bad for those interested in a healthy ecosystem and the seafood it produces.

It now appears that when oyster populations crashed in the late-1980s, the hard substrate required by the sea nettles' polyps fell into short supply. Combined with poorer water quality and warmer water temperatures, it seems the Chesapeake is no longer as hospitable to jellyfish as it once was . . . nor, alas, is it to most other living things.

Spectacular impacts on flora and fauna are resulting all over the world from jellyfish blooms. But how? It is not enough that perturbations occur in an ecosystem—perturbations large and small take place frequently, yet ecosystems are usually resilient and either manage to come through unscathed or quickly restore their equilibrium. Nor is it a sufficient explanation that jellyfish grow fast and have big appetites. Many cases must have occurred through the millennia in which jellyfish have bloomed and died without altering entire biota. Yet today we are witnessing multiple ecosystem shifts toward new stable states dominated by jellyfish, with few visible signs of return to their earlier fish-dominated stability.

The previous chapters have demonstrated many triggers that cause ecosystems to become unstable. Overfishing. Climate change. Pollution. Eutrophi-cation. Plague species. It is also important to understand several concepts that scientists talk about: the first, "Allee effect," describes reproductive success as a function of population density; the second, "trophic cascades," refers to the food web phenomenon that occurs when depletion of one level reverber-ates up and down the food chain; and the last, "shifting baselines," occurs when our interpretation of what is normal changes over generations due to our inherent short-term perspective.

The Allee Effect

Two quite opposite perspectives are used to explain why jellyfish bloom. One view holds that because the polyps are triggered at the same time by the same environmental stimuli, and the medusae ride the same currents together, they appear rapidly and in a group as a "bloom," appearing rapidly as a group like a meadow of flowers. The other explanation is that aggregating holds some benefit to the animals' survival or reproductive success. The truth is probably a blend of both explanations, but the second one is worthy of some discussion with regard to its ecological effects.

The Allee effect, named after W. C. Allee, who described this concept more than 60 years ago, refers to "the effects of population size, or density, on the fitness of individuals and the consequences that this may or may not have for population change" (Stephens, Sutherland, and Freckleton 1999, 188). In other words, the size of the aggregation and the distance between neighbors matters. This is often used in the field of conservation biology to describe the problem of finding mates at low population density, such as we have seen with white abalone, beluga sturgeon, and Atlantic swordfish.

On the flip side of the worries about dwindling abalones and fish, the Allee effect may help explain why jellyfish are doing so well in today's eco-systems. As more polyps bud off more medusae, those larger, denser blooms contain more individuals closer together. Many jellyfish species reproduce by "broadcast spawning," meaning that they release sperm and eggs freely into the water to make contact. In some jellyfish species, only the male is a broadcast spawner, with the female ingesting the sperm to fertilize her eggs internally. Either way, the denser the bloom, the closer the proximity of males and females, which increases fertilization success and in turn means more polyps.

Moreover, it appears that jellyfish may gain predatory and protective benefit from their aggregating behavior (Malej 1989). In particular, a prey item dodging one jellyfish in a swarm becomes more vulnerable to its neighbors. Furthermore, the predators of jellyfish larvae also serve as prey to jellyfish adults, so swarming may act as a protective hub for jellyfish larvae and juveniles.

Therefore, it appears that the mere fact of the presence of jellyfish blooms makes larger blooms more likely, and so on. Blooms therefore help ratchet jellyfish populations toward dominance.

Trophic Cascades

The word "trophic" comes from the Greek word meaning food or feeding; thus, the trophic level of an organism refers to the level it occupies on the food chain. Because of the interdependence between species in a community, when one species or functional group of species falters, others up and down the food chain are affected.

The canonical example of a trophic cascade is the assemblage of killer whales, sea otters, sea urchins, kelp forests in the Aleutian Islands (see chapter 2; Estes et al. 1998). While it might be hard to imagine killer whales having any effect on sea urchins or kelp forests, distant effects such as these occur quite often because of trophic linkages. Another well-studied example occurred along the Eastern Seaboard of the United States. Recall the issue of rapid worldwide shark depletion from chapter 6. We may think that sharks are of little use, or perhaps even better off eliminated. It is quite plausible that many fishermen who aren't fishing for shark fins or steaks may hold this view, such as cod fishermen wanting less competition for their catch and shellfish fishermen wanting safer working conditions among their clams and scallops. But what if the loss of sharks resulted in the unanticipated collapse of the scallop and clam industries?

Professor Ransom Myers and his colleagues studied the decline of sharks along the Atlantic Coast of North America (Myers et al. 2007). They found that 11 species of large sharks that eat rays and other sharks have become severely depleted since 1972. These sharks are now functionally eliminated: they have declined by 87–99 percent, and the few individuals that remain are generally half-grown. Their depletions have led to increases in most of their smaller shark and ray prey. One of these, the cownose ray, has benefitted from lack of predation to the point of a population explosion to over 40 million. These rays eat bivalves, such as scallops, clams, and oysters, to the tune of 210 grams (7 ounces) of shell-free wet weight per day. The rays migrate through the Chesapeake, spending about 100 days eating some 925,000 tons of bivalves per year. For comparison, the 2003 commercial bivalve harvest was only 330 tons—there, hyperabundant rays are consuming more than 2,500 times the commercial landings. The annual ray migration (and all-you-can-eat buffet) occurs prior to the scallops' spawning season, resulting in a rapid population crash of scallops and concomitant predation increase on clams and oysters. Once famous for rich harvests of bay scallops and clams, the Chesapeake is now best known for its pestilent jellyfish.

In this trophic cascade, the effects of shark depletion over thousands of miles of the Atlantic coast were felt in the collapse of the Chesapeake scallop and clam industries. Without doubt, the depletion of shark populations has had many other effects: even inconspicuous disruptions can perturb the ecosystem in ways that may not be recoverable. Thus, what may seem originally to be a small—or even justified—change can be magnified into massive havoc elsewhere in the ecosystem via the process of trophic cascade.

A third example, studied by Professor Terry Hughes of James Cook University in Australia, illustrates that sometimes events that we have no control over can push a teetering ecosystem over the edge . . . with catastrophic effects. Consider Jamaica. This island nation lies at the center of coral biodiversity in the Atlantic, but its exponential population growth has taken a heavy toll on its coral reefs. As its population rapidly expanded through the twentieth century, so too did the pressure to exploit its natural resources. By the late 1960s, artisanal fish trappers had already decreased fish biomass by 80 percent on the northern fringing reefs (Hughes 1994). By 1973, the number of fishing canoes deploying traps along the north coast was about 1,800—some 3.5 canoes per square kilometer (9 per square mile). Large predatory fish, such as sharks, snappers, jacks, triggerfish, and groupers, had virtually disappeared, as had turtles and manatees. The remaining species—herbivores like parrotfish and surgeonfish—were small. This pattern was mirrored on the south coast, as well.

As we have seen, sea urchins have voracious appetites for algae. In Jamaica, some species of fish prey on urchins, while herbivorous fish compete with them for the tender bits of new algae, keeping it from overgrowing the corals. As fish declined, the urchins flourished. Unlike the Aleutian example where the kelp forests support the entire ecosystem, the Jamaican ecosystem is based on the coral reefs. So algal reduction by urchins was a good thing. Sort of. Until 1983.

In 1983, a species-specific disease wiped out 99 percent of the sea urchin population. Without grazers, the algae flourished. First weedy species, then longer-lived species became established. Algal overgrowth shaded out existing corals, and new coral larvae had nowhere to settle. The scale of damage is enormous. Coral cover, which was previously 52 percent, has declined to 3 percent, and macroalgal cover has increased from 4 percent to 92 percent—the Jamaican ecosystem has shifted from healthy coral reefs to one dominated by algal beds. Although sea urchins have recolonized the reefs in small numbers, algae remain dominant (Scheffer, Carpenter, and Young 2005).

The Jamaican coral reefs are a striking example of synergistic effects triggering a spectacular trophic cascade. Overfishing had already weakened the ecosystem, resulting in the urchins being the only predator left to control the algae. The urchin blooms were a highly visible sign that something was wrong, but nobody noticed.

Trophic cascades generally result from disturbance to what is called a "keystone predator'—think of it like a Mafia boss who keeps all the others in line. The sea otters in the Aleutians. The great sharks in the western Atlantic. Cod in New England and Newfoundland. Algal grazers. The voracity of the keystone predator's appetite exerts chronic and continuous pressure on other species, keeping their numbers from growing out of control. As we have seen, disturbing this balance affects the entire food chain, essentially creating an ecological opening for jellyfish to exploit, or at the very least, removing the fish predators that keep jellyfish and other pests in check.

Shifting Baselines

Part of the overfishing problem, if not anthropogenic disturbance in the bigger sense, is what scientists call "shifting baseline theory." This concept refers to how our expectations of what healthy ecosystems look like have shifted over time, essentially lulling us into a false sense of security. Because we are familiar only with what we see during our lifetime, historical conditions are never apparent, so reference points of how things used to be shift toward a progressively more degraded state. "Shifting baselines results in an incremental 'lowering of standards' as each new human generation redefines what is considered 'natural' or 'normal' based on their own personal experiences. Each new generation lacks an understanding of how the environment 'used to be.' This lower standard is now the new baseline for the next generation" (Cudmore 2009, 12).

The concept was originally articulated by Professor Daniel Pauly. He explained it as follows:

> Essentially, this syndrome has arisen because each generation of fisheries scientists accepts as a baseline the stock size and species composition that occurred at the beginning of their careers, and uses this to evaluate changes. When the next generation starts its career, the stocks have further declined, but it is the stocks at that time that serve as a new baseline. The result is a gradual shift of the baseline, a gradual accommodation of the creeping disappearance of resource species, and inappropriate reference points for evaluating economic losses resulting from overfishing, or for identifying targets for rehabilitation measures. (Pauly 1995, 430)

In other words, the true baseline is the untouched state before human exploitation, perhaps hundreds of years ago, but this is very rarely the baseline available to us to work with. Our reference point of "natural" or "normal" has shifted. Part of the problem, Pauly demonstrates, is the scientific convention of ignoring anecdote as legitimate data—and even entirely disbelieving historical size or abundance information (Sàenz-Arroyo et al. 2005). In doing so, we deny ourselves access to centuries of information in the anthropological and historical record. "Thus, 'natural' means the way things were when we first saw them or exploited them, and 'unnatural' means all subsequent change" (Jackson 2001, 5411).

Paul Dayton and his colleagues eloquently summarize the scientific and social implications of shifting baselines in their paper, "Sliding Baselines, Ghosts, and Reduced Expectations in Kelp Forest Communities":

> The most important message of this paper is that, no matter how well one understands kelp populations, any current program will fail to discern the ghosts of missing animals. That is, any biologist studying the community now will see interesting biology of kelp and small animals, but the expectations of what is natural are much reduced and are likely to be an inappropriate basis for making fisheries and environmental decisions. (Dayton et al. 1998, 320)

A graphic example of a shifting baseline is that of large trophy fish in the Florida Keys (McClenachan 2009). As the larger fish were progressively fished out, the fishermen's expectations of what constituted a trophy fish became smaller and smaller (see plate 15).

We now perceive huge fish, and stories of huge masses of fish, as exaggerations or "fish stories," or, when credible, as aberrations. We take an up-close look at huge swordfish mounts—and maybe tap them just to be sure—because we simply don't believe that these behemoths could be real. Similarly, we laugh at stories from yesteryear that claim that fish were so abundant as to be caught in baskets near the shore, or that whales and manatees numbered in the hundreds of thousands . . . *or more*. These scenarios are so distant from our own experience that we simply don't believe they could possibly be true. Most large vertebrates suffered the effects of overhunting and overfishing long before anybody started keeping demographic details, and the few scientists who have seen pristine baselines are a dying breed. Our baseline of "normal" has shifted so far that we don't even try to include restoration of swordfish, sharks, sea turtles, or manatees in marine reserves, instead focusing on the species left after former top predators and grazers have long since been forgotten (Jackson 2001).

We see this all the time in the world around us and even in print. *Even here*. The limitations of publication inherently shift the baseline from the set of all knowledge on a subject to the subset that fits into limited space. A peculiarity of publication is that it is almost always shy of true numbers: not all occurrences are witnessed, not all those witnessed are documented, and not all documentation gets published. Alas, we are doomed to end up with an underestimation of the true occurrence of almost every phenomenon. That, too, is a shifting baseline. We are always fated to believe that an event is less common than it actually is, or that fewer species are struggling than actually are, or that the magnitude of jellyfish increase is less than it actually is.

In human history, two strong and chilling parallels exist with today's decline of big fish: (1) prehistoric decline of large land mammals, and (2) overhunting of whales. When *Homo sapiens* gains access to large animals, we eat them (Pauly, Watson, and Alder 2005). There was the rapid demise of marsupial megafauna in Australia 40,000–50,000 years ago following the arrival of aboriginal peoples. There was the rapid extinction of 30 species of big mammals in North America 12,000–13,000 years ago by Clovis hunters. The giant flightless moa was exterminated almost 1,000 years ago by the first Polynesians to arrive in New Zealand. As we spread across these land masses, large, slow animals that taste good roasted over a fire were hunted to extinction. Similarly, the once-mighty bison of North America fell victim to the more recent European spread, as have big game animals in Africa, though for different reasons.

Whales have been hunted by aboriginal peoples since time immemorial. But it was the industrialized and mechanized hunting that sent the world's whale populations into free fall. By the time commercial whaling was banned in 1986, all species of commercially important whales were close to extinction. Some have recovered. Many have not.

Like earlier examples of eating the big species to death, it appears that we are now again decapitating our food chains—this time it's the big fish. The sharks and the tuna, the swordfish and the grouper, the cod and the halibut and the sturgeon—the megafauna of the oceans. Ransom Myers and Boris Worm (2003) studied the worldwide depletion of predatory fish communities. In their 10-year study that used all available data from the beginning of exploitation, they found that 90 percent of all large fish have disappeared. It seems that we humans have a characteristic pattern of reducing community biomass by 80 percent within 15 years of a new fishery opening.

For example, as longline fisheries for tuna and billfish expanded in the 1950s and 1960s, newly fished areas showed high initial catch rates but declined to low levels after just a few years. As another example, 60 percent of large finfish, sharks, and skates in the Gulf of Thailand were lost during the first 5 years of industrialized fishing. Similarly, virgin communities of seamounts and continental slopes have declined rapidly over the first 3 to 5 years of fishing.

When a species is newly fished or an existing species is fished in a new region, we generally exhaust the resource in a few years flat. In most cases, these depletions occur before scientific monitoring has taken place, so there is no baseline to measure the loss by and the best we can hope to do is stabilize fish populations at low abundance levels.

The issue of shifting baselines was elegantly presented in Professor Callum Roberts's book, *The Unnatural History of the Sea*, which details the drivers and implications of overfishing throughout the history and spread of human habitation. Roberts says:

> In my work as a scientist, I find that few people really appreciate how far the
> oceans have been altered from their pre-exploitation state, even among profes-

sionals like fishery biologists or conservationists. A collective amnesia surrounds changes that happened more than a few decades ago, as hardly anyone reads old books or reports. People also place most trust in what they have seen for themselves, which often leads them to dismiss as far-fetched tales of giant fish or seas bursting with life from the distant, or even the recent past. The worst part of these 'shifting environmental baselines' is that we come to accept the degraded condition of the sea as normal. (p. xiv–xv)

And maybe it's more than that . . . maybe it is simply human nature.

Roberts further observes, "Where a resource is common property, shared by all, there is a tendency for individuals to take more of that resource than is sustainable. Individuals can obtain a private gain but at a cost to the rest of society by acting selfishly" (p. 217). Therefore, even though we kind of know there's a problem, it is easier on our soul not to think about it too much and better for us overall to get some before it's too late. And maybe in the back of our minds we hope a bit that the hullabaloo is wrong and some new population will be found or some new legislation will be enacted that makes others have to back off. In reality, this "don't worry, be happy" view may anesthetize us in the short term but does not serve us well in the long term. The current state of the oceans did not get this way overnight: we arrived here through cultural beliefs in birthrights and limitless bounty, through collective pursuit of progress and profit, and through many generations who ignored restraint. And like the party animal that imbibes without regard for credit card bills, drunk-driving implications, hangovers, or liver damage, there comes a time. . . .

A recent article demonstrated that while loss of large predators is generally presented as an ethical or aesthetic issue, it is, in fact, of much broader importance (Estes et al. 2011). Estes and his team showed that decapitation of the food chain, or "trophic downgrading" as they call it, leads to "unanticipated impacts of trophic cascades on processes as diverse as the dynamics of disease, wildfire, carbon sequestration, invasive species, and biogeochemical cycles" (p. 301). We have already seen how changes in phytoplankton can lead to changes in carbon sequestration, how decreases in biodiversity can lead to invasive species, and how losses of predators lead to jellyfish blooms.

A corollary to the shifting baseline concept is our intuitive confusion between ecological extinction (functionally gone) and biological extinction (totally gone). Ironically, with all the press about endangered species, many of us take comfort in thinking that they are not yet extinct, they can be saved. We are fooling ourselves. It's like keeping a brain-dead patient on life support in the hope that they awaken as normal as they once had been in their heyday. Many species have no real hope of survival; their numbers are so low that they cannot find mates, and their habitats are changing so fast that they cannot keep up. Even if we were to successfully save species in captivity, without suitable habitat to return to, they are doomed to exist only as show ponies. Where will we put the polar bears and penguins when the sea-ice is gone? And honestly, how many can we really save as captive-only species—our zoos are already nursing homes for the dying species of the planet. Certainly there will come a point where we stop trying to save every one.

Most large species on earth—terrestrial and marine, the elephants and tigers, the polar bears and penguins, the sharks and whales and sturgeon—now have "remnant populations" ranging from 10 percent to less than 1 percent of their prehunted abundance. These perilously low populations have lost their buffering capacity against disease, famine, or just a bad year.

But it's not just one or two species. In essence, we have reduced the whole upper half of the food chain to 10 percent of its population, so that most species still exist, but as mere ghosts of their formerly robust selves. It's like a dying town where only 1 in 10 houses still has a resident, the last few old-timers still holding on to something long ago lost as the tumbleweeds roll down Main Street.

These ghosts are functionally extinct, no longer performing their citizen functions in their ecosystems. Predator. Prey. Keystone regulator. The backbones of our ecosystems are disintegrating. Ninety percent of the sharks. Ninety percent of the big game fish. Ninety percent of the big mammals and big turtles and big seabirds. And like a fragile skeleton that has become brittle, when one leg breaks, the whole skeleton will begin to falter. Like a house of cards. Because a large number of species across ecosystems and food webs are all at the brink, it won't take much to trigger a domino effect to send ecosystems and food webs into freefall.

But that's just the endangered species, right? Well, maybe not. We tend to think a species must be okay if it is not listed as endangered, and that species not listed as extinct are better off than those that are. Well, not really. The process of getting a species listed is a lengthy and frustrating one and is generally impractical for noncommercial invertebrates. So the number of listed species, with its strong bias toward charismatic megafauna, is incredibly misleading in its incompleteness. More importantly, official labels of "endangered" or "extinct" are mere death certificates. Ecological extinction—the equivalent of a frail, forgetful human patient who has become physically and mentally incapacitated—occurs long before the final heartbeat of the species. Whales, sea turtles, bluefin tuna, cod, sturgeon, Atlantic halibut, sharks, river dolphins, Adélie penguins . . . their numbers have dropped so low that they are ecologically extinct in that they no longer contribute meaningfully to their ecosystem, either as predators or prey. These species are on life support.

One of the more vomitous implications of discussions on extinction, including this one, is that we are compelled to ignore the uncharismatic minor fauna, which comprises more than 99 percent of life in the oceans. Data are plentiful for the mighty humpback whales, the handsome penguins, the curious turtles, and tasty species like cod and tuna. But what of the vanishing *Scrippsia pacifica*, a stunningly beautiful jellyfish that used to be common throughout California? There's not enough data (or general sympathy) to get it listed as endangered. Or what of the lowly pycnogonids (sea spiders) or ascidians (tunicates)? Honestly, what would it take to convince most people that these unusual forms are threatened by trawling and hypoxia, and that that matters? Most invertebrates don't matter to us because we don't eat them . . . but they matter to those other species that do. And they matter to ecosystem integrity. Many thousands of invertebrates are, without doubt, already extinct due to anthropogenic pressures or well on their way. But when you look at the time, money, and effort spent tracking the declines of cod, bluefin tuna, sturgeon, and abalone, to name but a few, it becomes clear that most invertebrates will vanish without fanfare.

> The last fallen mahogany would lie perceptibly on the landscape, and the last black rhino would be obvious in its loneliness, but a marine species may disappear beneath the waves unobserved and the sea would seem to roll on the same as always.
>
> —G. CARLETON RAY, "Ecological Diversity in Coastal Zones and Oceans"

CHAPTER 12

The Jellyfish Double Whammy

What these jellyfish are eating are either the young of the next generation or the food of the next generation. We'll know the impact when
what they ate does not appear in the nets next year.

—DR. BELLA GALIL, National Institute of Oceanography in Israel

Lions Roaring in the Yangtze Estuary
(China, from 2003)

One of the most remarkable industrial feats of our time was the damming
of the Three Gorges in China's Yangtze River. Tourists flocked to the area,
eager to see the gorges one last time before they were filled with water. This
hydroelectric dam is the world's largest, forming a reservoir of 1,080 square
kilometers (420 square miles). The reservoir began to store water on 1 June
2003. That was also around the time when the jellyfish problems began (Xian,
Kang, and Liu 2005).

The rhizostome species *Rhopilema esculenta* is a favorite among jellyfish
connoisseurs—it has the right amount of "crunch"—and as a consequence
has been overfished as a food source. Once the most common jellyfish in the
Yangtze Estuary, *Rhopilema* has gradually been replaced by the lion's mane,
Cyanea. *Cyanea* is often described as looking like "a dinner plate with a mop

underneath," which is pretty accurate. And consistent with its raggedy, somewhat feral appearance, it is an opportunistic weed.

Cyanea responded rapidly to the increase in salinity, temperature, nutrients, and zooplankton caused by the damming of the Yangtze. In 1998, *Cyanea* comprised less than half a percent of the total samplings, whereas by November 2003, just five months after damming, *Cyanea* abundance had grown to a whopping 85 percent.

By May 2004, its abundance had grown to more than 98 percent of the total catches, causing bycatch problems for fisheries by clogging up the mesh of trawl nets.

May and November are the two key fish spawning seasons in the Yangtze Estuary; large blooms of jellyfish during these periods are detrimental to fish recruitment, both directly through predation of the fish eggs and indirectly through competition for the zooplankton as a food resource. There is also speculation that the jellyfish blooms drive the fish away. Because the discharge of water and sediments into the estuary are reduced by the dam, the saltwater intrusion to the estuary begins earlier and lasts longer, in essence, lengthening the peak season of *Cyanea* conditions.

Throughout the preceding pages, we have seen how jellyfish can be both predators and competitors of other species. This may seem like a small thing, but it is not. The implications are huge, especially for species higher on the food chain.

Jellyfish eating fish larvae and their prey is more or less the terrestrial equivalent of zebras eating lion cubs, as well as the birds and small mammals that the lion cubs most often prey on. Or the equivalent of grubs eating baby chickens as well as other grubs. Jellyfish not only kill fish directly by eating their eggs and larvae but also indirectly by outcompeting them for their food, leaving the fish to starve.

The predation and competition double whammy isn't the worst aspect of this. Through this process, a jellyfish-driven ecosystem may well become an alternative stable state to the long-established fish-driven ecosystem, making the latter difficult or impossible to reestablish.

Once jellyfish gain control, through this double whammy of predation and competition, they are able to keep control. Recall the decimation of the Black Sea fauna by *Mnemiopsis*—this is exactly what happened—and it was only through a radical alteration of multiple simultaneous factors that the ecosystem was able to begin recovery.

Not only are jellyfish both predators of and competitors with other species, they also respond to environmental fluctuations over shorter timescales than most other species. This allows them to reach bloom densities while other species are still developing as eggs or larvae. Furthermore, jellyfish do not reach satiety when feeding, that is, they don't become "full" even if their stomachs are full. Because they keep catching more food particles, jellyfish blooms carry the potential to kill vast quantities of prey.

Jellyfish as Predators of and Competitors with Fish

Jellyfish are predators of and competitors with more or less all species that have one or more life stages in the plankton. If a species spends any of its egg, larval, or adult stages drifting in the water column, it is likely that it is at the mercy of the jellyfish double whammy. While this technically applies to starfish and snails and tunicates, we tend to be more concerned about the effects on fish, particularly those species subjected to nonnatural impacts like overfishing and bycatch, because they are getting it from both sides so to speak.

Under normal circumstances, jellyfish eat larval fish, fish eggs, and a wide variety of copepods, invertebrate larvae, and other zooplankton. This broad diet overlaps with those of most other organisms feeding in the water column, particularly with the diets of forage fish like anchovies, sardines, herrings, pilchards, and menhaden. Under normal circumstances, jellyfish abundance is low enough for the ecosystem to stay in balance as the jellyfish and forage fish compete for these resources. Even when yearly seasonal blooms occur, the spike in predation quickly returns to normal.

Because jellyfish have few predators, control of their populations primarily comes from bottom-up forces like temperature, salinity, food availability, and availability of space for polyps, in those species that have them (Parsons and Lalli 2002). However, without predators and with diminishing competition from decreasing populations of fish and marine mammals, jellyfish populations can bloom without restraint. Because of their low metabolism, the total biomass of a jellyfish bloom can quite quickly and easily overtake the total biomass of other species in an ecosystem.

Think of jellyfish and fish populations as a sort of seesaw. When fish abundance is high, jellyfish abundance is low, and as fish abundance decreases, jellyfish abundance increases. So as we catch more and more fish, this leaves increasingly uneaten prey down the food chain, now available to be eaten by something else. Jellyfish take advantage of the available food, and in doing so, progressively tilt the competitive balance in their own favor. But jellyfish don't just eat the food that the fish eat, jellyfish also eat the eggs and larvae of the fish. And of course, as jellyfish numbers increase, they eat more and more, suppressing the ability of the fish population to recover from fishing pressure. With overfishing, the interaction between fishing and jellyfish becomes a self-enhancing feedback loop, whereby fewer fish leads to more jellyfish, which leads to fewer fish, and so on.

Fisheries in China are already suffering serious problems from the jellyfish double whammy, where the incidence of jellyfish blooms has dramatically increased since the late 1990s and is becoming an annual event (Dong, Liu, and Keesing 2010). Blooms of the refrigerator-sized *Nemopilema nomurai* in the East China and Yellow seas are blamed for a 20-percent decline in a commercially important species called the little yellow croaker. Similarly, the Bohai Sea fishery for the edible jellyfish *Rhopilema esculentum* declined about 80 percent due to a large bloom of the lion's mane jellyfish there in 2004. This resulted in economic losses estimated at $70 million.

Where Does the Feedback Loop Stop?

The implications of the jellyfish double whammy are bleak. It is difficult to imagine how fish might take back control and be able to flourish again, after a jellyfish-dominated ocean is established. Even if the original trigger, overfishing or pollution or whatever, is corrected, the jellyfish masses will still inhibit fish population growth via the double whammy of predation and competition. A jellyfish-controlled ecosystem will thus become the new stable state, the "new normal," and be very strongly resistant to change or correction.

A reasonable analogy would be where a prison guard gives the keys to an inmate, who then seizes the opportunity to revolt and run amok. The inmates take over. Some guards are killed, others are held hostage. Some inmates escape. Chaos reigns. Well, not for the inmates. Management can fire the guard who gave the keys in the first place, but that doesn't change the problem of the siege and the escapees. So how, then, do we take back the prison and round up those who have escaped? Generally, in a situation involving an actual prison revolt, military force would be used. Guns. Bullets. But jellyfish don't respond to guns and bullets. And of course, it's not just the problem of the jellyfish that we can see, but also the tiny polyps hiding heaven-only-knows-where out in that big, big ocean.

We discussed earlier the ratcheting effects of climate change. The jellyfish/fish dynamic can ratchet very rapidly in the direction favorable to jellyfish, but it's not so easy for the oceans to revert to a fish-dominated state.

Jellyfish as Top Predator

It is counterintuitive to think of jellyfish as a top predator. We normally picture big things with big teeth as the top predator. Sharks. Whales. Big fish. But in a scientific sense, a top predator is any species that has no predators. Size has nothing to do with it. Teeth have nothing to do with it. It's all about who eats whom.

Jellyfish are what can be considered a "trophic dead end"—a dead end on the food chain. They eat but are rarely eaten. In this way, they can become the dominant predator in an ecosystem, driving the food chain from the top.

Top-Down versus Bottom-Up Control

We generally think of community structure as being controlled by either top-down (predator-controlled) or bottom-up (resource-limited) forces. However, in an unexpected way, jellyfish are able to seize control and drive the ecosystem from both directions. While top-down control may appear more obvious than bottom-up control, both forces are at work to help jellyfish maintain control and keep other species sidelined.

Top-Down Control

Jellyfish prey directly on the eggs and larvae of fish and other species. This is top-down control. Jellyfish cripple species throughout the food chain—fast species, big species, species with big teeth—by preying out the smallest components of their life cycle. This truncates the contribution of these species' eggs and larvae to the next generation.

By amassing the predatory and competitive critical mass to take control, jellyfish can consume sufficient larvae and destroy the capacity of other species, such as fish, to regain their previous population size.

Jellyfish don't only limit fish populations—that's merely the part we can see the most easily, because it is reflected in fish catch statistics. As top predator, jellyfish also control the populations of just about every species in the ecosystem, either directly through predation on larvae or adults or indirectly through competition for food resources.

Bottom-Up Control

Copepods and other small plankton are the primary prey of adult schooling fish like anchovies and sardines, as well as the young of many other types of fish. Copepods are also the primary prey of most jellyfish. Jellyfish therefore compete with fish and other species by preying on their food sources, causing those other species to starve. This is bottom-up control.

In bottom-up control, we generally think in terms of physical factors like nutrients, nesting space, daylight length, and so on. Jellyfish cannot directly control these physical factors, but they can and do indirectly control how the nutrients are used and fed up the food chain. Even other physical factors like light penetration and physical access to space, are in some cases directly or indirectly influenced by jellyfish blooms to be favorable to themselves and unfavorable to their competitors.

Nutrient control. Due to mass consumption of copepods and other primary consumers (zooplankton) by jellyfish, the primary producers (phytoplankton) are released from grazing pressure. They bloom out of control, and their biomass often contributes to a dead zone beneath them near the seafloor. This in turn kills the various invertebrates that substantially contribute larvae to the next generation of zooplankton grazers, thus leaving yet more surplus phytoplankton to die uneaten and sink to the dead zone. Therefore, top-down control on one species can manifest as bottom-up control on another.

Light penetration. Recall the case of Norway's Lurefjorden. Over the past 40 years, the water in this fjord has become "darker" and more favorable to tactile predators, and recently at least 2 other fjords similarly have shifted to jellyfish-dominated ecosystems. Whether the jellyfish are actually driving it is not clear, but there can be no doubt that they are exploiting it.

Access to space. The Namibian case outlined in chapter 2 is a good example, where jellyfish have created a sort of "exclusion zone" around themselves in three-dimensional space. Beneath them, the dead zone prevents almost all species from surviving—except for jellyfish, which use their stored oxygen, and the bearded goby, which holds its breath. And with a "jellyfish curtain" spread across the width of the continental shelf like sentries guarding the golden chalice, their sheer stingy mass and the viscous stingy water that they produce combine to repel other species. Furthermore, by altering the physical conditions and food chain dynamics, jellyfish act as ecosystem engineers, ultimately altering the species composition in a community (Breitburg et al. 2010). This in turn is likely to be more favorable to jellyfish and less favorable to higher energy species. We examine this phenomenon in greater depth in the next chapter.

Through a combination of top-down control via fishing cascades and bottom-up control via eutrophication cascades, we are rapidly simplifying the marine ecosystems of the world to be more favorable to jellyfish blooms. And through a combination of top-down control via predation and bottom-up control via competition, jellyfish blooms can have serious implications for the marine community, which portends catastrophic consequences for fisheries. The ecosystem restructuring that leads to control by jellyfish has been called an "ecological climax" (Kirby, Beaugrand, and Lindley 2009). Whereas in small numbers, jellyfish represent an inconvenience to other species, in large numbers, they pose a formidable enemy too numerous and too well equipped to be defeated.

Seed Banks as Another Form of the Double Whammy

It is easy to imagine that as a jellyfish bloom expands, its larvae may settle in increasingly greater areas so that the next time a bloom occurs, it essentially picks up where it left off and continues to expand even more. In this way, the polyp stage acts as a sort of "memory" for the bloom.

Yes, that happens, but the problem is bigger than that. The periods of time between visible jellyfish blooms are an integral part of the process of bloom expansion. As other species occupying bottom space decline, polyps and hydroids spread. Think of them as real-estate moguls on the poor side of town, opportunistically grabbing space as soon as it comes available. They not only hold their ground, they rapidly populate within it so that the next bloom essentially starts with more "seeds."

As long as conditions are favorable for polyps, they can build their baseline populations, such that they are poised for vast explosive blooms when conditions become favorable for medusae. Then the medusa bloom is so dense that it has to expand, shedding larvae into previously unoccupied space, and so on.

Each stage acts as both a placeholder and a room-expander for the other. Polyps multiply within their geographical boundaries to add capacity to the medusa blooms, and medusae spread and shed their larvae to add more polyps. In this way, both stages of the jellyfish life cycle are able to ratchet each other toward increasingly larger abundances.

So we may breathe a sigh of relief in years when the refrigerator-sized *Nemopilema* or the spotted *Phyllorhiza* are less abundant, but it is a false sense of security. Beneath the cover of waves, their polyps are building the seedbank for the next big bloom. Similarly, *Mnemiopsis* multiplies in deeper waters into incredibly huge population numbers, waiting for favorable conditions. We saw this in the Mediterranean in 2009, when the species seemed to have emerged out of nowhere to simultaneously span the sea in superabundances from Israel to Italy to Spain.

> We can summarize the extent of human impacts on the oceans in stark terms. Humans have caused and continue to hasten the ecological extinction of desirable species and ocean ecosystems. In their place, we are witnessing population explosions of formerly uncommon species and novel ecosystems with concomitant losses in biodiversity and productivity for human use. Many of the newly abundant species, such as jellyfish in the place of fish and toxic dinoflagellates in the place of formerly dominant phytoplankton, are undesirable equivalents to rats, cockroaches and pathogens on the land. Moreover, there are good theoretical reasons and considerable empirical evidence to suggest that, once established, such newly established communities become stabilized owing to positive feedbacks among newly dominant organisms and their highly altered environments. (Jackson 2010, 3772)

CHAPTER 13

High-Energy and Low-Energy Ecosystems

In coastal areas, changes caused by eutrophication . . . appear to enhance jellyfish production and not fish production. This indicates that
a food chain favoring jellies occurs in these areas, rather than there
simply being a vacancy for an opportunistic feeder at a higher trophic
level.

—TIM PARSONS AND CAROL LALLI, "Jellyfish Population Explosions"

Late in the afternoon of 15 March 2011, the fire at the number 4 Reactor at
Japan's Fukushima Daiichi nuclear power plant had just been extinguished.
Earlier in the day, the third reactor building in 4 days had exploded. The death
toll from the previous Friday's tsunami was climbing while the temperature
in the badly devastated region was rapidly falling toward freezing. The world
nervously held its breath amid growing fears of radiation leaks and the first-
ever incident of multiple meltdowns. And yet, the number-one news story
on CNN's website at that time was "Powerful Neurotoxin Drove Millions of
Fish to Deaths."

There is something about nature that rivets the human mind. Edward O.
Wilson, one of the great scientific minds of the twentieth century, called it
"Biophilia"—the love of life, or the innately emotional connection of human beings to other living organisms. The neurotoxin in question is domoic
acid—but it should be called "demonic acid." It is produced by microscopic

single-celled marine "plants" (well, algae really) called diatoms. Diatoms are like a pair of speck-sized, opposing petri dishes fitted together, enclosing their photosynthesizing machinery within a glass case, like little tiny terrariums. Most diatoms don't produce neurotoxins, but species in the genus *Pseudonitzschia* do. This fact only became known following a mass poisoning episode in Canada in 1987, which left 4 people dead and over 200 ill after eating mussels. More recently, domoic acid was blamed for killing over 400 sea lions along the central California coast in 1998.

Domoic acid causes neurological damage, including permanent loss of short-term memory. It affects vertebrates, including fish, birds, and humans. This is apparently what caused the fish—sardines in this case—to, well, "pack themselves in like sardines" in the harbor at Redondo Beach near Los Angeles. There were so many fish in such a small area that they simply ran out of oxygen. Well, that's the official story anyway. Sounds fishy to me.

> The weaker tenants of the main
> Flee from their rage in vain,
> The vast menhaden multitudes
> They massacre o'er the flood;
> With lashing tail, with snapping teeth
> They stain the tides with blood.
> —Isaac McLellan, *The Bluefish*, 1886

When we think "top predator," we think of sharks. Bluefish. Big fish. Lots of teeth. Ferocious. Those vast menhaden multitudes flee indeed. . . . Or we think of a killer whale, with a dorsal fin like that of a shark, slicing through the water as smoothly as their teeth slice through flesh, their black-and-white color pattern stark in comparison to the blood-red water they leave in their wake. Or maybe we think of polar bears, or walruses, or sperm whales battling giant squid. We rarely, however, think of jellyfish.

Doctors Tim Parsons and Carol Lalli of the Institute of Ocean Sciences in British Columbia established a framework in which to explain the phenomenon of jellyfish blooms in terms of the perturbations discussed above. In particular, the oceans and seas appear to be shifting from high-energy food chains supporting large predators to low-energy food chains that support jellyfish (Parsons and Lalli 2002).

Food chains may be thought of as pathways of energy flow. Recall that each step in the food chain is called a trophic level. A familiar terrestrial example would be where energy from the sun is captured and used by grass to add biomass, that is, to grow. The energy stored in the grass as carbohydrate is eaten by cows, which use that energy to grow and add to their biomass. The energy stored by cows as meat and fat is then eaten by humans; we use that energy to power our bodies and to add to our own biomass. In this case, there are three trophic levels in the food chain: the grass (the primary producer), the cow (the herbivore), and us (the carnivores).

Historically, the marine food chain has been seen in rather traditional terms: phytoplankton (mainly diatoms) → zooplankton (mainly copepods) → small fish → big fish → large predators. The important, or perhaps romantic, parts are the more or less unlimited phytoplankton for grazers, and the top predators with big teeth and lots of muscle. But it has become clear in recent years that two quite different principal pathways exist for energy transfer from primary producer to apex predator.

The fundamental difference that drives these two food chains lies at their base with the phytoplankton and with the number of trophic levels each supports. If we wish to understand what drives population increases and decreases of fish, we must consider their food, and their food's food, and so on. Two quite different groups of phytoplankton can be delineated, corresponding to two groups of grazers and two groups of predators. The two food chains coexist, but one group or the other dominates in different ecological conditions. In general, "healthy" ecosystems, particularly upwelling systems and regions with a well-mixed water column, favor the high-energy food chain, whereas stratified water columns and ecosystems with too-high or too-low nutrient levels favor the low-energy food chain.

These two food chains have also been called the "muscle food chain" and the "jelly food chain," respectively, because of the types of top predators they support (Sommer et al. 2002).

Understanding Phytoplankton

Every second breath we take is thanks to phytoplankton.
—Author unknown

Phytoplankton are, in general, single-celled aquatic plants, and they are as diverse in their form and function as are animals or land plants (see plates 13, 14, and 16). The two most familiar types are diatoms and dinoflagellates. Other phytoplankton groups (including protozoans with symbiotic algae) include the foraminifera, the coccolithophores, the radiolarians, and a host of other very small creatures grouped as "flagellates." Foraminifera, or forams, are basically amoebae with ornate shells. Coccolithophores are single-celled spherical organisms covered in tiny plates that look like poker chips. Radiolarians are like tiny glass spheres covered in long, incredibly fine glass needles sticking out like a floating mine. Diatoms are a group of single-celled and chain-forming phytoplankton with silica cell walls—that is to say, glass. They come in a dazzling array of shapes and styles: discs, needles, stars, chains, some smooth, others spiky.

Dinoflagellates are among the most interesting creatures in the sea. Don't let their small size fool you—they are cool! Their name comes from the Greek *dinos*, which means "whirling," and the Latin *flagellum*, which means "whip." So literally, they are "whirling whips": microscopic, armor-plated creatures that are typically oddly shaped, resembling "Chinese hats, carnival masks, children's tops, urns, pots, and vessels of many kinds, the spiky knobs of medieval war clubs, balloons on strings, hand grenades or lances" (C. P. Idyll, in Rudloe and Rudloe 2010, 196). They have a groove down the middle into which they lay two long whips, or flagella, one for moving and the other for steering. They are able to swim to position themselves in the water column, aggregate, exploit sunshine and nutrients, or escape anoxic water zones. So they are essentially tiny motorized plants . . . er, um . . . armored motorized plants, like little World War II tanks powered by floppy helicopter blades . . . but some are naked. . . .

Using their flagella as their only source of locomotion, they are commonly able to swim up to 1 meter (3 feet) per hour—this is Formula One speed for a single-celled creature—and they migrate vertically through the water column up to 20–30 meters (60–100 feet) daily—imagine needing to swim in full body armor from California to Hawaii in molasses using only two floppy strands of spaghetti . . . oh, the glory of algae.

Not all dinoflagellates swim. Among the best-known dinoflagellates are the ones that have a symbiotic relationship with corals. The corals "farm" these temperamental algae inside their tissues but can lose them when conditions become unfavorable. It is these special symbiotic dinoflagellates, or rather their departure, that cause the worrying phenomenon "coral bleaching." Symbiotic dinoflagellates are not exclusive to corals; they also occur in many types of jellyfish, hydroids, sea anemones, and even other single-celled phytoplankton.

Many types of dinoflagellates produce bioluminescence, or biological light. Their light is a brilliant blue-green and comes as a bright flash; en masse, they put on a dazzling light show that rivals any rock concert. Glowing waves. Flashing footprints in the sand. Silhouettes of swimming dolphins. Dinoflagellates have an internal clock that prevents them from flashing during the day—even in a darkened room—but their clock can be re-set over a period of about 2 weeks and they can essentially be taught to think that day and night have switched. Curiously, dinoflagellates are the only photosynthetic organisms that are also bioluminescent.

But bioluminescence isn't the only way that dinoflagellates are entertaining. Many are also what you might call the Venus fly traps of the algal world: they are algae that eat animals. Yes, really. Instead of getting their energy in the normal plant and algal way of converting sunbeams into energy and body mass, they have to eat other plankton, primarily diatoms, other dinoflagellates, fish eggs, bacteria, and the occasional copepod larva. So as hard as it is to believe that an algal bloom could be *competing* with fish for food, and even preying on fish eggs, it's true. About half of the living species of dinoflagellates are "heterotrophic," or nonphotosynthetic, and must eat other creatures to survive.

Some types of dinoflagellates contain high levels of ammonia or other chemicals inside their body wall, like tiny water balloons. When these little ammonia bombs die (each individual dinoflagellate may only live a few days), the ammonia is released into the water. Together, they can change the water chemistry to be toxic.

One such noxious species, *Noctiluca scintillans*, Latin for "sparkling night light," appears to be increasing throughout the tropical and temperate waters of the world, producing vast blooms that result in severe fish kills, which cause problems for fisheries. And dinoflagellate blooms typically lead to jellyfish blooms, causing even worse problems. Ironically, while dreaded in most parts of the world, jellyfish blooms are actually used as a welcome indicator for improved fisheries conditions along the east and west coasts of India, because they signal that the dinoflagellate blooms have diminished (Aiyar 1936; Bhimachar and George 1950; Prasad 1958).

Diatoms and flagellates are, in essence, the "meadows of the sea." Well, by analogy with terrestrial meadows, diatoms would be the grasses, flagellates would be the weeds, and copepods would be the cows—okay, really, *really* small cows.

In the words of Professor Ted Smayda of the University of Rhode Island and a world expert on algal blooms, "Dinoflagellates behave as annual species, bloom soloists, are ecophysiologically diverse and habitat specialists, whereas diatoms behave as perennial species, guild members and are habitat cosmopolites" (Smayda 2002, 281). Diatom blooms are species-rich but relatively limited in suitable habitat, whereas dinoflagellates are quite diverse in their habitat requirements, but their blooms are usually dominated by single species.

Phytoplankton are the base of the marine food chains in the same way that grasses and other small weedy plants are on land. Where land-based herbivores primarily feed on grassy meadows, marine herbivores mainly feed on diatom and dinoflagellate blooms in the oceans. Phytoplankton photosynthesize the same way that terrestrial plants do. Each one acts like a tiny solar panel, capturing the sun's energy and converting it into useful carbohydrate energy—just like spinach, but a whole lot smaller.

Phytoplankton generally reproduce by simple asexual fission—a single cell divides into two, the two split into four, and so on. Exponential growth. By repeatedly splitting in half, one can produce 33 million offspring in only 25 divisions, exploding their populations into dense blooms in just a matter of days. One such bloom occurred in the Black Sea in the autumn of 1993, when the dinoflagellate *Gymnodinium splendens* bloomed to 3.8 million cells per liter (Zaitsev and Mamaev 1997). Thousands of these creatures were in each drop of seawater. In order for mass blooms to occur, phytoplankton need two things: plenty of nutrients and plenty of sunlight.

Phytoplankton, especially dinoflagellates, can rapidly create visually stunning but catastrophic blooms. The red tides that are normally associated with fish kills and toxic shellfish are developed in this way. But not all red tides are red: some may be fluorescent green, or vivid blue, or golden yellow, or bubblegum-pink, or black like oil. In addition to red tides, blooms of Cyanobacteria, or blue-green algae, can also lead to unpleasant foams and surface scums (Sommer et al. 2002).

The Carbon Cycle: Oxygen Production and Carbon Sequestration

Diatoms and other types of phytoplankton are vitally important to our oceans, and indeed, our survival. More than half of the carbon dioxide entering the atmosphere is "scrubbed" by plants, that is, it is absorbed and stored. The result of this carbon-sequestration process is that the effects of carbon dioxide emissions are buffered.

We've all heard of the potential problems with continued destruction of the Amazon rainforest, not only from a species point of view, but largely in terms of elimination of the carbon-scrubbing and oxygen-generating capacity that the Amazon holds. But what about the role that phytoplankton play? Of the atmospheric carbon dioxide that gets scrubbed, half is processed by land plants and the other half by phytoplankton. *Half.*

The earth's surface is 71 percent ocean and 6 percent rainforest. As we eliminate more and more rainforest, the plants in the ocean will become increasingly important to our survival. Of course, the opposite is also true: as we drive our oceans into increasingly poor condition, the rainforest becomes more and more important.

Toxicity and Other Problems with Dinoflagellates

Many dinoflagellates, and to a lesser extent other phytoplankton, produce powerful neurotoxins, which have often been associated with red tides. However, not all red tides are toxic or noxious, and not all noxious and toxic blooms produce red tides. These outbreaks are more accurately known as "harmful algal blooms." These appear to be on the increase in recent decades, with countless cases affecting humans, marine mammals, seabirds, and fish (Hallegraeff 1993; Anderson 1994). The syndromes have charmingly descriptive names like paralytic shellfish poisoning, diarrhetic shellfish poisoning, and neurotoxic shellfish poisoning. Ciguatera poisons its victims through the consumption of contaminated fish rather than shellfish, and is estimated to affect more than 50,000 people annually.

According to Jeremiah Hackett, a dinoflagellate expert from the University of Iowa, and his colleagues, saxitoxin, the cause of paralytic shellfish poisoning, is "1,000 times more potent than cyanide and 50 times stronger than curare" (Hackett et al. 2004, 1525). The type of toxin in neurotoxic shellfish poisoning, brevitoxin, not only accumulates in filter-feeding shellfish, but also is highly lethal in the water: "fish mortalities from [brevitoxin] blooms can be massive, involving tens of millions of wild fish of all types" (p. 1525). Furthermore, low, chronic doses of some dinoflagellate toxins have been shown to cause lung tumors in mice.

Exposure to estuarine waters containing the dinoflagellate *Pfiesteria*—recall the colloquially known "cell from hell"—produce symptoms in mammals, fish, and birds, including learning and memory deficiencies, skin lesions, acute respiratory problems, and eye irritation. Some toxins cause respiratory problems when their aerosolized components are inhaled in sea spray.

One of the most astonishing harmful algal blooms on record occurred in the Gulf of Oman in September to December 2008 (Bauman et al. 2010). The upper meter of water was exceptionally dense with dinoflagellates, with a single species comprising more than 90 percent of the biomass. The bloom occupied *more than 500 square kilometers* (200 square miles). Two species of corals that had covered more than 50 percent of the seabed experienced 100 percent mortality from the event. It is thought that reduced light penetration coupled with anoxia were likely to have rapidly decreased the efficiency of photosynthesis while increasing the respiratory rate. The algal bloom also resulted in massive decreases in fish abundance, believed to be from mortality rather than emigration.

Concentrating in the food chain. Toxic water. Aerosolized particles. Smothering corals. Mass fish kills. Diatoms are just plain boring in comparison. No bioluminescence. No whirling. No diarrhea.

High-Energy Food Chain (Diatom Based), a.k.a., the Muscle Food Chain

Diatoms in shells of glass,
reminders of earth's eons past.
Terrariums in single cells,
where tiny oxygenators dwell.

Bloom in nutrient's soupy mix,
in fantastic numbers carbon fix'd,
and tho' they're small to you and me,
all animals need them in the sea.

Diatoms grow rapidly, have short lifespans, are rarely a nuisance, and are the primary food of large copepods, krill, and filter-feeding fish, which in turn are the primary food of large fish, whales, penguins, and other high-energy consumers, including most of the commercially important fish species. This is what we call the high-energy food chain, because it can sustain high-energy predators (Parsons and Lalli 2002).

On land, most of the largest animals are vegetarians, primarily grass eaters, for example, cows, horses, buffalo, elephants, deer, and moose. The reason for this is that grasses require little effort to find and consume. They are also energy-rich food sources, so these primary consumers receive the full nutritional value without their food being processed by a middle-man consumer. In general, only about 10 percent of the energy value is retained and passed along each link in the food chain. For example, only 10 percent of the energy value obtained by a cow from a given amount of grass is passed along to its human predator in the form of meat.

In the sea, it works slightly differently. The largest animals are carnivores rather than vegetarians, and their food sources tend to be very plentiful and not very hard to find. Many of the whales, including the largest (the blue whale), and the largest fish (the whale shark) filter-feed on vast amounts of small zooplankton. Therefore, most primary consumers in the oceans, that is, the zooplankton that graze on the phytoplankton, have a large collective biomass rather than a large individual body mass.

But with that said, there are both large and small phytoplankton. As one might expect with big blades of lush grass versus small blades of weedy grass, different species graze on different types and sizes.

Diatom size is important for two key reasons. First, large diatoms are eaten by large copepods. Because copepods will generally take the largest food item they can handle, smaller copepods are restricted to smaller food. Furthermore, large diatoms are eaten by krill, and sometimes even by forage fish, which usually prey on large copepods. Large copepods, krill, and forage fish are all visual predators. Krill and forage fish, of course, are critical components of the high-energy food chain, being the primary prey of whales, dolphins, seals, penguins, and large fish.

A seasonally well-mixed water column generally produces an abundance of large diatoms during the mixing periods. Diatoms grow large in order to store energy as body mass during these pulsed bloom events. Moreover, many of the more abundant species of diatoms form colonies and aggregates that range in size from millimeters to centimeters (eighths of an inch or more). These masses of plant material are easily seen and eaten by large fishes.

The second reason diatom size is important is that large, heavy diatoms sink faster and are therefore more likely to reach the seafloor to bury their carbon with them when they die. In fact, diatoms are so abundant in the ocean that a quarter of the carbon dioxide in the atmosphere is sequestered in the sea sediments in this way. In contrast, smaller species fix more carbon overall but sink more slowly, thus decaying and releasing their carbon near the surface,

HOUSEHOLD TIP: *Diatomaceous earth* is the broken-glass cell walls of count-less numbers of diatoms; a sprinkling of this dust makes an outstanding natural boundary that ants and slugs won't cross, because the microscopic broken-glass edges are too sharp for them.

making them less efficient at carbon sequestration. This in turn could make the ocean less able to absorb carbon dioxide from the atmosphere, essentially setting up a feedback loop that we humans won't like.

Recent research demonstrates that excess nutrient conditions (i.e., eu-trophication) lead to smaller diatom body size, whereas seasonally pulsed nutrient conditions (i.e., normal conditions) lead to larger body size to store nutrients for times of need (Litchman, Klausmeier, and Yoshiyama 2009). Therefore, shifts in diatom size can lead to fundamental changes in carbon sequestration and the functioning of food webs.

Low-Energy Food Chain (Flagellate Based), a.k.a., the Jelly Food Chain

Like a coconut with two hokkein noodles, bless'd,
One for propulsion, the other steering, no less,

Whirling twenty thousand lengths a day,
Compelled to travel the seas this way.

Swim, swim, swim, li'l flagellar one,
Aggregate with your other mates until the day is done.

By night your spheres of light, blue the crashing tide,
And illuminate the shadowy dolphins hides.

And when the day is dawned in the crimson sea
All will suffer your toxicity.

The "other group" of phytoplankton is more or less "everything else": the nondiatom phytoplankton with such gloriously euphonous names as nano-plankton, coccolithophores, chrysophytes, chlorophytes, prasinophytes, and pelagophytes. Many of these other plankton have one or two flagella that they use for locomotion. While this nondiatom group is incredibly diverse and does not easily lend itself to a collective grouping, we shall refer to them as "flagellates" because that is the subgroup of most interest to jellyfish studies.

Like diatoms, flagellates grow rapidly, but they differ from diatoms in that they typically have longer lifespans, they are not consumed by most grazers, and they are often a nuisance. These include the red tides and other harmful algal blooms that can constitute serious toxic and anoxic pollution events, often linked with eutrophication. As we saw earlier, many flagellates can swim, respond to light, and concentrate themselves into astronomical densities. In addition to the flagellates, this other group also often contains large proportions of nonswimming green and blue-green algae (cyanobacteria, which can be smelly and harmful), as well as a strong microbial food web fueled by bacteria. Flagellates and microbes are often eaten by other carnivorous, predatory flagellates, or by small copepods, which are in turn the primary prey of jellyfish. This is the low-energy food chain.

Long thought to be of low nutritional content and therefore of little value to the food chain, flagellates were considered merely "extras" in the ecosystem, takers rather than givers. We now know that flagellates and the microbial community are neither a "link" nor a "sink" in the transfer of energy to higher levels on the traditional food chain, but are in fact, both a link and a sink as part of a separate food chain altogether. Of course, the two food chains are not completely separate: they are linked by tentacles.

Flagellates grossly outnumber diatoms in the ocean. But because they are very small, they are too costly in energy terms for the larger species like fish and krill to feed on under normal conditions. It would be like humans going fishing for dinner and chasing guppies rather than salmon: a single salmon is dinner, whereas a thousand guppies might be the protein equivalent, but are certainly a lot more work to catch and fillet.

In contrast to the well-mixed water column and high-energy food chain, a stratified water column can hasten eutrophication, leading to an ecosystem in which nutrients are very limited. Small phytoplankton dominate, particularly flagellates. One of the reasons for their success in these conditions is their ability to swim out of unfavorable water, whereas when diatoms encounter anoxic conditions, they simply sink and die, adding to the oxygen depletion. Small phytoplankton are consumed by small zooplankton, which are preferred by nonvisual predators.

In many conditions, jellyfish are superior competitors to fish. We have seen how turbidity can favor tactile predators like jellyfish over visual predators like fish. And because of the alternating clonal/sexual lifecycle of some jellyfish and incredibly rapid generation times of others, jellyfish can generally respond to subtle environmental changes and grow to bloom densities far faster than fish. Also recall that jellyfish have very broad diets—from copepods to fish eggs and larvae to invertebrate larvae—and even sometimes flagellates—and jellyfish are also able to "degrow" in times of low food availability. Finally, jellyfish feed continuously and do not satiate at natural food densities (Purcell, Bamstedt, and Bamstedt 1999). As our oceans and seas undergo the changes brought on by overfishing, pollution, and climate change, environmental conditions are becoming more favorable to ancient species, such as jellyfish, and less favorable to the more highly evolved species, such as marine mammals and fish (Parsons and Lalli 2002).

Understanding the Food Chains

The flagellate-based food chain is thought to support fewer marine mammals, turtles, seabirds, and fish than the diatom-based food chain in three fundamental respects. These major differences between the high-energy and low-energy food chains not only help explain them but also reflect our historical understanding of the low-energy food chain.

The Number of Trophic Levels Defines Food Chains

As discussed earlier, higher trophic levels (e.g., sharks) are higher on the food chain, while lower trophic levels (e.g., plankton) are lower on the food chain. John Ryther of the Woods Hole Oceanographic Institution in Massachusetts demonstrated that the flagellate-based food chain is longer than the diatom-based food chain, meaning that there are more trophic levels between the phytoplankton base and the top predator (Ryther 1969). Recall that each trophic step typically represents a 90 percent reduction in food energy—that is to say, only 10 percent of the energy consumed is retained and passed along every time food passes through the gut of a carnivore. The other 90 percent is burned up in reproduction, locomotion, daily living processes, and maintenance.

Picture the food chain as a pyramid, where, for example, the base level is the phytoplankton, the first level up is the copepods, the next level is the forage fish, and the top level is the top predator. Each level of the pyramid represents only 10 percent of the total energy passed along, so it is easy to see how even one or two extra trophic levels can make a huge difference in the food efficiency of an ecosystem. It becomes immediately evident in this example that a longer food chain is less efficient in terms of energy transfer.

For example, in the high-energy food chain, diatoms are eaten by krill, which are eaten by whales (3 trophic levels). Alternatively, diatoms are eaten by large copepods, which are eaten by forage fish, which are eaten by just about any larger vertebrate (4 trophic levels; but in many cases the forage fish eat diatoms directly, bypassing the copepods). To get to the same place with the low-energy food chain, flagellates are eaten by protozoans, which are eaten by predatory copepods, which are eaten by strange gelatinous creatures called chaetognaths (a.k.a., arrow worms) and various planktonic crustaceans, which are eaten by small to medium-sized fish, which are in turn eaten by larger animals (6 trophic levels).

Ryther hypothesized that the productivity in an ecosystem is governed by the number of steps in the food chain, that is, the fewer the steps, the higher the productivity. By eating close to the base of the food chain, energy transfer is more efficient—less food biomass is used to support the terminal link in the chain.

Prey Size Defines Food Chains

The second fundamental difference between the two food chains was articulated by Wulf Greve and Tim Parsons (1977). They proposed that productivity is also likely to be a factor of prey size, that is, visual predators, such as fish, are more efficient with larger prey, whereas tactile filter-feeders, such as jellyfish, require smaller prey that are less likely to rip tentacles and damage delicate gelatinous body parts.

Recall that in general, a copepod or other organism will take the largest particle size that it can handle (except for jellyfish, which target small prey). These differences in prey preference lead to differences in food web structure. Diatoms and flagellates typically differ in size by an order of magnitude or more, and the latter grossly outnumber the former, which effectively partitions the herbivores into those that take large diatoms and those that take small flagellates. This size partitioning leads to the two quite different (high-energy and low-energy) food chains.

Large phytoplankton are eaten by large copepods and krill, and sometimes anchovies. Large copepods are the primary food source of forage fish, such as anchovies, sardines, pilchards, and menhaden. Krill and forage fish are the primary food source for seabirds, seals, whales, and large fish. So from the phytoplankton to the whale or penguin, there are only two more steps in the food chain. Very little energy is being lost to other consumers along the way. So in ecosystems dominated by large phytoplankton, animals high on the food chain find food that has been quite efficiently produced, and jellyfish have a hard time competing.

Small phytoplankton are eaten by small copepods, as well as other types of small herbivorous planktonic creatures like radiolarians, tintinnids, and forams; these small copepods and other plankton are then eaten by larger carnivorous copepods or by jellyfish. The larger copepods are then eaten by the forage fish or krill, and so on, with a low percentage of food energy transferred at each step. Or if the smaller copepods are eaten by jellyfish, they are essentially wasted to the rest of the food chain, because not very many other organisms prey on jellyfish. So in ecosystems dominated by small phytoplankton, large predators either find food that has been inefficiently produced, or they starve in jellyfish-infested waters.

Energy Flow Defines Food Chains

The third fundamental difference between the two food chains was proposed some years later by Tim Parsons and Carol Lalli (2002), who built on the earlier insights and saw these food chains in terms of energy flow, that is, the energy requirements of the top predator. The diatom-based food chain supports high-energy predators like whales, seals, penguins, and large fish, which require a lot of energy to grow and maintain their large bodies and active lifestyles. In contrast, the flagellate-based food chain supports low-energy predators like jellyfish, which require very little energy to survive and grow. Jellyfish can thrive on the flagellate-based food chain because their overall energetic needs are lower than those of most other organisms, and because the total amount of food available that meets their needs is quite high.

The significance of this lies in the fact that this "other" food chain has still not been widely acknowledged. Textbooks don't teach it, ecological studies rarely consider it, and sustainability studies hardly ever account for it.

Professor Ulrich Sommer of the Institut für Meereskunde at Kiel, Germany, along with several colleagues, published an interesting paper a decade ago on this very subject (Sommer et al. 2002). They envisioned three different ecosystem scenarios based on nutrient availability.

Scenario 1 is typical of marine regions where upwelling on continental margins brings a high supply of all nutrients. This high-nutrient supply is also characteristic of seasonal mixing in temperate and polar seas. This nutrient regime favors the high-energy food chain.

Scenario 2 is modeled on the open ocean situation, where all nutrients are in low supply. This type of nutrient scenario is also characteristic of regions with stable or prolonged stratification and without land-based nutrient input. This type of nutrient regime favors the low-energy food chain.

Scenario 3 is modeled on coastal regions with eutrophication. Nitrogen and phosphorus are in high supply from land-based sources, but silicon supply is low due to rapid uptake by diatoms and slow natural replenishment by weathering processes. Nuisance algae are common and are not directly grazed. Rather, they enter the food chain as detritus being fed on by bacteria, which are in turn consumed by other microbes. This type of nutrient regime favors gelatinous predators.

Certainly, much more research needs to be undertaken to better understand this phenomenon, but the general pattern is loud and clear, underlined in bold italics, with flashing red lights, with sirens and whirling thingamajigs: as various anthropogenic disturbances alter our marine ecosystems away from diatoms, these systems are becoming increasingly favorable to flagellates, and therefore to jellyfish.

Recall Ric Brodeur, the scientist working on the Bering Sea jellyfish bloom problems discussed in chapter 2. Ric and his colleagues have been trying to better understand these alternate trophic pathways—high energy and low energy—by means of computer modeling (Brodeur et al. 2011). Their models have indicated that jellyfish have a major impact on lower food chain levels but translate relatively little energy to higher levels in the food web compared to forage fishes—they describe this pattern of energy transfer as a "large footprint" down the food chain and a "small reach" up the food chain. They also conclude that "a system dominated by jellyfish is not desirable and will actually decrease production of animals of interest to humans (fish, seabirds, marine mammals)" (p. 57).

We saw earlier that decaying jellyfish can be part of a positive feedback loop enhancing hypoxia in ecosystems, and that jellyfish goo and poo appear to promote climate change by favoring rapid carbon dioxide–producing bacteria over others. Similarly, a recent study on dead *Periphylla* in Norway found that some types of bacteria proliferated on the jellyfish tissues, while others were inhibited (Titelman et al. 2006). It therefore seems that jellyfish blooms not only promote certain phytoplankton through their predatory activity and enhance eutrophication and climate change which favor some species over others, but even in death, jellyfish chemically promote some microbes over others, thereby further regulating the species composition in ecosystems.

The millions of spotted *Phyllorhizas* in the Gulf of Mexico and Spain, the billions of moon jellies thwarting salmon farms and power plants, the zillions of sumo-wrestler *Nemopilemas* in Chinese and Japanese waters. . . .

Some jellyfish bodies float, while others sink. Most degrade rapidly, probably not even reaching the seafloor. Moon jellies disintegrate in 4–12 days, while *Periphylla* disappear in 4–7 days (Titelman et al. 2006). Either way, they act as smorgasbords for microbes and organisms that feed on them in the water column. And as some microbes flourish over others, these seemingly subtle changes at ground zero of the food web can cause huge ripples upstream.

It has been said that copepods act as a "switch" between the high-energy and microbial food chains (Stibor et al. 2004). In particular, copepods preferentially feed on large phytoplankton, whereas ciliates feed nearly exclusively on small phytoplankton. Copepods will feed on ciliates if large phytoplankton are unavailable. In this way, copepods are able to essentially convert the low-energy food chain to a useful tool for high-energy predators.

But jellyfish act in reverse: instead of converting low energy to high, they convert high energy to low. Back to the food pyramid, where each layer up represents a higher-energy food source that has eaten many times its own weight to bulk up its own nutritional value. Ounce for ounce, copepods are a higher-quality food source than the phytoplankton they eat, and forage fish are a higher quality food source than the copepods they consume. Jellyfish essentially turn the pyramid upside down by consuming many times their own body weight in high-value food to produce a lower-value food. Try to imagine the enormous number of copepods, fish eggs, fish larvae, and invertebrate larvae that a single jellyfish consumes to grow so large so fast. Recall that earlier jellyfish were described as a trophic dead end—that is, because they are of negligible food value to most higher organisms, they have few predators. Therefore, the food energy they consume is wasted to higher trophic levels—even if they do become prey, they provide considerably less food energy, ounce for ounce, than the fish eggs or zooplankton that they consumed in the growing process.

The animal and plant components of high- and low-energy food chains coexist throughout the year and in a variety of habitats. Large fish, whales, jellyfish, large and small copepods, diatoms, and flagellates can be found in most marine ecosystems throughout the world, more or less throughout the year. But in general, either the high-energy system or the low-energy one tends to dominate in space and time.

A good example of these two types of food chains was given by Greve and Parsons (1977). In temperate ocean waters, where vertical mixing and phytoplankton blooms tend to be seasonal, fish feeding and growth tend to be maximal in early spring when the diatoms bloom. Later in the summer, when the water column has stabilized and nutrients become limited and flagellates dominate, we see the large jellyfish blooms.

Another good example was given by Parsons and Lalli (2002). The highest biomasses of fish and whales are found off western continental coasts where significant upwelling occurs, such as inshore of the Humboldt, Benguela, and California currents. The phytoplankton in these systems are dominated by diatoms. Eastern continental coasts with convergent water masses, such as the Great Barrier Reef, the Caribbean, and the Florida Keys, by contrast tend to favor cnidarian ecology. The phytoplankton in these systems are dominated by flagellates.

While it is normal to have a mix of high-energy and low-energy components, and (seasonally) to have the low-energy community in dominance, we are increasingly seeing case after case where flagellate and jellyfish blooms are lasting longer, spreading larger, and becoming denser. And in many cases, as we have seen throughout this book, these jellyfish blooms are wreaking major havoc on fisheries and various other industries. How then does this occur, and what sets off the switch from one system to the other? Remember the lyrics to the old Peter, Paul, and Mary song, "The answer, my friend, is blowin' in the wind . . ." Yes, sand.

Silicate Limitation

The element silicon—better known to us in its mineralized form as silica, quartz, or glass—appears to be one of the keys to the shift from high-energy to low-energy food chains. Silicon is the second most common element in the earth's crust, comprising about 27 percent of the average rock. Silicon and oxygen link up to form a suite of common minerals called silicates. Silicate occurs naturally in seawater in its soluble form. Diatoms require dissolved silicate to make their tiny glass cases.

Diatoms require silicon in approximately a 1:1 ratio with nitrogen. As anthropogenic nutrients like nitrate and phosphate enter the oceans and seas in agricultural and urban runoff, sewage effluent, and industrial waste, they do so at a disproportionate rate to silicate, which enters the sea more slowly and seasonally through natural land erosion. Therefore, when nitrogen and phosphorus fertilizers are present to stimulate a bloom, diatoms are unable to take advantage of the lush conditions because of the relative scarcity of silicate. Flagellates don't use silicate and therefore have no such limitation, so they bloom to the point of dominating the phytoplankton community. One of the most graphic examples of this is in Tapong Bay in Taiwan (discussed in chapter 6), where the oyster rafts were removed from the highly eutrophic aquaculture bay. The local silicon:nitrogen ratio had decreased by almost half, which resulted in the proportion of diatoms being halved and the proportion of flagellates almost tripling (Lo et al. 2008).

Research strongly indicates that diatoms will outcompete flagellates when silicate is plentiful, but that flagellates will outcompete diatoms in a matter of days when silicate becomes limited. These outcomes apply across a broad range of environmental conditions (Schelske and Stoermer 1971; Officer and Ryther 1980; Egge and Aksnes 1992; Sommer 1994).

In a healthy (noneutrophic) coastal ecosystem where silicate is not limited, diatoms bloom in abundance. Copepods feed on them and quickly recycle their nutrients. The nitrogen and phosphorus are expelled as feces, which rapidly break down and become available again as nutrients in the water; the silicon glass cases that form the diatoms' "shells" are essentially spit out but dissolve back into the sea relatively slowly. Therefore, even in a normal diatom bloom, the low-silicate conditions favorable to flagellates will develop quite quickly through the limitations of the silicon regeneration cycle.

It further appears that enclosed or nearly enclosed bodies of water, such as the Baltic and the Black seas, are suffering from long-term depletion of silicate. It is thought that only a small fraction needs to be lost annually through preservation in sediments following large blooms in order to effect a major change over time (Conley, Schelske, and Stoermer 1993). This can become a self-enhancing feedback system, leading to flagellates representing a progressively higher percentage of the phytoplankton community.

While open coastal habitats, in theory, are less likely to suffer silicate depletion because of the mixing nature of the large ocean, this is not always as straightforward in practice as one might imagine. First, offshore water is notoriously low in nutrients, including silicate. Second, coastal silicate depletion can occur within days, leading to a sudden die-off of diatoms and a bloom of flagellates that can persist for months. Third, the eutrophication process itself enhances silicate depletion by reducing mixing and by stimulating mass blooms of diatoms that quickly use up the existing silicate faster than it can be recycled.

Intriguingly, while the sudden die-off of diatoms sends a massive pulse of organic matter to the seafloor, fueling the microbes and hypoxia that develop into a dead zone, the lower the concentration of dissolved silicate in the water, in theory, the better the oxygen condition of the bottom-waters (Conley, Schelske, and Stoermer 1993). Specifically, flagellates don't sink as quickly as diatoms when they die, and their bodies tend to decompose in the upper waters. Therefore, the earlier the diatoms exhaust the dissolved silicates, the less organic matter reaches the microbes, and the less anoxia occurs overall. However, this also leaves more nutrients in the water, stimulating larger flagellate blooms, decreasing water transparency and leading to a greater probability of toxic blooms and jellyfish blooms.

Some might consider it a chilling coincidence that the locations listed by Dan Conley and his colleagues in their 1993 review of silicate and eutrophication are the very same locations that are treated throughout this book as having severe jellyfish bloom problems. The Black Sea. The Baltic. The Kattegat. The Chesapeake. The Gulf of Mexico. The North Sea . Or perhaps it's no coincidence. "Environmental conditions that seem to favour jellyfish have high nutrients, but low Silicon:Nitrogen ratios, characteristic of eutrophic coastal waters" (Lo et al. 2008, 458).

Dams in rivers and agricultural diversion from rivers serve both to limit the amount of suspended sediments flowing into the oceans and to reduce the tumbling action and mechanical breakdown of sediments so that they come to rest prematurely. These factors substantially limit the volumes of dissolved silicate entering the oceans, which in turn changes the phytoplankton community composition (Harashima et al. 2006).

While harmful algal blooms, in a strict sense, are completely natural phenomena which have occurred throughout recorded history, in the past two decades the public health and economic impacts of such events appear to have increased in frequency, intensity and geographic distribution. (Hallegraeff 1993, 81)

By all accounts, dinoflagellate blooms are increasing in space and time. The low-energy food chain. Jellyfish. But dinoflagellates aren't all bad. Mike Schaadt, director of the Cabrillo Marine Aquarium in San Pedro, California, recounted the following story. During World War II, the Japanese soldiers were issued with packets of dried dinoflagellate powder. At night they would put a pinch or two in their palm, spit on it, and rub their palms together. This gave them enough light by which to read maps but couldn't be seen from a distance.

How Perturbations Change Ecosystems

While the two food chains naturally alternate in dominance under seasonally fluctuating environmental conditions, they are also driven into an unnatural relationship by overfishing and various forms of pollution, including eutrophication (Parsons and Lalli 2002). Any condition in which populations of copepods or krill increase unnaturally (e.g., release from predation by overfishing of their predators) will promote flagellate dominance through pressure on diatoms. Similarly, any condition which promotes a sequestration of nutrients in general, in particular, limitation of silicate relative to other nutrients, will also promote flagellate dominance; these conditions include tropical seas, reduction or absence of upwelling, and coastal eutrophication, to name a few.

We can regard the phytoplankton-zooplankton relationship as an oscillation. Any event or condition that favors zooplankton will exert heavy predation pressure on phytoplankton, while any event or condition that favors phytoplankton will lead to zooplankton blooms, which will eventually put pressure on the phytoplankton. So these lowest two links on the food chain are in a constant boom-and-bust struggle away from equilibrium, each side trying to get ahead of the other. But because copepods size-partition their prey, the struggle for dominance is not only between phytoplankton and zooplankton, but also between large and small.

If we keep in mind that a natural food chain is generally an oscillation around equilibrium, then overabundance of one link in the food chain must come at the cost of another. An increase in krill could result from either a relaxation of predation by whales, penguins, or fishermen, or from an excess of diatom prey.

Parsons and Lalli (2002) addressed some of the primary types of ecosystem perturbations and demonstrated that all can lead to shifts in phytoplankton community structure, that is to say, from large diatoms to small flagellates. As we have seen, these in turn lead from large copepods and krill to small copepods, which further lead from large fish and whales to expanding jellyfish populations.

Overfishing

In a simple three-step process, overfishing large quantities of forage fish initially leads to an overabundance of macrozooplankton, such as large copepods. However, heavy predation by the excess copepods soon leads to a decrease in diatoms. More nutrients are then available for the flagellate food chain, leading to jellyfish blooms. The same process can ensue from overfishing of whales or a die-off of penguins, from which krill populations initially benefit. However, an overabundance of krill soon leads to a diminishing diatom population, and hence, more nutrients available for flagellates, leading to jellyfish blooms.

Outstanding real-life examples of this process can be found in the shifts from fish to jellyfish in the Bering Sea, the Black Sea, and the Sea of Japan. In all three cases, significant overfishing occurred, followed by observed changes in phytoplankton size and community structure, resulting in out-of-control jellyfish blooms.

Pollution and Eutrophication

Pollution (e.g., by toxins, including heavy metals and petroleum hydrocarbons) and eutrophication (e.g., by agricultural runoff and sewage discharge) both tend to favor flagellates over diatoms. Low concentrations of hydrocarbons, such as may be present where oil exploration is in progress or over a very broad area following an oil spill, have been shown to enhance the growth of flagellates at the expense of diatoms (Parsons, Li, and Waters 1976; Greve and Parsons 1977).

Good examples of ecosystem shift due to eutrophication can be found in the studies of Tokyo Bay, the Gulf of Mexico, the Baltic, and the Gulf of

Maine, to name just a few. For example, two species of copepods in the highly eutrophic Tokyo Bay tell the story perfectly (Nomura and Murano 1992; Uye 1994). *Acartia omorii*, a diatom grazer, has historically predominated for most of the year. The much smaller species, *Oithona davisae*, feeds on flagellates; its abundance is 8–19 times as high in Tokyo Bay than in the Seto Inland Sea, which is less eutrophic. Furthermore, bottom layer hypoxia kills the eggs of native species, including *Acartia*, which are freely spawned and sink to the bottom, whereas *Oithona* carries its egg sacs attached to adults' bodies, so the young complete their life cycle in the oxygenated upper layers. As phytoplankton and grazing copepods switch from large to small in Tokyo Bay, the jellyfish community has also been growing. The moon jellyfish has exploded into plague proportions, increasingly interfering with fishing and industrial operations over decades. Today, little remains in the way of fisheries in Tokyo Bay.

Climate Change

Standardized oceanographic data on the concentration of chlorophyll in the ocean (as a measure of phytoplankton productivity) have been collected since 1899. While regional climate variability and coastal land runoff may induce local variation, analysis of this long-term dataset indicates unequivocally that the global concentration of phytoplankton has declined (Boyce, Lewis, and Worm 2010). Multiple lines of evidence link this decline to climatic variability, particularly increasing sea surface temperature over this timescale.

The effects of the chlorophyll decline "are particularly pronounced in tropical and subtropical oceans, where increasing stratification limits nutrient supply" (p. 595), the study concluded. These results support the general consensus that increasing ocean warming is leading to a restructuring of marine ecosystems and food webs, which in turn is likely to have substantial, if not catastrophic, effects on fishery yields, nutrient cycling, and biodiversity.

This decline in chlorophyll concentration over the past century is different from the shift in phytoplankton cell size from large to small. So while various ecosystems are experiencing a shift from high-energy to low-energy food chains, these changes are superimposed on a global decline in overall phytoplankton production.

Rising ocean temperatures create a feedback loop. Warmer surface water means greater stratification of the oceans, which reduces mixing, which is favorable to smaller diatoms and flagellates. Smaller diatoms and flagellates are less efficient at carbon sequestration. The glass cases of diatoms and the dense shells of calcifying organisms help them sink, but organisms with smaller cases and shells, as well as naked species, decay and release their carbon dioxide long before their remains reach the seafloor. Therefore, less carbon dioxide is sequestered in the deep ocean, even though photosynthesis may be higher overall. This could lead to more carbon dioxide remaining in the atmosphere, driving global warming.

A good example of climate change affecting the high-energy–low energy balance is found in the Bering Sea. Recall from chapter 2 that this regime shift brought on by sea surface warming led to coccolithophore blooms, which favored jellyfish and led to starvation of fish and seabirds. The reduction in fish and seabirds led to a decline in seals, driving the whales to shift to preying on otters. The reduction in otters lowered the predation pressure on sea urchins, which then exploded in numbers and began decimating the kelp forests, thus reducing habitat for many other species.

Global expansion of jellyfish blooms in space and time is not a good development for a growing human population. As we demand more food from the sea, the supply is simultaneously diminishing. As we demand more recreational opportunities on beaches and reefs and at island resorts, they will become less pleasant, both in terms of stings and reduced biodiversity.

> These enhancements of jellyfish populations may be indicative of major fundamental changes in marine ecosystems that are pushing the world's oceans into a less desirable state with respect to marine resources.
> —TIM PARSONS AND CAROL LALLI, "Jellyfish Population Explosions"

In Summary

Like the classic game Rock, Paper, Scissors, each of the perturbations vexing our oceans can be argued to overarch the others. But the reality is, they are severally and collectively part of an integrated system that both feeds and is fed, supplies and demands, gives and takes as part of the cycles of nature. Carbon dioxide in the air leads to warming and acidification . . . Acidification leads to flagellates and jellyfish . . . Warming leads to stratification, which leads to hypoxia . . . Hypoxia changes the water chemistry, which leads to flagellate blooms, which lead to jellyfish . . . Warming also enhances plant and animal growth, which leads to blooms, which lead to hypoxia . . . Warming also reduces oxygen in the water, making respiration more difficult, which leads to jellyfish . . . Overnutrification leads to flagellates . . . Pollution leads to flagellates . . . Flagellates lead to jellyfish. Rock, paper, scissors. . . .

It appears that, in many cases, jellyfish blooms are merely the most visible part of distressed ecosystems. In many cases, jellyfish are a symptom rather than a cause. As phytoplankton shift, so do copepods, and as smaller copepods dominate, fish decline. Perhaps in some cases jellyfish move into a vacant niche space, whereas in other cases, jellyfish are the last man left standing in an ecosystem where higher-energy animals can no longer thrive, while in still other cases, the ecosystem shifts to give jellyfish the competitive advantage. Anyway, jellyfish flourish.

As ecosystems shift away from diatoms toward flagellates, we can expect increasing frequency and duration of jellyfish blooms, along with decreasing population abundance of species on the high energy food chain.

Regions like the Bering Sea, the Gulf of Mexico, the North Sea, the Gulf of Maine, and Antarctica have traditionally provided a large fraction of the marine resources, such as finfish and shellfish, that feed the world. These too can be expected to further decline.

It probably won't happen slowly, where we can watch for signals and milestones to decide when "enough" has occurred and it's time to take action. Those signals and milestones have been all around us for decades and we have chosen to ignore them. It will probably happen more like a house of cards, where one card falls, causing others to fall, and so on. Pressure from above, unstable foundation below. We have, over the decades and through wanton disregard for consequence, turned our ecosystems into neighborhoods of card houses. So many populations of so many species are hanging on in dwindling numbers that it is hard to imagine that they will be resilient enough to endure a reshuffling of the food chain. And as one species collapses, so will those above it.

> The beauty and genius of a work of art may be reconceived, though its first material expression be destroyed; a vanished harmony may yet again inspire the composer; but when the last individual of a race of living beings breathes no more, another heaven and another earth must pass before such a one can be again.
>
> —WILLIAM BEEBE

PART IV

The Oceans Are Dying to Tell Us Something

It's a Friday night in 2050. It's been a long week at work and even if you could be bothered to cook, there's nothing in the fridge. So what fast food will you pick up on your way home? How about some squid and chips? Perhaps an algae burger? And don't forget the crunchy fried jellyfish rings on the side.

—CAROLINE WILLIAMS, "Jellyfish Sushi: Seafood's Slimy Future"

CHAPTER 14

Ocean Acidification: The "New" Problem

Aside from the dinosaur-killing asteroid impact, the world has probably never seen the likes of what's brewing in today's oceans. By spewing carbon dioxide from smokestacks and tailpipes at a gigatons-per year pace, humans are conducting a grand geophysical experiment, not just on climate but on the oceans as well.

—RICHARD KERR, "Ocean Acidification Unprecedented, Unsettling"

A colleague said to me, tongue-in-cheek, several years ago that sometimes it seemed to him like there was some grand balance at the Great Barrier Reef, a seesaw between corals and jellyfish, what he called the "conservation of cnidarian biomass." In good coral years, jellyfish were scarce, and the worst years for corals were the best for jellyfish. What a great paradox: jellyfish are essentially upside-down drifting corals without skeletons, and they *are* both in the same phylum. We both got a hearty chuckle over the yin and yang of it. But then I got to thinking. . . .

Perhaps he was right.

The two most severe coral bleaching events on record were 1998 and 2002, with 42 percent and 54 percent of the Great Barrier Reef affected, respectively (Done et al. 2003; Berkelmans et al. 2004). These were also very severe stinger seasons, with Queensland reporting 74 and 113 Irukandji hospitalizations, respectively, including 2 confirmed fatalities in 2002. The year 2006 was also among the worst mass coral bleaching events on record, as well as one of the worst stinger seasons, with 106 hospitalizations. Inexplicably, 1986 was a moderately severe stinger season, with 65 recorded hospitalizations, but seems to have been an unremarkable year for corals.

The factor that appears to determine which way the seesaw tips is climate. Since coral bleaching was first observed, there have been a total of eight mass bleaching events: 1980, 1982, 1987, 1992, 1994, 1998, 2002, and 2006. During this time, there have been six strong El Niño events: 1982, 1991–1992, 1994, 1997–1998, 2001–2002, and 2006. Routine data were not collected on Irukandji stings until 1985. However, since that time, in most years fewer than 20 people are hospitalized with Irukandji syndrome, whereas in some years there are more than 50. These severe years were 1985–1987, 1991–1992, 1994–1997, 2001–2004, and 2005–2006. You be the judge.

Regardless of any real or imagined seesaw between corals and jellyfish due to climate fluctuations, there is a more somber aspect to this relationship. Ocean acidification is a relatively new addition to today's parlance. Many people have heard of it but don't really understand it yet.

As we shall see in this chapter, ocean acidification is a special side effect of climate change with its own set of devastating consequences. Some people call it the sister problem of climate change, because the two go hand in hand. Think of it as global osteoporosis: shells and skeletons are simply vanishing. It is bigger than the other "whammies," and its teeth are already beginning to bite.

I began writing this book late on New Year's night, 1 January 2011. The following day, the atmospheric carbon dioxide was measured at 391.78 ppm. As I wrote this passage in March 2011, the carbon dioxide in the atmosphere was measured at 392.40 ppm. One year later, it was measured at 394.45. Just over 2 ppm may not sound like much for a year, but it's a lot. Because it's increasing every year. And it's not increasing at a steady rate, it's accelerating. Click, click, click . . . like a ratchet.

Let's look at atmospheric carbon dioxide over the last 50 years. Each decade, the *amount* of carbon dioxide has been increasing, just as the *rate of increase* has been accelerating:

1961–1970: increase of 8.77 ppm = rate of increase 0.88 ppm per year
1971–1980: increase of 13.00 ppm = rate of increase of 1.30 ppm per year

1981–1990: increase of 15.48 ppm = rate of increase of 1.55 ppm per year

1991–2000: increase of 15.24 ppm = rate of increase of 1.52 ppm per year

2001–2010: increase of 20.38 ppm = rate of increase of 2.04 ppm per year

A concentration of 450 ppm has been widely flagged as a "tipping point" beyond which global warming is irreversible and we are unlikely to be able to adapt. At our current rate of global emissions (>2 ppm per year), it is projected that atmospheric carbon dioxide will exceed 450 ppm within in the next 30 years (Hoegh-Guldberg 2010). Stabilizing carbon dioxide at 450 ppm requires cutting global emissions *to* less than 10 percent of today's levels by 2050. Don't get caught up in the numbers—the math is easy here: reducing to 10 percent requires that we cut by 3–4 percent per year for the next 30 years. But if the recent brouhaha in Australia about cutting *by* 5 percent by 2020 is any indication, controlling carbon dioxide emissions will be politically impossible. "To" or "by" . . . what a difference two little letters can make.

What Is Ocean Acidification?

Today's surface ocean is saturated with respect to calcium carbonate, but increasing atmospheric carbon dioxide concentrations are reducing ocean pH and carbonate ion concentrations, and thus the level of calcium carbonate saturation. Experimental evidence suggests that if these trends continue, key marine organisms—such as corals and some plankton—will have difficulty maintaining their external calcium carbonate skeletons. (Orr et al. 2005, 681)

Global warming science may be characterized by uncertainties, but this is not the case with acidification, which is a straightforward and predictable consequence of rising atmospheric carbon dioxide (Doney et al. 2009). Ocean acidification is, simply, the shift of the oceans toward acidity, that is, the lowering of their pH. What is uncertain is the effect that acidification will have on different species and how the inevitable reshuffling will affect ecosystem function.

The ocean's air-water interface is like the alveoli in our lungs, the thin surface where oxygen and carbon dioxide cross paths . . . but in reverse. Whereas we exhale carbon dioxide and breathe in oxygen, the ocean gives off oxygen and absorbs carbon dioxide.

The burning of fossil fuels adds more carbon dioxide into the atmosphere.

Much of it diffuses into the oceans, which helps buffer the effect of global warming. However, when seawater absorbs carbon dioxide, it lowers the pH, making the ocean more acidic.

Typically, marine organisms deal with the absorbed carbon dioxide in two ways, helping to buffer this acidifying effect. Organisms that photosynthesize, like phytoplankton macroalgae, convert carbon dioxide into plant carbohydrate in the same way that terrestrial plants make roots and leaves. Organisms with calcium carbonate shells and skeletons, such as pteropods and other mollusks, tropical corals, and starfish and their kin, use a derivative of carbon dioxide to make their hard parts. Some phytoplankton, called coccolithophores and foraminifera, both photosynthesize and calcify, helping scrub carbon dioxide through photosynthesis and calcification. When calcified organisms die and sink to the seafloor, this extra carbon is sequestered in their buried bodies. But the rate at which carbon dioxide emissions are increasing far exceeds the rate at which marine organisms are able to process it.

In becoming more acidic, the oceans won't turn to acid, but even a very small shift is enough to change seawater chemistry to become corrosive, leaching carbonate out of shells and skeletons. Try to imagine oysters and crabs without shells, or coral reefs with no hard parts, and you can quickly see that these organisms are unlikely to survive these processes.

Like climate that is naturally variable, the pH of seawater naturally varies with season, depth, productivity, and time of day. And just as climate change is a shift in the norm, ocean acidification is a shift in the average toward what may be considered a less desirable state. Even if all carbon emissions stopped today, it would still take thousands of years for the oceans to recover.

Can We Prevent Ocean Acidification?

The short answer is no, probably not. Scientists have not yet found a practical way to neutralize the extra acid. Dumping chalk or some other alkaline chemical to neutralize the world oceans would not be feasible as an engineering project, and probably wouldn't be effective as long as emissions continued.

Stopping or slowing global carbon dioxide emissions has proven problematical, and the drastic action required to reverse the current situation is unlikely to gain the global cooperation necessary for its success. For example, we are currently emitting about 7 billion tons of carbon dioxide per year, and the rate is increasing.

Professor Ken Caldeira of the Carnegie Institution in Stanford, California, thinks that the target needs to be zero emissions. "People laugh at this," he says, "until I point out a few simple facts." The oceans naturally absorb just 100 million tons more carbon dioxide per year than they release. Now they are soaking up an extra 2 billion tons a year, more than 20 times the natural rate, Caldeira says. "Even if we halve emissions, that will merely double the time until we kill off your favorite plant or animal" (Henderson 2006, 32).

> The concern of many scientists today is that the current episode of acidification is taking place more rapidly than anything that has transpired in the past, leaving oceanic species no time to adapt. Although the effects may be hidden from people's view, dramatic alterations in the marine environment appear to be inevitable. (Doney 2006, 65)

Dissolving Pteropods

Among the most interesting of all the sea's minibeasts are the pteropods: the sea butterflies and sea angels (see plate 14). They are related to snails, but they drift on currents instead of gliding across leaves and rocks. They spend their whole lives in the water column, never touching bottom. Their typically molluskan big foot is sculpted into little wings, which they use to flitter about like butterflies. Some pteropod species have shells; others don't. Yeah, you could call them slugs, but I wouldn't if I were you. They're mean.

A sea angel may look like the sweetest creature on the planet, or at least like something you could give a child for Christmas. Don't do it. They are only about an inch tall, some shorter. They have a little slit on top of the head, running between their pointy little "devil horns." Yeah, that's where the name starts to be misleading. That slit is its mouth. And inside that slit is some hardware that would make the producers of *Alien* swoon. Imagine a set of razor-sharp serrated hooks . . . now imagine them at high speed, shooting out, grabbing your insides, twisting really fast to shred everything, and then retreating back into its sweet little head to enjoy a snack of your guts. Welcome to the world of their prey—the sea butterflies—the pteropods with shells.

Probably as long as there have been pteropods, those without shells have preyed on the ones with shells: the angels eating the butterflies. And the violent way that the predator rips the guts out of the prey means that the latter has a pretty rough lot in life.

In recent years, however, scientists have become concerned about another threat to the sea butterflies. Their delicate shells, like many other calcified shells and corals, are in danger of disintegrating. You might think that losing a few snails—especially the mean ones—is no big deal, but sea angels and sea butterflies are a primary food source for salmon, baleen whales, seabirds, and many other species.

The plight of the unfortunate pteropods is immortalized in Sam Lardner's song "Pteropods," available on YouTube. It's worth a listen . . . and check out "Contaminated Clam" while you're there.

"Global warming is more than a matter of temperature. The villain of the piece is too much carbon dioxide, and carbon dioxide just like smoking, is damaging in many ways," says Robyn Williams, the esteemed host of *The Science Show* on ABC Radio National in Australia. The ocean is a natural sink for carbon dioxide in the atmosphere—the ocean absorbs and sequesters excess carbon dioxide. But the rate at which we now put carbon dioxide into the atmosphere via industrial smokestacks and automobile exhaust far exceeds the rate at which natural processes can neutralize it. And this excess carbon dioxide is changing the ocean's chemistry to be hostile to organisms with calcified parts, such as corals, shellfish, and plankton.

Environmental Hypercapnia

hypercapnia /hy·per·cap·nia/ (-kap´ne-ah) *n.* Excessive carbon dioxide in the blood. Also called *hypercarbia.* Etymology: Gk, *hyper* + *kapnos*, vapor.

The process of acidification is normal . . . well, sort of. In most marine animals and bacteria, carbon dioxide results naturally from the respiration process, while in most marine plants, carbon dioxide is used for photosynthesis to store energy as carbohydrate. However, in the deep sea, below the light penetration depth, photosynthetic plants cannot survive, so deeper waters tend to be more acidic than the shallower waters.

The depth in the ocean where water chemistry is exactly balanced between noncorrosive and corrosive, or the interface between undersaturation and supersaturation, is called the saturation horizon. Above this depth, organisms can extract calcium carbonate from the water to form their shells; below it, acidity leaches the calcium carbonate from their shells.

Pteropods have emerged not only as underdogs in the predator/prey dynamic of the open ocean, but also as prominent indicators for ecosystem health. Dr. Donna Roberts of the Antarctic Climate and Ecosystems Cooperative

A brief review of freshman chemistry—bear with me, no exam, I promise:

The chemistry of ocean acidification is straightforward. Excess atmospheric carbon dioxide (CO_2) diffuses into the oceans, where it reacts with water (H_2O) to form carbonic acid (H_2CO_3), the same weak acid that gives carbonated beverages their tang. The carbonic acid is unstable, so it dissociates into hydrogen ions (i.e., the "H" of pH) plus bicarbonate (HCO_3-) or carbonate ions (CO_3^{2-}). The hydrogen ions increase the ocean acidity, which, counterintuitively, is measured as a decrease in pH.

A pH unit is a measure of the acidity or alkalinity of a solution, or its relative concentration of hydrogen ions ranging from 0 to 14. A pH of 7 is neutral, and the lower the pH, the more acidic the solution. The pH scale is logarithmic, meaning that there is a tenfold change between units of pH, for example, a change from pH 8 to pH 7 means that 7 is ten times more acidic than 8.

For many millions of years, the ocean's pH was 8.2, slightly alkaline, but this varies by season and geographical location. Since the beginning of the Industrial Revolution, the ocean's pH has decreased by 0.1 unit (to 8.1), which represents about a 30 percent increase in hydrogen ions with the logarithmic scale (Royal Society 2005).

To make their calcium carbonate hard parts, most organisms with shells or skeletons combine a negatively charged carbonate ion with a positively charged calcium ion (Ca^{2+}). Organisms must absorb both of these ions from seawater. Corals, mollusks, echinoderms, and other organisms make their hard parts in this way. By volume, the main contributors to this process are the marine plankton with carbonate shells.

Excess hydrogen ions react with carbonate to form bicarbonate, decreasing the availability of carbonate for shell-making; excess hydrogen also lowers the pH, making the oceans more acidic. Normal ocean processes over hundreds of thousands of years can weather enough carbonate to neutralize dissolved carbon dioxide, but not fast enough to keep up with its current increase. The 22 million metric tons of CO_2 that the ocean now absorbs from the air *each day* is overwhelming its buffering capacity.

Experts predict that under our "business-as-usual" CO_2 emissions—that is, assuming no major actions to reduce emissions—oceanic pH is likely to decrease by 0.3–0.4 units by 2100—this would mean that the abundance of hydrogen ions in seawater will increase by 150 percent (UNEP 2010).

Kitchen Experiment #1: Water Temperature and CO_2

Part 1. Put one bottle of carbonated beverage in the refrigerator and keep another bottle at room temperature. Once the refrigerated bottle is cold, open both. The cold one gently bubbles, while the warm one bubbles so violently that it may erupt and overflow. Take a taste: the room-temperature soda seems to have more "fizz," whereas the colder one seems flat in comparison. That fizz is carbon dioxide.

Part 2. Now put the cold soda back in the fridge, uncapped (but keep the cap). Leave overnight. The next day, look at the bottle of soda—it has stopped bubbling. Remove the bottle from the fridge and replace the cap; leave at room temp for another day. Then open the bottle of warm soda. What's going on? It's bubbling again—it's gotten its fizz back. You have just demonstrated that cold water holds more carbon dioxide than warm water.

Research Centre at the University of Tasmania led a 9-year study to better understand the role of acidifying waters on pteropods (Roberts et al. 2011). The researchers placed cups to trap sinking sediments above (1,000 meters, or 3,000 feet) and below (2,000 meters, or 6,000 feet) the aragonite saturation horizon (currently at 1,200 meters, or 3,600 feet) in the sub-Antarctic waters of the Southern Ocean off Tasmania. Alarmingly, while 50 percent of the shells collected above the aragonite saturation horizon were in pristine condition, only 3 percent of the shells collected below the horizon were. Those collected in acidified waters had become opaque and were quite pitted and brittle (see plate 14).

Pteropods don't live in a world of their own. In fact, their biomass sometimes exceeds that of krill, making them a significant link in the food chain. But as they dissolve into extinction, the fate of the species that feed on them will rely on whether or not they can find suitable prey among species less vulnerable to acidification. Corrosive waters created by undersaturation of carbonate will eventually progress toward the equator. This is not just theoretical; there is evidence that the corrosive effect of acidification is already occurring.

We have been adding carbon dioxide to the atmosphere faster than the ocean can suck it up, scrub it, and sequester it. In the northeastern Pacific, the saturation horizon has shoaled upward by 50 to 100 meters since the pre-

---

Kitchen Experiment #2: Acidification Effect

If you want to see for yourself what happens to calcium carbonate in weak acid, try putting a piece of chalk or an egg (in its shell) in a glass of vinegar or cola for a few days. Go ahead—don't feel intimidated by science—it's actually pretty cool!

---

industrial era, where it is now captured by upwelling and brought to the surface. The saturation horizon in other oceans has risen too: 80–150 meters in the eastern South and North Atlantic, 100–200 meters in the Arabian Sea and Bay of Bengal, and 30–80 meters in the South Pacific (Feely et al. 2004).

Dr. Richard Feely of the National Oceanic and Atmospheric Administration's Pacific Marine Environmental Laboratory in Seattle has been studying carbon dioxide in the oceans for over 35 years. Feely and his colleagues conducted a study off the west coast of North America from Canada to Mexico in May and June 2007, where they demonstrated that along the coast near the Oregon-California border, upwelling was bringing corrosive waters to the surface, a condition not predicted to occur until around 2050 (Feely et al. 2008). The big question now is how these upwelling pulses of corrosive water will affect the marine life there.

Feely and his colleagues' modeling experiments have examined the behavior of carbon dioxide in the atmosphere and oceans over time. At current emission rates, their experiments predict that by midcentury, all Arctic waters will corrode aragonite; while by 2100 the Southern Ocean and parts of the North Pacific will be similarly corrosive, with a projected increase in the surface ocean's acidity of about 150 percent. Laboratory experiments with pteropods at these predicted pH levels are frightening, indicating that pteropod shells will dissolve in a matter of weeks.

Acidification of the oceans as a mechanism of extinction is nothing new. About 55 million years ago, the largest mass extinction of deep-sea organisms in earth's history divided the geological periods we now recognize as the Paleocene from the Eocene.

From a still-mysterious source, nearly 4,500 billion tons of carbon were released into the atmosphere, leading to catastrophic global warming. "Sea-surface temperatures rose by 5°C at the equator and by up to 9°C near the poles," wrote Jeff Hecht in the June 2005 issue of *New Scientist*, a sobering article highlighting the work of James Zachos of the University of California at Santa Cruz and his colleagues. In a geological instant of just 10,000 years, the acidity of the oceans rose to catastrophic levels, and then took more than 100,000 years to return to where they had been (Zachos et al. 2005). And through no fault of the creatures with shells, their hard parts turned to clay.

While the source of the carbon and the degree of acidification remain controversial, the apocalyptic consequences are not. This period, known as the Paleocene-Eocene Thermal Maximum (PETM), triggered the extinction of 35–50 percent of bottom-living forams and caused the worldwide disappearance of coral reefs. This quick belch of carbon led to massive changes in species composition and the restructuring of food webs. The effect of PETM acidification on calcareous nannofossils is demonstrated in plate 14.

Interestingly, the rate of atmospheric carbon dioxide increase during the PETM was 200 million tons of carbon per year, whereas the present rate of carbon dioxide emissions is some 16 times the PETM rate, or 8 billion tons of carbon per year (Gibbs et al. 2006). In fact, our carbon dioxide emissions today are altering the chemistry of seawater faster than the earth has previously experienced. Ever.

Ken Caldeira has written nearly 100 scholarly publications on carbon dioxide and acidification in the last 10 years alone. One of his recent articles compares today's carbon emission and acidification rates with the catastrophic meteorite impact that brought the Cretaceous to an abrupt halt 65 million years ago (Caldeira 2007). It no doubt began as a normal day—dinosaurs munching away on each other, flying lizards screeching, gigantic pterosaurs, four-foot clams, fifty-foot mosasaurs, and enormous turtles—just a normal day . . . and then, whammo! A 6-mile-wide comet slammed into the Yucatán Peninsula. Bad luck. The debris ejected into the atmosphere by the comet left soot and dust that blocked out the sun for months or more. Plants died. Animals that ate plants died. In fact, every land animal bigger than a cat died.

The seas didn't fare well, either. The Yucatán Peninsula is essentially calcium carbonate—a big block of chalk—and gypsum (calcium sulfate). The impact of the comet would have superheated the chalk and gypsum, releasing sulfur dioxide, a poisonous gas. And just as carbon dioxide dissolves into the ocean, so do other gases. Carbon dioxide reacts with water and oxygen to become carbonic acid, and sulfur dioxide reacts with water and oxygen to become sulfuric acid. Bad news. . . .

Surface and shallow waters became corrosive, and species with calcium carbonate shells or skeletons simply vanished. Because this process occurred in one big pulse, organisms died rapidly and the buffering capacity of the large global ocean was able to neutralize the acidic water over probably just a year or two. But for most calcified species living in the surface waters, it was too late.

Volcanoes, hydrothermal vents, and other natural sources add about a half billion tons of carbon dioxide per year to the atmosphere. Over the eons, the atmosphere stays in balance through the natural process of carbonate uptake and sequestration by marine and terrestrial organisms. It is thought that during the Cretaceous, volcanic emissions doubled, but so did the weathering rate, so the whole carbon cycle essentially sped up to twice as fast as normal.

But we have accelerated the rate of carbon dioxide being added to the atmosphere to rates way beyond the capacity of natural processes to handle it. We now emit about 25 billion tons of carbon dioxide per year—about 100 times more than natural emissions.

In Caldeira's words, "If we were to cut carbon dioxide emissions to just 2 percent of current emissions, this would still be enough to double the natural sources of carbon dioxide. Just 2 percent of current carbon dioxide emissions, if sustained over a long period of time, would be enough to produce the warmer climate experienced on our planet 100 million years ago" (Caldeira 2007, 191). *Two percent.* But honestly, we as a global community don't seem willing to cut emissions *by* 2 percent, let alone *to* 2 percent. . . .

These mass extinctions of the PETM and Cretaceous may seem extraordinary and far away from our lives today, but in fact, they offer a good comparison for future climate change. The figures are a reasonable estimate of what we can expect if we continue to burn known fossil fuel reserves over the next few hundred years. Furthermore, many of the types of organisms likely to be most affected are of socioeconomic importance to us as food sources and our recreation: mollusks like oysters, clams, scallops and mussels; crustaceans like prawns, crabs, lobsters, and krill; and corals and the diving and fishing habitats associated with them.

But How Fast—Not in My Lifetime, Right?

The conclusion that, even if we act promptly and resolutely, the world is on a path to reach 650 ppm is almost too frightening to accept. That level of greenhouse gases in the atmosphere will be associated with warming of about 4°C by the end of the century, well above the temperature associated with tipping points that would trigger further warming . . . These conclusions are alarming, to say the least, but they are not alarmist.

—CLIVE HAMILTON, *Requiem for a Species*

How fast? A reasonable question. A recent study has suggested that some polar and subpolar surface waters are likely to begin dissolving the shells of pteropods and other creatures with aragonite shells by 2050 (Orr et al. 2005). Another study using a "business-as-usual" scenario concluded that pteropod shells in the Southern Ocean should begin dissolving by 2030, and no later than 2038 (McNeil and Matear 2008).

To look at it in another way: human industrial and agricultural activities since 1750 have increased carbon dioxide in the atmosphere by about 100 ppm, to about 395 ppm now. It may not sound like much, but it is the highest that atmospheric carbon dioxide has been in more than a half million years. Our species has never known these conditions before. About 50 percent of the carbon dioxide has remained in the atmosphere, 30 percent has diffused into the oceans, and 20 percent has been taken up by terrestrial plants. Under scenarios developed by the Intergovernmental Panel on Climate Change, atmospheric carbon dioxide concentrations are expected to rise at an increasing rate, and could exceed 500 ppm by 2050 and reach 800 ppm toward 2100 (Feely et al. 2008). This increase would reduce oceanic carbonate ion concentration by about 50 percent. Most organisms simply cannot evolve quickly enough to keep up with such a rapid rate of change.

Furthermore, there is an inherent lag time of decades to centuries between the carbon dioxide we pump into the air today and the effect it will have on temperature and ocean acidity due to physical processes governing the diffusion of molecules across distances and interfaces. This means that we won't see the biological effects of today's carbon dioxide emissions until decades from now, whereas the corrosive and warming effects that we are currently witnessing are from emissions in earlier decades. Similarly, recovery takes tens to hundreds of thousands of years, as it requires the slow and steady weathering of rocks and sediments to flush more carbonate into the oceans.

The Effects of Ocean Acidification on Ecosystems

Australia has a pretty good reason for being interested in ocean acidification: models and projections suggest that its effects are most likely to be first and hardest felt in the Southern Ocean, right on Australia's doorstep. But it's not just speculation—it is already happening.

Dr. Richard Matear of CSIRO Marine and Atmospheric Research in Hobart, Tasmania, sums up 3 primary effects of ocean acidification.

- **Effect 1**: the physical aspects of seawater become more acidic and the saturation horizon shoals into shallower water.
- **Effect 2**: the process of organisms manipulating their environment becomes more costly in terms of energy required to remove carbonate from the water to build and repair shells and skeletons.
- **Effect 3**: metabolic processes of organisms change in ways that may not be favorable to the organism.

Effect 1: Seawater becoming Corrosive

Why Chemistry Matters

Organisms in the ocean use two common mineral forms of calcium carbonate to build their shells and skeletons: aragonite and calcite. Aragonite is used by most mollusks and corals, whereas the more prevalent calcite is found in the shells of plankton, such as coccolithophores and foraminiferans, calcified red algae, some sponges, brachiopods, echinoderms, and bryozoa. Aragonite is less stable than calcite, so is therefore more likely to dissolve first.

Scientists talk about the aragonite saturation state and calcite saturation state. These refer to the state of the ocean as being at equilibrium where aragonite and calcite are fully saturated, that is, no more can go into solution. At the saturation point, the carbonate concentration in water is in equilibrium with shells, the limiting factor on its carbon-scrubbing ability being the speed at which animals and plants can process carbonate into their shells. Below the saturation point, the less-saturated water acts like a magnet, drawing carbonate out of shells and skeletons molecule by molecule—first aragonite and then calcite. Think of it as being like osmosis, where, in an effort to stabilize, molecules pull across a gradient from where they are saturated toward where they are undersaturated.

As excess carbon dioxide lowers pH and raises the carbonate saturation point, carbonate dissolves out of shells, thus making them weaker and lighter. As water acidifies, its saturation point rises meaning that the ocean can hold more carbonate than before. But carbonate has a stronger affinity to bond with hydrogen than with calcium, so less carbonate remains available overall, thereby rendering the process of extracting carbonate from the water for shell building and repair ever more difficult.

Recall that colder water holds more dissolved oxygen than warmer water; it's the same with carbon dioxide. Therefore, as more carbon dioxide is diffused into the oceans from the atmosphere, polar seas absorb more carbon dioxide and become corrosive first. Absorption also rises in wintertime, when the temperatures are lowest and wind-driven upwelling and mixing occur. And sadly, pteropods are largely polar and subpolar, so their aragonite shells are the first to suffer the ill effects of ocean acidification (Orr et al. 2005). They are tiny, squishy, fluttering canaries in the coal mine. Matear explains the real-life effects of ocean acidification in an interview on ABC's *The Science Show*:

> If I took a shell and stuck it in the Southern Ocean, it would just slowly dissolve away. This is a dramatic change . . . we're not just reducing it, we're actually driving the system to such a state where organisms that form calcareous hard parts will actually experience dissolution in the upper ocean, so they won't be able to retain their shells. So I think that the consequence of that is any organism that forms aragonite that lives in the Southern Ocean will probably disappear by the end of the century. (Matear 2006)

But it's not just in faraway Antarctica and it's not something that may happen 90 years from now. Tropical pteropods are already showing signs of disintegration. Liza Roger of the University of Western Australia and her colleagues examined pteropods collected across northern Australia between 1963 and 2009 and found a decline in shell thickness and an increase in shell porosity over this time (Roger et al. 2011). The time and location of this study coincided with a decline of about 10 percent in aragonite saturation level.

Decalcifying Corals

Temperature-induced mass coral bleaching causing mortality on a wide geographic scale started when atmospheric carbon dioxide levels exceeded ~320 ppm. When carbon dioxide levels reached ~340 ppm, sporadic but highly destructive mass bleaching occurred in most reefs world-wide, often associated with El Niño events. Recovery was dependent on the vulnerability of individual reef areas and on the reef's previous history and resilience. At today's level of ~387 ppm, allowing a lag-time of 10 years for sea temperatures to respond, most reefs world-wide are committed to an irreversible decline.(Veron et al. 2009, 1428)

Corals, like pteropods, use aragonite to build and repair their hard parts. They too are among the hardest hit by the effects of ocean acidification.

Corals are very special organisms. Osha Gray Davidson called them the "enchanted braid," because corals are animal, vegetable, and mineral at the same time. The animal part is related to jellyfish, and indeed, from a structural perspective, a coral polyp is essentially an upside down jellyfish stuck to the bottom and encased in limestone. The vegetable part is algae that live symbiotically inside the coral tissues, giving corals not only their dazzling fluorescent colors but also supplying a major proportion of their daily nutritional carbon requirements. The mineral part, of course, is the corals' limestone skeleton.

But corals are more than just an enchanted braid—they also create the three-dimensional habitats that act as the world's hotspots of biodiversity. Although representing only a tiny proportion of physical area of the ocean, about a third of all marine fish species are associated with corals and coral reefs. Corals and other creatures with calcified structures normally build and sustain their calcification because seawater near the ocean surface is saturated with calcium carbonate. Essentially, all the hydrogen ions in it are partnered with carbonate. But as acidification progressively changes the hydrogen-carbonate balance, near-surface water drops below its carbonate saturation point. And because carbonate has a higher affinity to bond with free hydrogen ions than with calcium, the carbonate will begin dissolving out of shells and skeletons. As skeletal density decreases, the corals become increasingly brittle, making them less resilient to storm damage and predation.

Long before acidification was a recognized concern, many coral reefs were already in decline because of the actions of destructive fisheries catering to the aquarium trade, increasing sedimentation from coastal development, and nu-trient pollution. Stressed-out corals have fallen ill with coral bleaching and a host of diseases and fungal infections. Furthermore, cyclones, squalls, and even normal wave action continually erode and destroy corals, which need a continuous supply of calcium carbonate not only to grow but also to repair and recover from such events.

Independent teams of researchers have shown that corals slow their rate of calcification in conditions of increased acidity (i.e., lowering of pH). Projec-tions based on current carbon dioxide emissions indicate that by 2100, corals' ability to calcify is likely to decline by 25–30 percent. Whether corals actually die or simply grow more slowly, is currently a matter of debate among scien-tists. But in many regions, corals will soon be eroding faster than they can rebuild (Hoegh-Guldberg et al. 2007).

Another point of debate is about whether or not slowing growth due to acidification will be offset by stimulation of growth by rising ocean tempera-tures. Carol Turley, director of the Plymouth Marine Laboratory, argues per-suasively that this is unlikely, partly because many studies have demonstrated that even a small rise in temperature—even over a small period of time (e.g., just 1°C (1.8°F) higher than average for just a month)—can result in coral bleaching. So the synergistic effects of warming and acidification are likely to have deleterious consequences for corals in particular.

The debate today is not about whether acidification is a good or bad thing—the debate is whether acidification is a bad thing or a *really* bad thing. If cor-als grow more slowly, they will reach sexual maturity more slowly. And the young that they do have will find it harder to extract carbonate from seawater in order to begin making their skeletons in the first place. Moreover, without recruitment of new young each season, older, dying corals will not be replaced and the competition for space is hampered.

Some of these questions may already have an answer. Dr. Glenn De'ath and his colleagues from the Australian Institute of Marine Science studied 328 colonies of massive corals in 69 reefs across the Great Barrier Reef and found that calcification has declined more than 14 percent since 1990 (De'ath, Lough, and Fabricius 2009). They attributed this sudden decline in calcifica-tion to stress from rising temperatures and decreasing aragonite saturation of seawater.

Different species respond in different modes and tempos to different stim-uli. And factors like symbiotic relationships and evolution can buffer popula-tions from environmental change or accelerate species decline (Pandolfi et al.

2011). So the big outstanding question is who will live and who will die—which species will be most resilient, and which will succumb to pressures too great, and what levels of environmental variables will be the "tipping point" for different species.

Recall the cold-water corals from the deep sea that were discussed in chapter 6. An article in *New Scientist* in August 2006 raised ocean acidification as another interesting problem faced by cold-water coral reefs (Henderson 2006). Because ocean acidification is of greatest concern in colder waters, deeper waters, and waters closest to the poles, it is likely that colder reefs will suffer corrosion and destruction even earlier than their tropical cousins (Doney et al. 2009). Numerous independent experts have concluded that at current carbon dioxide emission levels, corals are likely to become increasingly rare through this century, and by 2100 there may well be nowhere left on earth with the type of water chemistry that corals need.

Osteoporotic Forams and Confused Coccos

Phytoplankton—diatoms and flagellates—play a fundamental role in ecosystem food chain dynamics. However, these are not the only types of phytoplankton that affect and are affected by environmental changes.

Two more types of phytoplankton, the foraminifera and coccolithophores, are primary contributors to photosynthesis while alive, and to chalky sediments when dead. The famed White Cliffs of Dover are the fossilized deposits of trillions of foraminifera and coccolithophores that rained down the water column through time, many millennia ago.

Foraminifera, or forams, are very strange little creatures. Neither animal nor vegetable nor mineral, they nevertheless can eat animals, cultivate plants, and build minerals. They are single-celled protists, like amoeba, but with shells, which are often chambered. They range in length from about a tenth of a millimeter to nearly 20 centimeters long. But mostly they are small. Really small.

Like their amoeba relatives, forams have thin extensions of their cytoplasm which they use to catch food and move around. Some forams farm symbiotic algae in much the same way as corals do, while others graze phytoplankton, or even hunt copepods. Because of their prevalence over time and across the globe, forams are important in the fossil record. Their abundance and distribution patterns inform us about all kinds of useful and interesting phenomena: petroleum deposits, extinction events, climate patterns, and so on.

Forams build their skeletons from calcite. As their ability to calcify is impacted, their little bodies will vanish by the billions. No more carbon dioxide sequestration in vast limestone graveyards. No more White Cliffs of Dover. This process may already have started. A recent study found that in one Southern Ocean foram species, shell weights today are 30–35 percent lighter than those from the same species in Holocene sediments (i.e., from the past few thousand years), which is consistent with today's acidification causing reduced calcification (Moy et al. 2009).

Far more perplexing are the coccolithophores, those strange little spherical phytoplankton covered in ornate limestone plates like tiny poker chips (see plate 14). Coccolithophores are key players in the ocean acidification story. Historically major producers of calcium carbonate in the world's oceans, they account today for approximately a third of total marine calcium carbonate production.

Coccolithophores, like forams, are part of the flagellate food chain. They are the survivors. They are the weeds. They date back to the mid-Mesozoic, 200 million years ago. Because they form calcium carbonate plates, it is widely believed that they will suffer badly the effects of ocean acidification (Riebesell 2004). But recent studies have demonstrated that different coccolithophore species respond to acidification with different calcification behaviors (Iglesias-Rodriguez et al. 2008). One such study examined a range of carbon dioxide concentrations, from preindustrial levels (280 ppm) up to the level predicted for the year 2100 (750 ppm) using a coccolithophore known for producing truly massive blooms in polar and subpolar regions, *Emiliania huxleyi* (see plate 14). This is the species discussed earlier, whose blooms in the Bering Sea have been photographed from space. Whereas carbon metabolism remained stable between 280 and 490 ppm, between 490 and 750 ppm, body size and calcification mass, as well as population growth rate, all increased significantly.

Furthermore, the sedimentary record since 1780 reveals an increase in calcification of about 40 percent, with the rate of increase accelerating over recent decades. It therefore seems that coccolithophores have been responding to increasing carbon dioxide levels and are likely to flourish in a high–carbon dioxide world. In particular, from the findings of the above study, it seems likely that at 750 ppm, polar and subpolar coccolithophores are likely to double their rate of calcification and photosynthesis. Likely effects on tropical and temperate coccolithophores are unknown.

Effect 2: Manipulating the Environment becomes More Costly

While some species' skeletons and shells are leaching away, a close corollary is that the process of capturing carbon dioxide from seawater and precipitating it into limestone for one's own use becomes harder and requires more energy. Think of it like trying to warm up your home on a cold winter's night with the window open, or like trying to save your change when there's a hole in your pocket.

Osteoporosis of existing shells and the inability to build new ones are essentially two sides of the same coin, both becoming worse as carbon dioxide increases. One startling example of what happens when organisms find it harder to build shells is found in the lucrative (and tasty) American oyster industry. . . .

Oysters without Shells

My attention is naturally drawn to anything to do with Tillamook, Oregon, because, as a child, Tillamook cheddar was the only cheese I would eat. So when I read about mass mortality of oyster larvae near Tillamook (Kerr 2010), I was intrigued. The Whiskey Creek Shellfish Hatchery in Tillamook usually supplies 50–85 percent of the young oysters to West Coast oyster operations each year. But their larval oysters are failing to thrive.

Just up the coast at Willapa Bay, where oysters have been farmed for over 100 years, they're also having problems (Kennedy 2009). The region provides one-sixth of the nation's oysters—but not in 2008 . . . or in 2007 . . . in fact, for 5 years running, the oyster larvae have failed to settle and begin growing.

These billions of oysters dying in Oregon and Washington share another feature aside from early mortality. The water upwelling along their coastline used to bring huge quantities of nutrients from the ocean depths, stimulating dense phytoplankton and zooplankton blooms—ample food for hungry, growing oysters . . . and for a $111 million shellfish industry. But nowadays, the upwelling water is bringing something besides food: excess hydrogen ions. It's corrosive.

Recall the work of Richard Feely, who confirmed that acidified waters were already reaching near the surface off the California–Oregon border. He and his colleagues believe that this is the most likely explanation for the oyster dearth. Deep water is naturally lower in pH because bacterial decay of dead organisms releases carbon dioxide, and light can't penetrate the deep sea, so phytoplankton can't photosynthesize to use up the carbon dioxide. Therefore, deeper water is more acidic than well-lit upper coastal waters.

In the past, the acidic deeper waters lay too deep to be captured by upwelling currents. But in recent years, the acidic horizon has expanded upward to the point where corrosive water is now being captured by the upwelling (Welch 2009).

The earliest layers of a baby oyster's shell are composed of aragonite, whereas later layers are made of calcite. The period of upwelling corresponds with the period when the oyster larvae are trying to build their first shells. And it is no longer working.

Effect 3: Metabolic Processes Change

There is the obvious issue of shells and skeletons losing carbonate. But there is also the subtler issue of the water being "not quite right" for reproduction and survival of noncalcified species of animals and plants. Many organisms are highly sensitive to subtle chemical changes, particularly larval and juvenile stages. Vital functions, such as gonad maturation, sperm motility, speed of embryonic development, larval settlement, and growth, as well as mate choice in fishes and photosynthesis in plants, are also affected by changes in pH. Many of these processes are governed by enzymes and other cellular mechanisms that may perform differently in environments with even subtle changes in pH. The fact is that for most species, we don't really know what will happen.

In most marine invertebrates, larval mortality rates exceed 90 percent and can exceed 30 percent the first day of life, particularly in bottom-dwelling marine invertebrates like bivalves, snails, barnacles, ascidians, bryozoans, and echinoderms (Gosselin and Qian 1997). Therefore, even an apparently small extra threat can make a big difference to an already struggling population.

In general, larvae are more sensitive to environmental conditions than adults. As demonstrated by the few experiments thus far on larval response to conditions with elevated carbon dioxide, reproduction and survival of many invertebrates are likely to decline with changes in seawater pH. Recent experiments and observations on the effects of acidification include the following.

- **Deformed Krill.** Krill hatched in different levels of carbon dioxide under experimental conditions showed ill effects from carbon dioxide. In the worst-case scenario for carbon dioxide levels predicted for the end of this century, the krill were deformed and listless; with additional carbon dioxide, the eggs failed to hatch (Darby 2008).

- **Acidosis in mammals and fish.** A more acidic environment may increase carbonic acid in body fluids, leading to lowered resistance, metabolic and behavioral depression, physical activity and reproduction, and asphyxiation (Hood 2004).
- **Sensory ability.** Olfactory cues for homing and predator avoidance in larval clownfish and damselfish are increasingly disrupted by increasing acidification (Munday et al. 2009; Munday et al. 2010).
- **Fertilization.** Sperm swimming speed and motility were significantly reduced even by slight carbon dioxide–induced acidification, reducing fertilization success by 25 percent (Havenhand et al. 2008). A similar effect was found in another study on a coral and a sea cucumber (Morita et al. 2010), and is likely for a wide range of free-spawning marine invertebrates.
- **Development.** Scallops slow their development rate with acidification, as do sea urchins, apparently because of decreased protein synthesis. In copepods, both hatching and larval survival decrease with increasing carbon dioxide, although it appears that this may only apply at much higher carbon dioxide levels.
- **Settlement.** Coral larvae don't seem to be affected by high carbon dioxide, but once they settle and become polyps, their skeletons become deformed. So too, settlement of bivalves and prawns appears to be negatively affected by increased carbon dioxide.
- **Reproduction.** In sea urchins and prawns, gonad growth and production were delayed or reduced in conditions of elevated carbon dioxide, possibly as a result of physiological stress.

Furthermore, even extremely minor changes in pH can force organisms to spend more energy on restoring and maintaining their internal pH balance, diverting energy away from important processes like growth and reproduction (Hardt and Safina 2010).

Some species will fail to mature, while others will mature but the sperm won't find eggs. Still others will fertilize, but their larvae will perish. It is highly likely that community structures will be reshuffled, as occurred after past mass extinction events. Some species adapt while others go extinct. Species least affected by change, or that are able to quickly adapt to it, will inherit the niches vacated by those that cannot.

Our scientific understanding of plunging pH is still in its infancy, and its catastrophic effects are only just beginning to be appreciated.

We can look to an ecosystem off the island of Ischia near Naples, Italy, to get a feel for what our ecosystems may look like. Here, undersea volcanoes release carbon dioxide, causing substantial local acidification (Hall-Spencer et al. 2008). "Although surrounded by a diverse rocky shore community with abundant calcareous organisms, the carbon dioxide venting site is impoverished in sea urchins and coralline algae, and is bare of stony corals. The shells of snails found in this area are weakened, and snail juveniles are completely absent. Are these changes a foretaste of the fate of the oceans in general?" (Riebesell 2008, 46).

Moreover, because the larval stages are the ones most vulnerable to acidification, as well as the stages most likely to become food for predator species, a decrease in overall planktonic food supply will affect the food chain higher up, as food becomes scarcer and harder to find. Those species with more versatile feeding habits and able to endure longer periods of starvation will be more likely to survive. You guessed it: jellyfish.

It may be that organisms routinely experience low pH conditions in their habitats for brief intervals, so may be better adapted to periodic acidic conditions than we realize (Winans and Purcell 2010). Fairly large fluctuations in pH occur over the course of daily cycles due to plants photosynthesizing during the day and absorbing carbon dioxide, but respiring at night, so giving off carbon dioxide.

If so, it may be that some organisms have a built-in buffer, allowing them not only to tolerate fluctuating conditions but perhaps even to rapidly adapt, a process that has been dubbed "contemporary evolution" (Stockwell, Hendry, and Kinnison 2003). We generally think of evolution in, well, evolutionary time scales. But many recent studies have reported "rapid evolution" in populations facing environmental changes, such as habitat loss and degradation, overharvesting, and introduced species. In response to strong pressures, evolution in heritable traits has been documented in some organisms even in a matter of decades.

In general, species with longer generation times will evolve more slowly than species with shorter generation times. In some ways, this is intuitively obvious, because change in heritable traits takes place over generations in populations and species, not within individuals. But the implications of this are somewhat disturbing. Slow-growing species like mammals and large fish, as well as invertebrates that need to calcify their shells and skeletons probably won't fare as well as species with very rapid growth rates and short generation times . . . yep, jellyfish.

Using future climate scenarios and projections of species' distributions, researchers have concluded that at midrange warming scenarios for 2050, 15–37 percent of species in their sample regions and taxa will be "committed to extinction" (Thomas et al. 2004, 145). But these studies were only of terrestrial vertebrates, invertebrates, and plants. Given the added complications of ocean acidification, dead zones, and lower dissolved oxygen retention in warmer water, it seems plausible that marine extinction rates may be similar or worse.

Possible Good News

There is, however, some speculation that perhaps not all is doom and gloom. There may be a positive side to adding carbon dioxide into the seas (Henderson 2006). Research suggests that under more acidic conditions, microbes may produce organic compounds which act as seeds for cloud formation, thereby helping to cool the atmosphere, thereby potentially slowing global warming. Furthermore, since carbon dioxide is used by phytoplankton in photosynthesis, more may act as a "fertilizer" to boost their activity, potentially mopping up excess carbon and thus helping to slow global warming.

The Effect of Ocean Acidification on Jellyfish

As we have seen with overfishing, climate change, eutrophication, pollution, and plague species, jellyfish can take the punches and keep on standing or, more often than not, duck, weave, and miss the punches altogether. But what about acidification?

Most studies on acidification effects naturally have focused on organisms with obvious hard parts, such as mollusks, corals, echinoderms, forams and coccolithophores. Currently only a small handful of studies exist on acidification and jellyfish. One of these, by Martin Atrill of the University of Plymouth and his colleagues, concluded from their 43-year dataset that declining pH was linked to increasing abundance of gelatinous organisms (Attrill, Wright, and Edwards 2007). The authors were unable to determine from their study exactly *why* the decrease in pH and increase in jellyfish were correlated, but the correlation itself allowed them to model future patterns under different scenarios. Not surprisingly, they concluded, "The evidence available from our study, however, suggests that projected climate change, including reducing ocean pH, may increase the frequency, and thus influence, of gelatinous predators in the North Sea over the next 100 yr [*sic*]" (p. 484). However, a second study examined rising acidity and jellyfish population patterns in the North Sea and North Atlantic and found no significant correlation (Richardson and Gibbons 2008). So it will be interesting to see how other research pans out.

Jellyfish Sensory Organs

Some have wondered whether jellyfish might be negatively affected by ocean acidification, because they too have hard parts that may disintegrate (Richardson et al. 2009). Statoliths are the microscopic "ear bones" in the balance organs of jellyfish, allowing them to sense gravity. Statoliths are the only hard part in an otherwise very soft body. Statoliths are *really* small, like grains of fine sand. Scyphozoans (moon jellies, sea nettles, lion's manes, and blubbers) typically have 8 sensory ganglia (called "rhopalia"), and each one has a dozen or more microscopic statoliths. Cubozoans (box jellies and Irukandjis) have 4 sense organs, and each has a somewhat larger (thought still quite small) statolith.

But alas, the quick answer is no. Jellyfish statoliths are made of a different chemical compound than are the hard parts of most other marine organisms. Instead of calcium carbonate, jellyfish statoliths are composed of calcium sulfate hemihydrate (Becker et al. 2005; Tiemann et al. 2006), a form of gypsum, better known as plaster of Paris.

In comparison to calcium carbonate, which is quite soluble in acidic solution, calcium sulfate is only slightly soluble. Furthermore, calcium sulfate is not carbon based, so should not be affected by the calcium carbonate saturation states, that is, the level at which the seawater becomes corrosive. Finally, sulfate is virtually unlimited in the ocean, so statoliths should not be inhibited from forming. In light of the chemical construction of jellyfish statoliths and seawater chemistry, it would therefore appear that jellyfish may have functional statoliths long after their calcium carbonate-dependent prey and competitors have fallen victim to ocean acidification.

Curiously, in contrast to most other salts, the solubility of calcium sulfate decreases with increasing temperature. However, in seawater, its solubility is largely independent of temperature (Shaffer 1967). Therefore, one could surmise that in the unlikely event that warming waters have any effect on statoliths, it would be to make them more stable.

A recent experiment by Amanda Winans and Jenny Purcell of Shannon Point Marine Center in Anacortes, Washington, examined the effects of ocean warming and acidification on jellyfish (Winans and Purcell 2010). They set up different combinations of water temperatures and pH levels, mimicking a range of conditions predicted for the next 100–300 years, then monitored the development of moon jellyfish polyps for 17 weeks. All the polyps survived and reproduced, indicating that low pH is tolerated by these jellyfish.

However, while the polyps thrive in such conditions, life may be more difficult for their offspring, that is, juvenile jellyfish. When raised at low pH, their statoliths were significantly smaller than normal, suggesting that while overall reproduction and growth might be unaffected, jellyfish raised in conditions of acidification may suffer from impaired balance.

> Ocean acidification is yet another red flag being raised, carrying planetary health warnings about the uncontrolled growth in greenhouse gas emissions. It is a new and emerging piece in the scientific jigsaw puzzle, but one that is triggering rising concern . . . The phenomenon comes against a backdrop of already stressed seas and oceans as a result of over-fishing to other forms of environmental degradation. Thus the public might quite rightly ask how many red flags do governments need to see before the message to act gets through.
> —ACHIM STEINER, UN undersecretary-general and
> UNEP executive director, December 2, 2010

CHAPTER 15

The Rise of Slime

Some scientists, such as Jeremy Jackson at Scripps Institution of Oceanography, have suggested the oceans are reverting back to primeval seas of millions of years ago, when algae, bacteria and jellyfish ruled the oceans. He playfully dubs this the "rise of slime."

—*Los Angeles Times*, 8 March 2011

We've seen how jellyfish blooms can cause immense problems for industrial and recreational activities and for ecosystems. Many species have clogged fishermen's nets and the intake pipes of nuclear power stations and desalination plants. So too, many jellyfish species have caused mass mortalities and monetary losses in salmon farming operations, and some species even transmit bacterial disease. Many others are causing stinging problems at tropical and temperate beaches around the world, from Sydney to Spain, from Thailand to Trinidad, from the Florida Keys to Waikiki. *Mnemiopsis* has invaded the seas of Eurasia, leaving collapsed fisheries in its wake and apparently building its populations for the next opportunity to bloom out of control.

We have also seen how the weedy lifestyle of many jellyfish species enables them to exploit disturbed ecosystems. They can eat just about anything, from fish eggs and larvae to copepods, from microbes to mud, and even each other—and if times get really tough they can degrow for a while and regrow later. They grow fast, breed early, rapidly regenerate lost parts, and can clone in 13 different ways. They can easily tolerate hypoxia by using oxygen stored in their own jelly tissues. *Mnemiopsis*, at least, is a self-fertilizing simultaneous hermaphrodite. And *Turritopsis*, at least, is immortal.

The vast majority of jellyfish treated throughout these pages are the large and colorful ones, the ones whose blooms you couldn't miss unless you completely closed your eyes. Thick red bands of sea tomatoes. Cerulean armadas of man-o'-war and blue bottles. Ghostly plagues of moon jellies or *Mnemiopsis*. Spotted *Phyllorhiza*. Stripey sea nettles . . . but what about the little ones, the ones that aren't so obvious? Over 90 percent of the jellyfish worldwide are smaller than a dime, transparent, colorless, and essentially off the radar . . . which is why a recent paper is so startling. Small jellyfish (less than 2 inches in diameter) were monitored in Jiaozhou Bay, China, from 1991 to 2009 (Sun, Li, and Sun 2012). The authors found an abrupt shift in dominant species after 2000, along with an almost fivefold increase in biomass of small jellies after this point. The authors blamed the changes on eutrophication, aquaculture, and coastal construction.

This Chinese study is of fundamental importance for two reasons. First, because this is another stream of independent evidence demonstrating that 2000 was some kind of tipping point for numerous different species of jellyfish in quite different habitats around the world: the giant Nomura's jelly in Japan, the huge burgundy sea nettle off southern California, sea tomatoes off Western Australia, and so on. The second alarming thing about the Chinese study is that it's not just the big species that are the problem . . . it's even the ones we can't see. It is one thing to look down from a spotter plane and see a jellyfish bloom stretching for hundreds of miles; it is quite another to see nothing and have no idea of the true extent of the problem.

They're dangerous. They're beautiful. They decimate fisheries and alter ecosystems. They provide medical science with important diagnostic tools. Love them or hate them, jellyfish are in their renaissance as the world's oceans head toward a more gelatinous future. There is overwhelming evidence that (A) jellyfish are behaving badly, and (B) they are probably going to get worse as long as we continue to give them ideal conditions to do so.

Jellyfish superabundances are a symptom that something is seriously wrong with the oceans. They thrive in conditions where other species cannot, and they are now thriving as perhaps never before. Even before the Cambrian explosion, a time in deep history when diversity of living things was on an exponential growth curve, jellyfish weren't as well off as they are today. In many of today's oceans and seas, the phytoplankton bases of food chains are changing to favor jellyfish over fish, and jellyfish are out-competing fish for control of shared food resources. But they are also outcompeting the human race, because we depend on the oceans' fish for our own food.

Loss of Ecosystem Goods and Services

Beyond food, we depend on healthy oceans for a whole variety of vital life support systems. The ocean is, in many ways, the "heart and lungs" of our planet (Hoegh-Guldberg 2010). The sea carries and stores nutrients, filters impurities from the air, tempers local weather patterns, buffers the planetary effects of climate change, and provides half of the oxygen we breathe. As if that's not enough, hundreds of millions of people rely on the ocean for their protein, and millions rely on it for their livelihood.

Marine ecosystems provide many important goods and services to humankind (Worm et al. 2006), including food supply and employment for many millions of people; nursery habitats for commercially important species and the species that support them; filtering and detoxification of our wastewater through filter-feeders and wetlands; buffering from flooding events in coastal regions; also pharmaceutical and genetic resources that we have only just begun to tap as well as carbon sequestration and oxygen production. As marine systems degrade through eutrophication and other types of pollution, a cascade of events is set into motion, whereby the death of large beds of filter-feeders reduces filtration → which leads to poor water quality → which leads to algal blooms and jellyfish blooms → which leads to reduced biodiversity. Similarly, invasive species can exploit disturbed ecosystems → leading to increased predation on, or competition with, native species → which leads to reduced biodiversity.

Four categories of ecosystem services have been identified that contribute to human well-being, each underpinned by biodiversity (TEEB 2010):

- **Provisioning services**—for example, wild foods and pharmaceuticals;
- **Regulating services**—for example, climate regulation through carbon sequestration and storage, and filtration of pollutants by wetlands;
- **Cultural services**—for example, recreation, spiritual and aesthetic values, education, and tourism;

- **Supporting services**—for example, photosynthesis and oxygen production, and nutrient cycling.

As anthropogenic pressures degrade the ocean's productivity, these goods and services are vanishing. Species composition at the base of the food web is being altered by warming trends, increasing carbon dioxide, and excess nutrients. Conversely, the top of the food web is being truncated by overfishing. Changes in phytoplankton composition affect absorption and sequestering of carbon dioxide, as well as oxygen production. Phytoplankton changes are reverberating up the food chain, affecting fish populations.

Biodiversity appears to be the key factor in the ability of ecosystems to buffer against disturbance and to recover afterward (Worm et al. 2006). Diverse food sources are vital for meeting the requirements of marine organisms' life cycle stages both as larvae and as adults, including growth and reproduction. Therefore, species-rich ecosystems are more resistant to loss of species and the services they provide to us. However, in today's rapidly changing marine environments, with dead zones, habitat destruction, stock depletion, regional extinctions, homogenizing harbors, and trophic cascades, many locally adapted populations and species are simply unable to continue to play their normal roles in the communities to which they belong.

The problem is in the multitude of simultaneous disturbances, each easily capable of changing the oceans as we know them, but in concert, threatening the planet with unfathomable consequence. The idea that we are in a real marine crisis is not a minority view. Let's see what some of the big cheeses say. . . .

- Professor Ferdinando Boero (2012) of the Università del Salento in Lecce, Italy: "Gelatinous plankton blooms occurred also in the past, of course, but they lasted for one or two seasons and then disappeared, and they were mostly local. Now, apparently, the jellies are here to stay."
- Professor Jim Carlton (2001, ii), director of the maritime studies program of Williams College and Mystic Seaport: "Introduced species are a growing and imminent threat to living marine resources in the United States. Hundreds of species arrive in US waters from overseas each day, playing a game of ecological roulette with ecosystem and economic stability . . . These introductions lead to vast alterations in species interactions and

changes in nutrient cycling and energy flow, which results in cascading and unpredictable effects throughout entire communities."

- Professor Paul Dayton, of the Scripps Institution of Oceanography, and colleagues (1995, 224): "Fishing exerts a profound effect on almost all components of associated communities and ecosystems . . . It appears as though most continental shelf and coastal habitats are already heavily disturbed by fishing impacts of many types . . . In almost all cases the situation is so desperate that we cannot afford to wait for more research but must begin strong risk aversion management now."
- Professor Robert Diaz (2001, 275) of the Virginia Institute of Marine Science: "No other environmental variable of such ecological importance to estuarine and coastal marine ecosystems around the world has changed so drastically, in such a short period of time, as dissolved oxygen . . . It appears that many ecosystems that are now severely stressed by hypoxia may be near or at a threshold of change or collapse (loss of fisheries, loss of biodiversity, alteration of food webs)."
- Dr. Scott Doney (2010, 1512), chief scientist at the National Oceanic and Atmospheric Administration and geochemist at the Woods Hole Oceanographic Institution: "Most of these perturbations, tied either directly or indirectly to human fossil fuel combustion, fertilizer use, and industrial activity, are projected to grow in coming decades, resulting in increasing negative impacts on ocean biota and marine resources."
- Professor Paul Ehrlich, population biologist at Stanford University and author of *The Population Bomb*: "Of course a new emerging disease or toxic problem could alone [also] trigger a collapse. My pessimism is deeply tied to the human failure to do anything about these problems, or even recognise or talk about them" (Jowit 2011).
- Professor James Estes of University of California at Santa Cruz and colleagues (2011, 301): "Our planet is presently in the early to middle stages of a sixth mass extinction . . . characterized by the loss of larger-bodied animals in general and of apex consumers in particular . . . but also highlights the unanticipated impacts of trophic cascades on processes as diverse as the dynamics of disease, wildfire, carbon sequestration, invasive species, and biogeochemical cycles."
- Professor Richard Feely, senior scientist at the National Oceanic and Atmospheric Administration Pacific Marine Environmental Laboratory in Seattle, Washington: "Ocean acidification could ultimately threaten a reorganization of the entire marine food chain, which could lead to major

changes in the distributions of marine species . . . Many valuable marine organisms such as crabs, lobsters, and shrimp are expected to suffer. Many coral reefs, which bring in millions of dollars in revenue for the tourism industry, are expected to be lost. The loss of planktonic species—an important food source for many commercial and recreational fish species—could have significant economic impacts on the multi-billion dollar US seafood industry" (quoted in Kennedy 2009, 3).

- Professor Tim Flannery (2005, 112), environmentalist author, paleontologist, Australian of the Year 2007, and the Australian climate commissioner: "So what is the prognosis for the world's coral reefs? . . . The damage already sustained is a strong indication that reefs are sensitive to the perturbations climate change brings, leading me (and many other scientists) to believe that the future for reefs under the emerging new climate is bleak . . . Visitors travelling to Queensland by 2050 may see the Great Stumpy Reef."

- Professor Ove Hoegh-Guldberg (2010, 2), director of the Global Change Institute at the University of Queensland: "Planetary emergency . . . It may seem far-fetched to some that human activities are causing such large-scale and fundamental changes to the world's oceans. However, the rapidly accumulating evidence and understanding of these changes is largely irrefutable."

- Professor Jeremy Jackson (2010, 3765), a marine ecologist and paleontologist at the Scripps Institution of Oceanography: "Today, overfishing, pollution and increases in greenhouse gases are causing comparably great changes to ocean environments and ecosystems. Some of these changes are potentially reversible on very short time scales, but warming and ocean acidification will intensify before they decline even with immediate reduction in emissions. There is an urgent need for immediate and decisive conservation action. Otherwise, another great mass extinction affecting all ocean ecosystems and comparable to the upheavals of the geological past appears inevitable."

- Professor Jane Lubchenco, administrator of the National Oceanic and Atmospheric Administration, and colleagues (2003, S3): "A broad spectrum of land and oceanbased activities, coupled with continued growth of the human population and migration to coastal areas, is driving unanticipated, unprecedented, and complex changes in the chemistry, physical structure, biology and ecological functioning of oceans worldwide. Symptoms of complex and fundamental alterations to marine ecosystems abound, including increases in: coral bleaching, zones of hypoxic or anoxic water, abrupt changes in species composition, habitat degradation, invasive species, harmful algal blooms, marine epidemics, mass mortalities, and fisheries collapses."

- Professor Daniel Pauly, director of the University of British Columbia Fisheries Centre and the Sea Around Us Project, and colleagues (2005, 5): "With the development of industrial fishing, and the resulting invasion of the refuges previously provided by distance and depth, our interactions with fisheries resources have come to resemble the wars of extermination that newly arrived hunters conducted 40 000–50 000 years ago in Australia, and 12 000–13 000 years ago against large terrestrial mammals in North America."

- Dr. Jenny Purcell (2012, 28) of Western Washington University: "Human population and demands from coastal areas will continue to increase. In the conflict of interests between human economics and preservation of the environment, economics usually prevails. Therefore, all of the anthropogenic forcing factors discussed here probably will escalate. Despite efforts to restore the environment, earlier conditions have not returned. We need to understand how present ecosystems, including jellyfish, will change under continued anthropogenic forcing."

- Professor Nancy Rabalais, executive director of the Louisiana Universities Marine Consortium, and colleagues (2009, 1535): "The overall forecast is for more eutrophication and for hypoxia to worsen, with increased occurrence, frequency, intensity, and duration."

- Professor Callum Roberts (2007, xv), conservation biologist at the University of York, England, and author of *The Unnatural History of the Sea*: "Those charged with looking after the oceans set themselves unambitious management targets that simply attempt to arrest declines, rather than rebuild to the richer and more productive states that existed in the past. If we are to break out of this spiral of diminishing returns and diminished expectations of the sea, then it is vital that we gain a clearer picture of how things have changed and what has been lost."

- Professor Boris Worm, marine ecologist and conservation biologist at Dalhousie University in Nova Scotia, and colleagues (2006, 787, 790): "We conclude that marine biodiversity loss is increasingly impairing the ocean's capacity to provide food, maintain water quality, and recover from perturbations. . . . This trend is of serious concern because it projects the global collapse of all [species] currently fished by . . . 2048."

Outlook for the Future

When you choose the behavior, you choose the consequences.
—Dr. Phil McGraw

We, the gung-ho *Homo industrialis*, have chosen as a species to write off whales, coral reefs, penguins, sea turtles, tuna, and cod, to name but a few, so that we do not have to suffer the inconvenience of industrial and gastronomic restraint. If we had known all this ahead of time, would we have made the same decisions? Yes, probably, because in truth, we have known the facts for quite a while and yet we have turned a blind eye, made excuses, found exceptions, questioned the obvious reality all around us, attacked the messengers, and continued to sacrifice nature instead of our own lusts and greed. Living in a state of delusional optimism, we have ratcheted our ecosystems—click by click—toward a state of freefall.

A typical illustration is the sad saga of the Yangtze River dolphin, which went extinct in 2006 while all the environmental groups were busy haggling over which had the best idea of how to save it. We are missing the big picture by bickering about little details like precise figures of fisheries decline and sustainability, precise wording of global emissions treaties and carbon reduction schemes, and precise longitudes and latitudes of polar ice melt. There is a time to haggle and a time to question, but there is also a time to act. And I believe the evidence herein demonstrates that we have well and truly missed that window of opportunity. It came and went while our leaders were debating over which direction to point the finger of blame.

But what about now, who should we blame? The fishermen . . . the farmers . . . the government . . . the big industrial companies with big spewing factories . . . people who use plastic . . . people who drive big cars . . . people who drive little cars . . . ? The reality is that we are all to blame. Like some ultimate *Twilight Zone* version of Garrett Hardin's *The Tragedy of the Commons*, we all want our fair share of fish and seafood, of cheap electricity, of industry, of progress. We see these as our birthright . . . but alas, so are their consequences.

Our outlook is worrying. The severity of marine perturbations is expected to increase in the foreseeable future. The six major threats to our seas, summarized by the global expert organizations, make this all too clear:

- **Overfishing.** According to the Food and Agriculture Organization of the United Nations (FAO 2005), for the 441 wild fish stocks globally for which population status information is available, 77 percent are fully fished *or worse*. Of these, 52 percent are fully exploited, 17 percent are overexploited, 7 percent are depleted, and 1 percent are recovering from depletion; only 23 percent of the oceans' wild fish stocks are less than fully exploited. No information is currently available on the status of the other 143 stocks.

- **Climate change.** According to the Intergovernmental Panel on Climate Change (IPCC 2007, 45, 46), "For the next two decades a warming of about 0.2°C per decade is projected for a range of SRES [reference] emissions scenarios. Even if the concentrations of all GHGs [greenhouse gases] and aerosols had been kept constant at year 2000 levels, a further warming of about 0.1°C per decade would be expected . . . Anthropogenic warming and sea level rise would continue for centuries due to the time scales associated with climate processes and feedbacks, even if GHG concentrations were to be stabilised."

- **Eutrophication.** According to the United Nations Environment Programme (UNEP 2006, 33), "The potential seriousness of this problem was not foreseen only a few decades ago, when it was first emerging. Over the past few years, the magnitude and intractability of the problem has become apparent. Increased demands for food for an expanding global population, intensified agriculture and an estimated 2.4–2.7-fold increase by 2050 in nitrogen and phosphorus-driven eutrophication of terrestrial, freshwater and near-shore marine ecosystems are all elements of a worrying picture of the future."

- **Other pollution.** According to the Ocean Conservancy and the United Nations Environment Programme, "The hardest truth about the state of our marine environment is that we've trashed our ocean, the source of much of the food, water, and oxygen we need to survive. Marine debris is now considered one of the most pervasive pollution problems plaguing our ocean and waterways, and our growing population is generating more of it than ever before" (Ocean Conservancy 2010, 16). "The degradation time for plastic in the marine environment is, for the most part, unknown. Estimates are in the region of hundreds of years" (UNEP 2011, 26).

- **Plague species.** According to the Global Invasive Species Programme (GISP 2010, 5), "Climate change will have direct and second order impacts that facilitate the introduction, establishment and/or spread of invasive species. Invasive species can increase the vulnerability of ecosystems to other climate-related stressors and also reduce their potential to sequester greenhouse gasses."

- **Ocean acidification** According to the Intergovernmental Panel on Climate Change (IPCC 2007, 52), "The uptake of anthropogenic carbon since 1750 has led to the ocean becoming more acidic with an average decrease in pH of 0.1 units. Increasing atmospheric CO_2 concentrations lead to further acidification. Projections based on SRES [reference] scenarios give a reduction in average global surface ocean pH of between 0.14 and 0.35 units over the 21st century . . . The progressive acidification of oceans is expected to have negative impacts on marine shell-forming organisms (e.g., corals) and their dependent species."

It is now clear that our marine ecosystems are in freefall due to multiple stressors, and there is no easy fix. Hell, there isn't even a hard fix. The startling truth is we screwed up. It's not a matter of cutting back a bit on fishing . . . or reducing a bit on carbon emissions . . . or keeping an eye out for introduced species. Even if we could somehow convince the whole world to do all of these things . . . *it's too late.*

Our oceans are the equivalent of a patient with cancer that has metastasized to six different organs. There are some things that we, as wise as we are, still can't make right.

Amid ever-increasing demand for fish, and vanishing supply, our commercial wild fisheries are becoming bankrupt. We have decapitated the food chain by overfishing, leaving small species, short-lived species, and species with less nutritional bang for the buck. We have taken the good ones and left behind those we didn't want as the breeders for the next generation. We have turned coastal shelf waters anoxic and compressed the liveable space through shoaling of the oxygen minimum layer of the deep sea. And we continue to add more and more carbon dioxide and fertilizers and industrial waste every single day.

We have made seawater corrosive so that the small species, the short-lived species, the species lower on the food chain, such as snails and corals, are losing their shells and skeletons to leaching. Microplastics and oil spills and pesticides are clogging the stomachs and the feathers and the hormones of innumerable marine species, from birds to clams to seaweeds.

So then, what is left? After the big fish and the marine mammals have vanished, after the clams and the worms have suffocated in the bottom hypoxia and the snails and the corals and the calcified plankton have disintegrated, after the birds and the mussels and the sea cucumbers have choked on plastic bullets, and the macroalgae have succumbed to the shading of the dinoflagellates, what is left?

Jellyfish.

The state of the oceans may now seem so bleak that "where to from here?" seems too overwhelming to think about, or perhaps an impotent exercise. Maybe we simply cannot fathom what this all really means, so we relegate it to the "too-hard basket" to try to face another day.

Professor David Jablonski, a paleontologist at the University of Chicago, says it best: "A lot of things are going to happen that will make this a crummier place to live—a more stressful place to live, a more difficult place to live, a less resilient place to live, before the human species is at any risk at all" (Quammen 1998).

So we are confronted now with a choice. When we look at the challenges we are facing, shall we throw up our arms and party like there's no tomorrow or try to find ways to help the system be as resilient as possible by reducing multiple pressures as much as we can? I don't know: you tell me. Do we hasten our journey to Jablonski's "crummy place," or do we decide that there is still value in our efforts at stewardship of the environment . . . and in making it less crummy?

Communities in many regions around the world are already adapting to a more gelatinous way of life, for example:

- Protective clothing is increasingly being worn in jellyfish-infested areas around the world, following the impressive success in sting reduction demonstrated in the Whitsunday region of the Great Barrier Reef.
- Community-action programs have begun in many Mediterranean regions, such as Spain, Italy, and France, with hotlines and jellyfish alerts. The French even recently launched a 48-hour Internet forecast for jellyfish blooms at beaches in the Alpes-Maritimes region (Samuel 2012).
- Swimmers in the Chesapeake can now check a smartphone app for the local jellyfish forecast (Basch 2012).
- Chefs in Japan and Europe have become creative with gourmet uses for jellyfish, including marketing jellies as the "ultimate diet food," with only 36 calories per 3-ounce portion (CalorieKing 2011).
- South Korea has released hundreds of thousands of juvenile filefish along

the coasts of Busan and Gwangalli Beach, in an effort to combat stings in these regions (Park 2009). When not being dried and made into a sweet-salty jerky as a popular snack food with beer, filefish are natural predators on jellyfish.

- In Spain and the Canary Islands, loggerhead turtles are being reintroduced in an attempt to control the jellyfish now plaguing these regions ("Marine Turtles in Fuerteventura" 2011; Torné 2009).

Full Steam Ahead to the Past

Such tales of doom and gloom—ecosystem collapse without re-covery—have become the norm. Those of us who study the health of the ocean sometimes feel less like PhDs and more like MDs. But medi-cal doctors do not spend their careers writing ever-more-refined obitu-aries of their patients, and neither can we.

—DR. NANCY KNOWLTON, founder of the Center for Marine Biodiver-sity and Conservation at the Scripps Institution of Oceanography

When I began writing this book, I had the idea that contributing to public understanding of the role jellyfish play in our deteriorating ecosystems would be kind of groovy. At the time, I had a naïve gut feeling that all was still sal-vageable and that by really "getting it," we would still have time to act. I even thought that we could hand our children a better world, with all the perks we enjoy plus a bit of extra wisdom gleaned along the way.

But I think I underestimated how severely we have damaged our oceans and their inhabitants. I now think that we have pushed them too far, past some mysterious tipping point that came and went without fanfare, with no red cir-cle on the calendar and without us knowing the precise moment it all became irreversible. I now sincerely believe that it is only a matter of time before the oceans as we know them and need them to be become very different places indeed. No coral reefs teeming with life. No more mighty whales or wobbling penguins. No lobsters or oysters. Sushi without fish.

In their place, we shall see blue-green algae, emerald green algae, golden algae, flashing blue algae, red tides, brown tides, and jellyfish. Lots of jellyfish. The seas were dying for us to notice their distress, but we collectively chose to overlook the red flags. Of course, we don't need to listen, we know better: we are *Homo sapiens* . . . the *wise man.*

Throughout the history of life on earth, major macroevolutionary events, such as mass extinctions and periods of intense evolutionary diversification, have been linked to global-scale changes in environmental conditions. Even relatively small changes in climate have driven major ecosystem change through sea-level fluctuations leading to the opening of straits or the emergence of isthmuses. Today's overfishing, pollution, and greenhouse gas emissions are comparable to the intense global warming, acidification, hypoxia, and mass extinctions throughout history . . . all at once.

It is unlikely that all life in the oceans will disappear, or that photosynthesis will simply cease; it is, however, likely that our marine ecosystems will undergo radical simplification. This is already occurring in many locations around the world, as outlined in the preceding pages. Diatoms are being outcompeted by flagellates. Large copepods are being replaced by small copepods. Fish-dominated systems are flipping to jellyfish-dominated systems. High-energy food chains are being replaced by low-energy food chains.

Jeremy Jackson explains it best:

> Synergistic effects of habitat destruction, overfishing, introduced species, warming, acidification, toxins, and massive runoff of nutrients are transforming once complex ecosystems like coral reefs and kelp forests into monotonous level bottoms, transforming clear and productive coastal seas into anoxic dead zones, and transforming complex food webs topped by big animals into simplified, microbially dominated ecosystems with boom-and-bust cycles of toxic dinoflagellate blooms, jellyfish, and disease. (Jackson 2008, 11458)

By fishing down our food webs and initiating trophic cascades, we are removing the large predators—the things with teeth. Ocean acidification is dissolving the hard parts of corals, mollusks, echinoderms, and crustaceans, sparing only the soft and squishy species. And hypoxia is killing off even the species we can't bring ourselves to care about.

Jackson has been quoted a lot lately in the popular press for his view that we are creating a "rise of slime." And he appears to be right. For a glimpse of what our oceans may look like in the future—and maybe not all that far in the future—in my lifetime and yours, our children's lifetimes—we can look to a time in the past when jellyfish and other soft, squishy organisms dominated.

Recall the opening words to this book about the seven successive bedding planes of fossilized jellyfish blooms. Jellyfish evolved in a world devoid of predators, and their only competitors were each other.

Long before plants and animals became abundant, the seas were anoxic or hypoxic. After about two billion years of low oxygen—only 10 percent of today's levels—oxygen shot up around 750–800 million years ago. The oldest unambiguous jellyfish fossils are from the Ediacaran period of the Late Precambrian, about 565 million years ago.

The Ediacaran ocean had small phytoplankton, such as cyanobacteria and flagellates, whereas diatoms didn't evolve for another 400 million years. Big predators, such as fish and whales, didn't evolve until much later when a higher energy food chain was possible, 300 million years ago for fish and 100 million years ago for whales. Diatoms had become abundant by the time marine mammals evolved, but the Ediacaran ecosystem appears to have been based primarily on jellyfish-type ecology. The climate and the ecosystem were a jellyfish heaven, in which they ruled the seas for nearly 100 million years as top predators.

The ancient seas were dominated by flora and fauna similar to those that today's seas appear to be shifting toward. No spectacular coral reefs. No vast filtering mussel beds. No sharks slicing through the water as menhaden multitudes flee.

Just jellyfish . . . lots of jellyfish. It might seem outlandish and farcical to think that jellyfish could rule the seas. But they've done it before, and now we have opened the door for them to do it again. Jellyfish are weeds. They are opportunists. When they have the opportunity, taking over is probably, to some extent, just what jellyfish do.

If you are waiting for me to offer some great insight, some morsel of wisdom, some words of advice . . . okay then. . . .

Adapt.

[**ACKNOWLEDGMENTS**]

This work was inspired by the work of Dr. Claudia Mills, who has been and remains one of my heroes since my earliest days in science, and Professor Tim Parsons, whose work on high-energy and low-energy ecosystems has infiltrated my waking hours and REM sleep for more than a decade. This book was probably always destined to be, developing through more than nineteen years of late-night, wine-fuelled debates with many colleagues and interested friends, but yes, a long time in coming. It was an evening with my dear friend Liz Turner (the godmother of this book) and her neighbors, Julie Bollen, Melagueta Mattay, and Scott Mollison, that finally inspired the putting of pen to paper (well, ok, fingers to keyboard).

The book finally came to fruition through painstaking word-by-word and thought-by-thought editing by Tom and Tina McGlynn, Patrick Filmer-Sankey, Scott Condie, Jennifer Lavers, Ellen Lund Jensen, Rudy Kloser, and Julian Uribe-Palomino, for whom words just do not seem to be enough to communicate my gratitude. Professors Paul Dayton and Ferdinando Boero gave valuable and thought-provoking review comments as well as profoundly wonderful encouragement; I thank them sincerely and this book is better for their contributions.

Ultimately, however, it was the brilliant editors, designers, and other production wizards at the University of Chicago Press who have collectively transformed *Stung!* into something that I am so very proud of. Christie

Henry, editorial director, thank you, thank you, thank you, for giving my book a chance. Mary Gehl, the best editor anyone could ask for, you have been an absolute joy to work with! Ryan Li, your design work is simply splendid! Amy Krynak and Joan Davies, thank you for all the bits and bobs that have made for such a smooth and enjoyable process. And Carrie Adams, my publicist, thank you for your contagious excitement, enthusiasm, and commitment! Sincere thanks also to Alex Lubertozzi for the short-notice indexing, and Sylvia Earle for spinning a last-minute foreword into solid gold.

Early conversations with, and recent publications by, the doyennes and doyens of the jellyfish bloom world have been informative and formative: Nando Boero, Dave Cargo, Bella Galil, Mark Gibbons, Jacqueline Goy, Monty Graham, Bill Hamner, Claudia Mills, Jenny Purcell, Anthony Richardson, Tamara Shiganova, and Shin-ichi Uye.

To the countless others who inspired me, enriched my thoughts, and kept me sane along the way, I am grateful to all, but I would be a very bad girl if I failed to thank the following by name: Phil and Ann Alderslade; Larry Allen; Brad Armstrong; Ross Atkinson; Ian Barr; Pam Beesley; Jeanne Bellemin; Franklin and Audrey Bridgewater; Avril Brown; Joe Burnett; Lyndy Burt; Dale Calder; Marnie Campbell; Jim Carlton; Kate Charlton-Robb; Rose Clark; Philip Clarke; Andy Cohen; Peter Davie; Sue and John Davis; Paul Dayton; Charlotte Doctor; Mae Downs; Stef Evans; Scott Farley; Peter Fenner; Beth Fulton; Chuck Galt; Dennis Gordon; Karen Gowlett-Holmes; Michael Gross; Peter Haaker; Chad Hewitt; Judith Handlinger; Geoff Harris; Bob Hill; Meg Hoggett; Rosemary Hooper; Graham Hosie; Ashley Wong Hoy; Ellen Lund Jensen; Andrew Jones; Tom Jones; John Keesing; Michael Kingsford; John Kitchener; Rudy Kloser; Lyn Koller; Ron and Kathy Larson; Deb Lewis; Mark Lewis; John Lippmann; Loisette Marsh; Duncan Massey; Richard Matear; Felicity McEnnulty; Nick McGowen; Kirrily Moore; Kim Moss; Nick Murfet; Ken Nelson; Dave O'Meara; David O'Sign; John Pearn; Al Pegler; David Pepper; Martine Plakalovic; Ingrid Radkey; John Rees; Anthony Richardson; David Ritz; Dennis Robson; Lindi Routledge; Andrew Rozefelds; Michael Schaadt; Stuart Scoon; Scoresby and Anna Shepherd; Grant Small; Robin Smith; Lynn Strefford; Jane Taylor; Peter Thompson; Ian Tibbles; Judi Tibbles; Madeleine van Oppen; Raymond Wells; Albert Wertheim; Sally, Nick, Ben, and Mitchell White; Ian Whittington; David Wilkes; Caroline Williams; David Williams; Brett Williamson; John Williamson; Ken Winkel; Cathy Young; and Wolfgang and Lynn Zeidler. I offer a very sincere and heartfelt thank you to Torsten Lund Jensen and Ngouja, who helped me enormously with obtaining literature. And with a grateful heart and a bowl of cashews, I thank my very special friends, Charlee, Nimh, Pickles, and Yellows.

With deep gratitude I thank the many people who generously shared their own spectacular photographs or helped me obtain those of others: Bruce Barker, Louise Bell, Pru Bonham, Joana Cubillos Castillo, Nicholas Eagar, Gary Florin, Karen Gowlett-Holmes, Jacky Graham, Jana Guenther, P. J. Hahn, Gustaaf Hallegraeff, Amanda Hamilton, Elizabeth Havice, Arlo Hemphill, Tomoko Ikawa, Glenn Jones, Kazuhiko Kashiwagi, Masato Kawahara, Ron Larson, Jenn Lavers, Carol Li, Derek Lohuis, Ed Mastro, Makoto Miyazaki, Sue Morrison, David Pepper, Jenny Purcell, Nancy Rabalais, Anne Layton Rice, Dan Richards, Callum Roberts, Donna Roberts, Mike Schaadt, Alexander Semenov, Robin Smith, Captain Jamie Snediker, Leif Magne Sunde, Peter Thompson, Shin-ichi Uye, Kyle van Houtan, Alan Williams, Caroline Williams, Neil Woolmer, Dave Wrobel, Marsh Youngbluth, and Patrizia Ziveri.

My work leading to this book was facilitated and funded by the Australian-American Fulbright Foundation; Australian Biological Resources Study; Australian Institute of Marine Science; Australia-Thailand Institute; Broome Shire Council; Cabrillo Marine Aquarium; California State University, Long Beach; California State University, Northridge; CRC Reef Research; CSIRO Centre for Research on Introduced Marine Pests; CSIRO Marine and Atmospheric Research; Darwin City Council; Darwin Lagoon Corporation; Divers Alert Network Asia-Pacific; Friday Harbor Laboratories; Howard Hughes Medical Institute; James Cook University; Lions Foundation; Los Angeles Pierce College; Museum and Art Gallery of the Northern Territory; Museum of Tropical Queensland; Paspaley Pearling Company; Pearl Producers Association; Queen Victoria Museum and Art Gallery; Robert King Memorial Foundation; Scripps Institution of Oceanography; Smithsonian Institution; South Australian Museum; South Australian Research and Development Institute; Surf Life Saving Australia; Surf Life Saving Queensland; Tasmanian Museum and Art Gallery; and University of California Berkeley.

Finally, I thank the Launceston City Council for giving me the time off work and the inspiration to write this book.

[APPENDIX]

Appendix Table 1. Reported events of jellyfish ingress interfering with power plant operations around the world. Each emergency shutdown event is not only costly but, in the case of nuclear plants, is potentially dangerous.

Locality	Dates	Nuisance type	Source
Arabian Gulf, Saudi Arabia	Annually between March and July	"The incidence of jellyfish was reported to be about 100 tons/h during its peak incidence in a power plant intake on the Arabian Gulf coast."	Azis et al. 2000, 9
Tokyo, Japan	1963	*Aurelia aurita* clogged the intake for power plants along the coast of Tokyo Bay.	Yasuda 1988, in Uye 1994
Tuggerah Lakes, New South Wales	1967–1971	The jellyfish *Catostylus mosaicus* was so abundant as to block the cooling water screens at the Munmorah Power Station.	Pulley 1971, in Scott 1999
Tokyo, Japan	1972	A dense patch of medusae forced a suspension of the supply of electricity in the Tokyo area.	Yasuda 1988, in Uye 1994
South of Gothenburg, Sweden	1974, 1975, 14–15 July 1976, 1977, 1978	The Ringhals nuclear power plant is the largest power plant in Scandinavia. Two of the reactors have suffered numerous shutdowns due to high concentrations of jellyfish in the cooling water intakes. Parts of the traveling screens have been destroyed on at least three occasions.	Verner 1983; Andermo 1977

Location	Date	Event	Reference
Karratha, Western Australia	1977	The cooling pipes of the Cape Lambert power plant were clogged by large numbers of red jellyfish, resulting in a plant shutdown.	L. Marsh, personal communication
Ringhals, Sweden; Kiel, Germany; and Peru	various, pre-1983	*Chrysaora* sp. was implicated in the clogging of cooling water inlets in Peru, and *Aurelia aurita* was implicated in incidents in the Baltic region and Japan	Möller 1984
Barsebäck, Sweden	Pre-1983	Intake problems due to jellyfish have been reported from this nuclear power plant.	Verner 1983
Denmark	Pre-1983	Intake problems due to jellyfish were reported from Danish power stations.	Verner 1983
Germany	Pre-1983	"Intake problems due to jellyfish have . . . been reported from . . . German Federal Republic power stations."	Verner 1983, 206
Hutchinson Island, Florida	1983	The operators manually tripped the unit 1 reactor at the St. Lucie nuclear plant because of jellyfish blocking the cooling water supply.	Lochbaum 2011
New Delhi, India	November 1983	"Jellyfish closed a nuclear power plant in India by blocking pipes bringing coolant from the sea. (*West Australian,* 9 November 1983)"	Ludlam 2012

Locality	Dates	Nuisance type	Source
Hutchinson Island, Florida	August 1984	The operators manually tripped the unit 2 reactor at St. Lucie when the jellyfish returned en masse.	Lochbaum 2011
Hutchinson Island, Florida	1 September 1984	Both nuclear reactors at the St. Lucie facility were closed due to thousands of jellyfish clogging the filtering screens for the reactors' cooling systems. Both reactors remained closed for several days.	Reuters 1984
Florida City, Florida	3 September 1984	Both nuclear reactors at the Turkey Point nuclear power plant were shut down when jellyfish clogged the flow of cooling water to the main condensers, causing a metal filtering screen to be bent inward nearly two feet. Both reactors remained shut down for eleven days.	Lochbaum 2011
Tanagwa, Japan	Pre-1986	Large quantities of jellyfish (150 tons per day) were removed from the Tanagwa Power Station in one instance.	Kawabe and Traplin 1986
Takasago, Japan	Pre-1989	A record quantity of 165 tons of jellyfish was removed from the seawater cooling intakes of the coal-fired power plant in a single day.	Rajagopal et al. 1989

Location	Date	Description	Source
Hunterston, Clyde Sea, Scotland	21 August 1991	"This advanced gas-cooled nuclear power station requiring up to 40 million gallons of seawater per hour for turbine condenser cooling was shut down when thousands of jellyfish were sucked into the screens of the seawater cooling intake. (*Nuclear News*, October 1991)"	*Nuclear News* 1991
Hunterston, Clyde Sea, Scotland	1992	The nuclear power station was shut down due to a *Rhizostoma* bloom that was blocking the cooling system.	Houghton et al. 2006
Hutchinson Island, Florida	18 and 20 September 1993	On September 18, the operators tripped the unit 1 reactor at St. Lucie because of jellyfish intrusion. On September 20, while making another attempt to restore the reactor to full power, the operators again had to trip the reactor because of the jellyfish.	Lochbaum 2011
Kashiwazaki, Japan	7 July 1999	At the Kashiwazaki Kariwa nuclear power station, "on July 7, 1999, an unusually large flow of jellyfish arrived all at once, shutting down the filtering equipment for reactors 1, 2, and 3. The resulting lack of cooling water forced us to reduce power output."	Takizawa 2005, 36
Gyeongsang-buk-do Province, South Korea	1 May, 11 August, and 26 August 2001	The Uljin nuclear power plant had numerous problems with jellyfish blocking the seawater inflow to the cooling system. On 26 August, reactor 1 was completely stopped and reactor 2 was affected but still operating.	*Chosun Ilbo* 2001

Locality	Dates	Nuisance type	Source
		The reactors each produce 950,000 kilowatts of electricity per hour, with a combined output of 3.7 percent of Korea's national production.	
Hadera, Israel	Summer 2001	Tons of jellyfish had to be scooped out of the cooling pool and seawater intake pipes of one of Israel's largest power plants, at an estimated cost of $50,000.	Brahic 2008; Waldoks 2010; and Galil et al. 2010
Gulf of Oman	2002	A large number of jellyfish blocked Oman LNG's seawater cooling system intake; a shutdown could cost the company about $7 million a day.	Vaidya 2003
Oscarshamn, Sweden	29 August 2005	One of three plant reactors was shut down due to jellyfish in the cooling water from the Baltic Sea hindering its flow. The Oscarshamn plant supplies about 10 percent of the electricity used in Sweden.	Environmental News Network (AP 2005)
Chalk Point, Maryland	June–July 2006	Maryland's largest power plant, the Chalk Point Generating Station on the Patuxent River, had so many jellyfish clogging the nets that protect the cooling water intakes that jellyfish had to be dumped from the outermost line of nets twice a week.	Delano 2006

Location	Date	Description	Reference
Calvert Cliffs, Maryland	July 2006	Jellyfish clogged the cooling water intake pumps at least three times in July (6th, 7th, and 12th). On 7 July, unit 1 was forced to reduce power to 41 percent capacity.	Delano 2006
Hamaoka, Japan	20 July 2006	Two reactors at Chubu Electric Power Company's nuclear plant were affected by jellyfish clogging a cooling system filter, causing the intake system to shut down automatically. For about three hours, output for the two reactors was reduced by 60–70 percent.	BBC 2006
Niigata, Japan	pre-2007	Jellyfish at the Higashi-Niigata Thermal Power Station have been such a problem that engineers developed an under-water image analysis system to detect and remove jellyfish automatically.	Matsuura et al. 2007
San Luis Obispo, California	October 2008	At the Diablo Canyon nuclear power plant, north of Los Angeles, jellyfish caused a total shutdown of one reactor and reduction to half power of another for three days	Di Savino 2008
Perak, Malaysia	Pre-2009	The coal-fired power plant at Manjung, in the state of Parak, has had ingression problems with massive swarms of jellyfish. Other power plants in the area have had problems too.	Raj 2011 and Wan Maznah 2011

Locality	Dates	Nuisance type	Source
Ashdod, Israel	5 July 2010	Israel Electric issued a statement that "an enormous amount of hundreds of tons a day" of jellyfish were affecting its power plants adjacent to the coastline: Ashdod, Hadera, Ashkelon, Haifa Aridnge power plant site in Tel Aviv.	Weiss 2010
Lothian, Scotland	28 June 2011	Both reactors at the Torness nuclear power station had to be shut down for two days when it became swamped with moon jellyfish; fishermen on three trawlers helped clear the nearby waters of jellyfish.	SkyNews 2011; BBC (Miller 2011)
Hutchinson Island, Florida	August 2011	The St. Lucie nuclear power plant was shut down for two days; the pulverized jellyfish swarm also killed about 75 critically endangered goliath grouper, weighing an average of 200 pounds each.	Huffington Post ("Moon Jellyfish Kill Tons of Goliath Groupers" 2011)
Chubu, Japan	July 2012	About 24,000 tons of jellyfish swarming in Ise Bay—about twice the usual level—threatened 9 thermal power plants, and caused reductions in power at 3 plants for 9 days.	*Chunichi Shimbun* 2012

Appendix Table 2. Reports of jellyfish ingress events interfering with desalination plant operations around the world. This is likely to be a gross underestimation of the actual scale of the problem.

Locality	Dates	Nuisance type	Source
Saudi Arabia	March–April 1998	Periodic ingress of jellyfish clogging the intake screens and "seriously affecting the pumping of seawater needed for the plant."	Azis et al. 2000, 3
Kuwait	Pre-2003	A plant larger than 5 million gallons per day had to be shut down on an emergency basis, due to jellyfish having blocked the travelling band screens.	Gille 2003
Gulf of Oman	March 2003	More than 300 tons of jellyfish blocked and damaged the seawater intake screens at the Al Ghubra Desalination Plant, causing a 50 percent reduction in output.	Vaidya 2003
Gulf of Oman	Mid-March 2003	Swarms of jellyfish blocked the seawater intakes at the Birka Power and Desalination Plant, badly affecting the water supply to Muscat City.	Vaidya 2003
Kuwait	Pre-2008	Intake basin 1 in Kuwait Bay "has suffered from many incidents of excessive entrainment of jelly fish . . . In some cases this results in the closure of some sections of the plant."	Rakha et al. 2008, 1
Israel	3 March 2009	A swarm clogged the filters of a 100 million liter-per-day desalination plant on the Mediterranean coast, reducing water production by more than a third.	Galil et al. 2009a; Faris 2009

Appendix Table 3. Published reports of jellyfish blooms interfering with trawling operations around the world. No doubt there are many hundreds or thousands more interference problems that have not been reported.

Species	Locality	Dates	Nuisance type	Source
Class Scyphozoa				
Periphylla periphylla	Norway	Late 1940s; since 1973 (October–November, and April–May)	Clogging fishing nets. Impeding trawling operations during heavy infestations.	Fosså 1992
Aurelia sp.	Black Sea	1976–1988	Clogging fishing nets and trawls.	Zaitsev and Mamaev 1997
Rhopilema nomadica	Israel and the Mediterranean	Since mid-1980s (June–September)	Clogging fishing nets. Impeding trawling operations.	Lotan et al. 1992; Galil and Zenetos 2002
Chrysaora melanaster	Bering Sea	1990–1999 (June–September)	Clogging fishing nets. Highly infested regions avoided.	Brodeur et al. 2002
Phyllorhiza punctata	Gulf of Mexico	2000 (May–September)	Clogging fishing nets. Shrimp losses to US$10 million.	Graham et al. 2003
Crambionella orsini	Gulf of Oman and Persian Gulf	2002	Decreased catches. Damaged fishing gear.	Daryanabard and Dawson 2008

Species	Location	Date	Impact	Reference
Cyanea sp.	Yangtze Estuary, China	Since 2003 (November)	Clogging fishing nets.	Xian et al. 2005
Nemopilema nomurai	Japan	2002–2006 (August–December)	Clogging/bursting fishing nets. Reduced catch of finfish. High finfish mortality and reduced value. Increased labor to remove jellyfish from nets. Increased risk of boat capsizing. Stings to fishermen.	Kawahara et al. 2006
Nemopilema nomurai	Japan	November 2009	Capsized a 10-ton fishing trawler	Ryall 2009
Aurelia sp.	Gulf of Mexico	Fall 2009 and 2011	Clogging nets for fishers and shrimpers	Dugan 2011
Lychnorhiza lucerna	Southern Brazil	September–November (annually)	Clogging fishing nets. Reduced trawling time. Further from ports. Temporarily switch to other fish.	Nagata et al. 2009

Species	Locality	Dates	Nuisance type	Source
Lychnorhiza lucerna	Northern Argentina	December–May (annually)	Clogging of/damage to fishing nets. Reduced quantity and quality of fish. Unable to operate.	Schiariti et al. 2008
Chrysaora hysoscella	Namibian Benguela	August–September	Clogging/bursting fishing nets. Collapse of pilchard fishery.	Brierley et al. 2001
Aurelia sp.	UK and the Baltic	Late fall and winter	Clogging/bursting fishing nets. Displace fish to other areas. Cod-end is open due to weight. Fish move away from jellyfish swarm.	Russell 1970
Rhizostoma octopus	Black Sea	N/A	Clogging fishing nets.	Netchaerff and Neu 1940, in Russell 1970
Catostylus mosaicus	New South Wales, Australia	October–May	Restrict trawling.	Broadhurst and Kennelly 1996

Species	Location	Time	Impact	Reference
Aurelia sp.	Inland Sea, Japan	Summer (annually)	Clogging/bursting fishing nets. Declining catches of fish. Reduction in catch quality. Stings to fishermen. Increased labor to remove jellyfish from nets.	Uye and Ueta 2004; Uye and Shimauchi 2005
Unspecified jellyfish	Turkey	Ongoing	"Large quantities of jellyfish are discarded in the anchovy, horse mackerel, bluefish and bonito fisheries in Turkish waters."	Özdemir 2007,9
Class Cubozoa				
Chiropsalmus quadrumanus	Texas, USA	1955 and 1956 (August–September)	Stings to fishermen.	Guest 1959
Class Hydrozoa				
Olindias sambaquiensis	Southern Brazil / northern Argentina	July–October	Clogging fishing nets. Shrimp move away from jellyfish. Stings to fishermen.	Nagata et al. 2009

Species	Locality	Dates	Nuisance type	Source
Tima bairdii	Sweden	1966–1967	Severe decline in prawn stocks. Clogging/bursting fishing nets. Cod-end is open due to weight. Unable to operate for days at a time.	Dybern 1967
Nanomia cara	Gulf of Maine	Fall and winter, 1975	Clogged trawl nets. Give confusing sound-scattering signals to equipment. Considerable losses of time and money to commercial fishermen at several New England ports.	Rogers 1978

Source: Modified from Nagata et al. 2009.

[GLOSSARY OF TERMS]

amphipod—A type of small crustacean that generally appears "squished" side-to-side; compare with *isopod*, which is "squashed" flat.

anoxic—Without oxygen; compare with *hypoxic* (low oxygen) and *normoxic* (normal oxygen).

anthropogenic—Of humans; in this case, the making of disturbances to the environment, such as overfishing or pollution.

aragonite—A form of calcium carbonate that corals, pteropods, and some other organisms use to make their shells and skeletons; aragonite is the form most vulnerable to ocean acidification.

artisanal fisheries—Either for subsistence use or for small-scale commercial enterprise, usually local and with a small boat.

ascidian—A type of marine invertebrate in the phylum Urochordata, also called "sea squirts"; see also *tunicate*.

asexual reproduction—Cloning, i.e., production of progeny from a single parent.

ballast water—Water used as weight to improve the stability of transoceanic ships.

benthic—Pertaining to the sea floor or to organisms that live on the sea floor; contrast with *pelagic*.

biodiversity—"The variability among living organisms from all sources including, *inter alia*, terrestrial, marine and other aquatic ecosystems and the ecological complexes of which they are part; this includes diversity within species, between species and of ecosystems," as defined by the Convention on Biological Diversity (United Nations 1992, 3).

brachiopod—A type of invertebrate in the phylum Brachiopoda, resembling bivalves (clams and oysters) but structurally quite different.

bycatch—The incidental killing or damaging of organisms when fishing for targeted species.

Cnidaria—The phylum (very high level grouping of organisms) containing the animals with stinging cells, e.g., stony corals, soft corals, hydras, jellyfish, sea anemones, and sea fans.

coccolithophore—A type of spherical single-celled phytoplankton covered in tiny calcified discs that look like poker chips.

community—An association of living organisms with mutual relationships among themselves and with their environment, functioning to some extent as an ecological unit.

copepod—A type of small (usually) planktonic crustacean that typically acts in the food chain between the phytoplankton and small-to-medium–sized animals.

Ctenophora—The phylum containing the comb jellies, strange creatures without stinging cells but that usually have eight rows of cilia used for locomotion.

diatom—A type of single-celled phytoplankton encased in a silica (glass) case; the base of the high-energy food chain; compare with *dinoflagellate*.

dinoflagellate—A type of single-celled phytoplankton characterized by having (usually) two flagella; often associated with toxic algal blooms and red tides; the base of the low-energy food chain; compare with *diatom*.

echinoderm—A member of the phylum Echinodermata, e.g., a starfish or sea urchin.

ecology—The study of relationships between and among organisms and their environment.

ecosystem—Communities of animals and plants together with their physical environment.

ENSO—El Niño Southern Oscillation, a basin-wide climate pattern that oscillates over multiyear time scales in the South and Equatorial Pacific.

envenomation—The process whereby venom is injected from a venom gland into the recipient, e.g., by jellyfish, stonefish, stingrays, etc.

environment—The physical and biological conditions in which an organism or group of organisms exists.

ephyra (plural, ephyrae)—The juvenile medusa, produced asexually from a polyp; looks like a little snowflake.

eutrophication—A type of water pollution, usually caused by excessive fertilizer runoff or sewage effluent causing an accelerated growth of phytoplankton, algae, and other plant life, which causes an undesirable disturbance to the natural balance of the ecosystem.

exotic species—A species that has been transported by human activities into a region where it does not naturally occur, either intentionally or by accident.

flagellate—A single-celled organism that uses one or more whiplike flagella for propulsion and steering; most contain chlorophyll and are photosynthetic, whereas others feed wholly or partly on other organisms.

forage fish—Small pelagic fish that are preyed on for food by larger predators, such as larger fish, seabirds, and marine mammals; they generally feed on small zooplankton, such as copepods, thereby transferring energy up the food chain to top predators, and they compensate for their small size by forming schools, e.g., anchovies, sardines, herrings, menhaden, sprats, and halfbeaks.

foraminifera—A type of single-celled phytoplankton, essentially amoeba with shells.

gonad—The reproductive organs of an animal, i.e., the organs that make the sperm and eggs.

ground fish—Fish that live on, in, or near the ocean or lake bottom, e.g., halibut, sole, and flounder.

harmful algal bloom—a mass proliferation of toxic phytoplankton species.

herbivorous—Plant-eating, or vegetarian, e.g., herbivorous copepods and krill graze on diatoms.

hydroid—The (benthic) polyp stage of hydrozoans, often in alternation with *hydromedusae.*

hydromedusae—Small, often transparent, jellyfish belonging to the class Hydrozoa; compare with *Hydrozoa* and *Scyphozoa.*

Hydrozoa—The class of organisms containing the (benthic) hydroids, (solitary pelagic) hydromedusae, and (colonial pelagic) siphonophores.

hypoxic—Low oxygen; compare with *anoxic* (no oxygen).

ichthyoplankton—Fish eggs and larvae, which drift in the water before hatching or settling.

introduced species—A nonnative species entering an ecosystem via human action; may become invasive and pestilent.

isopod—A type of small crustacean that generally appears "squashed" flat; compare with *amphipod,* which is "squished" side-to-side.

jellyfish—A nontechnical term collectively applied to the pelagic cnidarians and ctenophores.

juvenile—A sexually immature animal that resembles the adult form; compare with *larva.*

larva—An early life stage in many marine animals that is different in form and appearance from the juvenile and adult stages; compare with *juvenile.*

medusa (plural, medusae)—The "jellyfish" part of the jellyfish life cycle, generally spending its time drifting on the currents rather than stuck to the substrate like a polyp.

menhaden—A type of small pelagic schooling fish that occurs in vast abundance in southeastern US waters, caught by the billions for processing into fish oil and fish meal.

NAO—North Atlantic Oscillation; a basin-wide climate pattern that oscillates on a multiyear scale; the North Atlantic version of El Niño.

oligotrophic—Water that is very low in nutrients; about 90 percent of the ocean surface (i.e., the open sea) is essentially a biological desert, and flagellates dominate in these regions, but even they can be scarce.

overfishing—Fishing a population faster than or beyond its reproductive capacity to replenish itself.

pelagic—Pertaining to the water column or the organisms that live there; some organisms spend their entire life in the pelagic zone, whereas others spend only their larval stages; organisms that drift at the mercy of currents are called *plankton*, whereas those that swim against a current are called *nekton*.

perturbation—Disturbance or stressor to the environment; often, a relatively small perturbation can cause major problems through a cascade effect.

phytoplankton—Microscopic single-celled, plantlike algae that drift in the ocean, e.g., diatoms, dinoflagellates, and coccolithophores.

piscivorous—fish-eating, e.g., sharks, dolphins, and larger fish; compare with *planktivorous*.

plague species—A species that becomes pestilent in an ecosystem, whether introduced or native.

planktivorous—plankton-eating, e.g., jellyfish and many types of small fish; compare with *piscivorous*.

plankton—Drifters at the mercy of currents; compare with *nektonic* swimmers.

planula—The larval stage of a cnidarian.

podocyst—A small encapsulated fragment of tissue left behind by a jellyfish polyp, which can "hatch" into a new polyp; may be used as an "overwintering" form during unfavorable conditions.

polyp—A small, benthic (bottom-dwelling) stage in the jellyfish life cycle that is produced through sexual reproduction of sperm and egg, or asexually (clonally) by a parent polyp; also sometimes called a scyphistoma or a hydra.

primary productivity—Carbohydrate production by phytoplankton in growing their bodies and replicating into vast blooms; phytoplankton are consumed by herbivorous zooplankton, and energy is transferred up the food chain by organisms that eat the zooplankton, and so on.

protist—A member of the kingdom Protista, containing the slime molds and algae.

pteropod—A type of "winged" mollusk, e.g., the sea butterflies and sea angels.

recruitment—Survival of young fish to adulthood and being added into the reproductive population.

rhizostome—A blubber jellyfish, e.g., the spotted *Phyllorhiza* or the refrigerator-sized *Nemopilema*.

salp—A pelagic tunicate, may be solitary or colonial; resembling jellyfish, but more closely related to humans than to jellyfish.

Scyphozoa—The class of organisms containing the sea nettles, moon jellies, lion's manes, blubbers, and other typically large and conspicuous jellyfish species.

siphonophore—A type of colonial pelagic hydrozoan, e.g., a Portuguese man-o'-war.

strobilation—A process of metamorphosis that some types of jellyfish polyps undergo in order to bud off juvenile medusae.

symbiont—An organism living in symbiosis with another, usually internally, e.g., algal symbionts in the tissues of corals, or gut bacteria in cattle.

tunicate—A type of marine invertebrate in the phylum Urochordata, also called a "sea squirt"; see also *ascidian*.

upwelling—Bottom water brought to the surface by wind blowing surface waters seaward. The combined total of the world's upwelling regions comprises no more than about one-tenth of 1 percent of the surface of the ocean, but produces about half of the world's fish supply.

zooplanktivorous—A species that eats zooplankton, such as the anchovy or the whale shark.

zooplankton—Literally, drifting animals, often small; this includes species such as copepods, which spend their entire life in the plankton (called *holoplankton*), as well as the larvae of bottom-dwelling animals (called *meroplankton*). The term *zooplankton* also technically includes jellyfish, although they are often referred to as *gelatinous zooplankton* to differentiate them from crustaceans, mollusks, fish, and other zooplankton.

[SOME PRACTICAL CONVERSIONS]

Length

Meters and Yards

- A meter and a yard are about the same (39 inches, compared to 36).
- A square yard is 80 percent of a square meter, whereas a square meter is about 20 percent more than a square yard.
- A cubic yard is about ¾ of a cubic meter, whereas a cubic meter is about 1⅓ of a cubic yard.

Centimeters and Inches

- An inch is 2½ centimeters.
- A centimeter is about ⅓ of an inch.
- A millimeter (⅒ of a centimeter) is about the thickness of a dime.
- There are 100 centimeters to a meter, and 1,000 meters to a kilometer.

Kilometers and Miles

- A mile is about 1½ kilometers; a kilometer is about ⅔ of a mile.
- Driving at 60 miles an hour is the same as about 95 kilometers per hour; at this speed, you'll travel a mile every minute, or 1½ kilometers.

Volume

Gallons, Quarts, and Liters

- A liter and a quart are about the same, with 4 of them to the gallon.

Weight

Pounds and Kilograms

- There are 2.2 pounds in a kilogram, or just over ½ a kilogram to a pound.
- An average person weighs 150 pounds, or 70 kilos.

Temperature

- The freezing point is at 32° Fahrenheit or 0° Celsius; each degree Celsius is equal to 1.8 degrees Fahrenheit—so, for example, 10°C would be 50°F: 32 (because 32°F equals 0°C) + 18 (i.e., 10 × 1.8).

[REFERENCES]

Abbriano, R. M., M. M. Carranza, S. L. Hogle, R. A. Levin, A. N. Netburn, K. L. Seto, S. M. Snyder, and P. J. S. Franks, with the students of Biological Oceanography Class SIO280. 2011. "Deepwater Horizon Oil Spill: A Review of the Planktonic Response." *Oceanography* 24:294–301.

ABC. 2005. "Japan's scientific whaling a sham: WWF." *ABC News Online*, 13 June. http://www.abc.net.au/news/2005-06-13/japans-scientific-whaling-a-sham-wwf /1591324.

Abend, L. and G. Pingree. 2006. The Sea Stings Back. *Time*, 13 August. http://www .time.com/time/magazine/article/0,9171,1226063,00.html.

Aiyar, R. G. 1936. "Mortality of Fish of the Madras Coast in June 1935." *Current Science* 4:488–89.

Alessi, E., G. Tognon, M. Sinesi, C. Guerranti, G. Perra, and S. Focardi. 2006. *Chemical Contamination in the Mediterranean: The Case of Swordfish*. Rome: WWF—World Wide Fund for Nature.

Algalita. 2009. "Ocean Protection Comments." Algalita Marine Research Foundation, 22 September. http://www.algalita.org/pdf/ocean-protection-comments.pdf.

AMCS. 2011. "Threats to Sharks." Australian Marine Conservation Society. Accessed 3 May 2011. http://www.amcs.org.au/default2.asp?active_page_id=516.

Amstrup, S. C., I. Stirling, T. S. Smith, C. Perham, and G. W. Thiemann. 2006. "Recent Observations of Intraspecific Predation and Cannibalism among Polar Bears in the Southern Beaufort Sea." *Polar Biology* 29:997–1002.

Andermo, L. 1977. "Report on Safety Related Occurrences and Reactor Trips, July 1, 1976–December 31, 1976." Technical report 1977-1. http://www.iaea.org/inis/ collection/NCLCollectionStore/_Public/08/343//8343443.pdf.

Anderson, D. M. 1994. "Red Tides." *Scientific American*, August, 62–68.

Anderson, P. J., and J. F. Piatt. 1999. "Community Reorganization in the Gulf of Alaska following Ocean Climate Regime Shift." *Marine Ecology Progress Series* 189:117–23.

AP. 2005. "Jellyfish Cause Shutdown of Swedish Nuclear Reactor." *Environmental News Network*, 30 August. http://www.enn.com/top_stories/article/2454.

———. 2010. "Dead Jellyfish Washing Up along Gulf Coast." *CBSNews.com*, 3 May. http://www.cbsnews.com/stories/2010/05/03/national/main6456715.shtml.

Arai, M. N. 2001. "Pelagic Coelenterates and Eutrophication: A Review." *Hydrobiologia* 451:69–87.

"Asia: Dark Days for Estrada." 1999. *Economist* (London), December 18, 36–37.

Atkinson, A., V. Siegel, E. Pakhomov, and P. Rothery. 2004. "Long-Term Decline in Krill Stock and Increase in Salps within the Southern Ocean." *Nature* 432:100–3.

Attrill, M. J., J. Wright, and M. Edwards. 2007. "Climate-Related Increases in Jellyfish Frequency Suggest a More Gelatinous Future for the North Sea." *Limnology and Oceanography* 52:480–85.

Australian AP, with Australian Geographic staff. 2011. "Radiation Risk to Japan Marine Life?" *Australian Geographic*, 5 April. http://www.australiangeographic.com.au/journal/radioactive-water-from-fukushima-affect-marine-life.htm.

Azis, P. K. A., I. Al-Tisan, M. Al-Daili, T. N. Green, A. G. I. Dalvi, and M. A. Javeed. 2000. "Effects of Environment on Source Water for Desalination Plants on the Eastern Coast of Saudi Arabia." *Desalination* 132:29–40.

Azzarello, M. Y., and E. S. V. Vleet. 1987. "Marine Birds and Plastic Pollution." *Marine Ecology Progress Series* 37:295–303.

Bader, H. R., and J. Guimond. 2005. Tundra Travel Modeling Project. Alaska Department of Natural Resources, Division of Mining, Land and Water.

Bahnsen, C. J. 2006. "Saving Sharks in Baja California?" Center for Shark Research, 1 September. http://www.thefreelibrary.com/Saving+sharks+in+Baja+California%3f-a0151548874.

Baker, L. D., and M. R. Reeve. 1974. "Laboratory Culture of the Lobate Ctenophore *Mnemiopsis mccradyi* with Notes on Feeding and Fecundity." *Marine Biology* 26:57–62.

Båmstedt, U., J. H. Fosså, M. B. Martinussen, and A. Fosshagen. 1998. "Mass Occurrence of the Physonect Siphonophore *Apolemia uvaria* (Lesueur) in Norwegian Waters." *Sarsia* 83:79–85.

Banks, M. R., and D. Leaman. 1999. "Charles Darwin's Field Notes on the Geology of Hobart Town—A Modern Appraisal." *Papers and Proceedings of the Royal Society of Tasmania* 133:29–50.

Barham, E. G. 1963. "Siphonophores and the Deep Scattering Layer." *Science* 140:826–28.

Barker, R. 1998. *And the Waters Turned to Blood: The Ultimate Biological Threat.* New York: Touchstone.

Barnes, D. K. A. 2005. "Remote Islands Reveal Rapid Rise of Southern Hemisphere Sea Debris." *Scientific World Journal* 5:915–21.

Basch, M. 2012. "App Lets Swimmers Check the 'Jellyfish Forecast.'" *WTOP*,

28 June. http://www.wtop.com/968/2921944/App-lets-swimmers-check-the -jellyfish-forecast.

Baum, J., and R. A. Myers. 2004. "Shifting Baselines and the Decline of Pelagic Sharks in the Gulf of Mexico." *Ecology Letters* 7:135–45.

Baum, J. K., R. A. Myers, D. G. Kehler, B. Worm, S. J. Harley, and P. A. Dohert. 2003. "Collapse and Conservation of Shark Populations in the Northwest Atlantic." *Science* 299:389–92.

Bauman, A. G., J. A. Burt, D. A. Feary, E. Marquis, and P. Usseglio. 2010. "Tropical Harmful Algal Blooms: An Emerging Threat to Coral Reef Communities?" *Marine Pollution Bulletin* 60:2117–122.

Bax, N. J., R. Tilzey, J. M. Lyle, S. Wayte, R. Kloser, and A. D. M. Smith. 2005. *Providing Management Advice for Deep Sea Fisheries: Lessons Learnt from Australia's Orange Roughy Fisheries.* Deep Sea 2003: Conference on the Governance and Management of Deep-sea Fisheries. Part 1: Conference Reports, 1–5 December 2003, 259–72. Queenstown, New Zealand.

Baxter, E. J., M. M. Sturt, N. M. Ruane, T. K. Doyle, R. McAllen, L. Harman, and H. D. Rodger. 2011. "Gill Damage to Atlantic Salmon (*Salmo salar*) Caused by the Common Jellyfish (*Aurelia aurita*) under Experimental Challenge." *PLoS One* 6:e18529.

Bayha, K. M., G. R. Harbison, J. H. Mcdonald, and P. M. Gaffney. 2004. "Preliminary Investigation on the Molecular Systematics of the Invasive Ctenophore *Beroe ovata*." In *Aquatic Invasions in the Black, Caspian, and Mediterranean Seas*, edited by H. Dumont, T. A. Shiganova and U. Niermann, 167–75. Dordrecht: Springer.

BBC. 2006. "Nuclear Plant Struck by Jellyfish." *BBC News*, 20 July. http://news.bbc .co.uk/2/hi/asia-pacific/5197846.stm.

Beaugrand, G., K. M. Brander, J. A. Lindley, S. Souissi, and P. C. Reid. 2003. "Plankton Effect on Cod Recruitment in the North Sea." *Nature* 426:661–64.

Becker, A., I. Sötje, C. Paulmann, F. Beckmann, T. Donath, R. Boese, O. Prymak, H. Tiemann, and M. Epple. 2005. "Calcium Sulfate Hemihydrate Is the Inorganic Mineral in Statoliths of Scyphozoan medusae (Cnidaria)." *Dalton Transactions* 2005:1545–50.

Bennett, B. 1999. "Healing the Derwent's Murky Blues." *Ecos* 100:10–17.

Benningfield, D. 2006. "Attack of the Killer Jellyfish." *Science and the Sea*, 15 October. http://www.scienceandthesea.org/index.php?option=com_content&task=view&id =29&Itemid=10.

Berkeley, S. A., C. Chapman, and S. M. Sogard. 2004. "Maternal Age as a Determinant of Larval Growth and Survival in a Marine Fish, *Sebastes melanops*." *Ecology* 85:1258–64.

Berkeley, S. A., M. A. Hixon, R. J. Larson, and M. S. Love. 2004. "Fisheries Sustainability via Protection of Age Structure and Spatial Distribution of Fish Populations." *Fisheries* 29:23–32.

Berkelmans, R., G. De'ath, S. Kininmonth, and W. J. Skirving. 2004. "A Comparison of the 1998 and 2002 Coral Bleaching Events on the Great Barrier Reef: Spatial Correlation, Patterns, and Predictions." *Coral Reefs* 23:74–83.

Bhimachar, B. S., and P. C. George. 1950. "Abrupt Set-Backs in the Fisheries of the Malabar and Kanara Coasts and 'Red Water' Phenomenon as Their Probable Cause." *Proceedings of the Indian Academy of Sciences (B)* 31:339–50.

"Big Old Fat Fecund Female Fish: The BOFFFF Hypothesis and What It Means for MPAs and Fisheries Management." 2007. *MPA News* 9:1–2.

Bigot, J. 2002. "Les stocks ne diminuent pas, ils se déplacent." L'Usine Nouvelle 12:10.

Bilio, M., and U. Niermann. 2004. "Is the Comb Jelly Really to Blame for It All? *Mnemiopsis leidyi* and the Ecological Concerns about the Caspian Sea." *Marine Ecology Progress Series* 269:173–83.

Birkeland, C., and P. K. Dayton. 2005. "The Importance in Fishery Management of Leaving the Big Ones." *Trends in Ecology and Evolution* 20:356–58.

Bishop, J. W. 1967. "Feeding Rates of the Ctenophore, *Mnemiopsis leidyi.*" *Chesapeake Science* 8:259–61.

Bizzarro, J. J., W. D. Smith, R. E. Hueter, and C. J. Villavicencio-Garayzar. 2009. "Activities and Catch Composition of Artisanal Elasmobranch Fishing Sites on the Eastern Coast of Baja California Sur, Mexico." *Bulletin of the Southern California Academy of Sciences* 108 (3): 137–51.

Black, R. 2011. "Arctic Sea Routes Open as Ice Melts." *BBC*, 25 August. http://www.bbc.co.uk/news/science-environment-14670433.

Bobko, S. J., and S. A. Berkeley. 2004. "Maturity, Ovarian Cycle, Fecundity, and Age-Specific Parturition of Black Rockfish (*Sebastes melanops*)." *Fishery Bulletin* 102:418–29.

Boero, F. 2012. Evaluation of "From Fish to Jellyfish in the Eutrophicated Limfjorden (Denmark)," by H. U. Riisgård et al. Faculty of 1000, 20 February. http://f1000.com/715498001.

Boero, F., M. Putti, E. Trainito, E. Prontera, S. Piraino, and T. A. Shiganova. 2009. "First Records of Mnemiopsis leidyi (Ctenophora) from the Ligurian, Thyrrhenian and Ionian Seas (Western Mediterranean) and First Record of *Phyllorhiza punctata* (Cnidaria) from the Western Mediterranean." *Aquatic Invasions* 4:675–80.

Boersma, M., A. M. Malzahn, W. Greve, and J. Javidpour. 2007. "The First Occurrence of the Ctenophore *Mnemiopsis leidyi* in the North Sea." *Helgoland Marine Research* 61:153–55.

Bolton, T. F., and W. M. Graham. 2004. "Morphological Variation among Populations of an Invasive Jellyfish." *Marine Ecology Progress Series* 278:125–39.

Boyce, D. G., M. R. Lewis, and B. Worm. 2010. "Global Phytoplankton Decline over the Past Century." *Nature* 466:591–96.

Brahic, C. 2008. "Alien Species Named and Shamed on European List." *New Scientist*, 6 February. http://www.newscientist.com/article/dn13284-alien-species-named-and-shamed-on-european-list.html.

Branch, T. A., R. Watson, E. A. Fulton, S. Jennings, C. R. McGilliard, G. T. Pablico, D. Ricard, and S. R. Tracey. 2010. "The Trophic Fingerprint of Marine Fisheries." *Nature* 468:431–35.

Breitburg, D. L., B. C. Crump, J. O. Dabiri, and C. L. Gallegos. 2010. "Ecosystem Engineers in the Pelagic Realm: Alteration of Habitat by Species Ranging from Microbes to Jellyfish." *Integrative and Comparative Biology* 50:188–200.

Brierley, A. S., B. E. Axelsen, E. Buecher, C. A. J. Sparks, H. Boyer, and M. J. Gibbons. 2001. "Acoustic Observations of Jellyfish in the Namibian Benguela." *Marine Ecology Progress Series* 210:55–66.

Brierley, A. S., P. G. Fernandes, M. A. Brandon, F. Armstrong, N. W. Millard, S. D. McPhail, P. Stevenson, M. Pebody, J. Perrett, M. Squires, D. G. Bone, and G. Griffiths. 2002. "Antarctic Krill under Sea Ice: Elevated Abundance in a Narrow Band Just South of Ice Edge." *Science* 295:1890–92.

Broadhurst, M. K., and S. J. Kennelly. 1996. "Rigid and Flexible Separator Panels in Trawls that Reduce the By-Catch of Small Fish in the Clarence River Prawn-Trawl Fishery, Australia." *Marine Freshwater Research* 47:991–98.

Brodeur, R. D., M. B. Decker, L. Ciannelli, J. E. Purcell, N. A. Bond, P. J. Stabeno, E. Acuna, and G. L. Hunt Jr. 2008a. "Rise and Fall of Jellyfish in the Eastern Bering Sea in Relation to Climate Regime Shifts." *Progress in Oceanography* 77:103–11.

Brodeur, R. D., C. E. Mills, J. E. Overland, G. E. Walters, and J. D. Schumacher. 1999. "Evidence for a Substantial Increase in Gelatinous Zooplankton in the Bering Sea, with Possible Links to Climate Change." *Fisheries Oceanography* 8:296–306.

Brodeur, R. D., J. J. Ruzicka, and J. H. Steele. 2011. "Investigating Alternate Trophic Pathways through Gelatinous Zooplankton and Planktivorous Fishes in an Upwelling Ecosystem Using End-to-End Models." In *Interdisciplinary Studies on Environmental Chemistry-Marine Environmental Modeling & Analysis*, edited by K. Omori, X. Guo, N. Yoshie, N. Fujii, I. C. Handoh, A. Isobe, and S. Tanabe, 57–63.Tokyo: Terrapub.

Brodeur, R. D., C. L. Suchman, D. C. Reese, T. W. Miller, and E. A. Daly. 2008b. "Spatial Overlap and Trophic Interaction between Pelagic Fish and Large Jellyfish in the Northern California Current." *Marine Biology* 154:649–59.

Brodeur, R. D., H. Sugisaki, and G. L. Hunt Jr. 2002. "Increases in Jellyfish Biomass in the Bering Sea: Implications for the Ecosystem." *Marine Ecology Progress Series* 233:89–103.

Brotz, L. 2011. "Changing Jellyfish Populations: Trends in Large Marine Ecosystems." Master's thesis, University of British Columbia.

Brown, C. 2007. "A Scorching Future." *Los Angeles Times*, 24 January. http://articles .latimes.com/2007/jan/24/food/fo-wine24.

Browne, M. A., A. Dissanayake, T. S. Galloway, D. M. Lowe, and R. C. Thompson. 2008. "Ingested Microscopic Plastic Translocates to the Circulatory System of the Mussel, *Mytilus edulis* (L.)." *Environmental Science & Technology* 42:5026–31.

Bruno, D. W., and A. E. Ellis. 1985. "Mortalities in Atlantic Salmon Associated with the Jellyfish, *Phialella quadrata*." *Bulletin of the European Association of Fish Pathologists* 5:64–65.

Buck, E. H. 2005. "Hurricane Katrina: Fishing and Aquaculture Industries—Damage and Recovery, a CRS (Congressional Research Service) Report for Congress." 7 September. http://www.fas.org/sgp/crs/misc/RS22241.pdf.

Caddy, J. F. 1993. "Toward a Comparative Evaluation of Human Impacts on Fishery Ecosystems of Enclosed and Semi-Enclosed Seas." *Reviews in Fisheries Science* 1:57–95.

———. 2000. "Marine Catchment Basin Effects versus Impacts of Fisheries on Semi-Enclosed Seas." *ICES Journal of Marine Science* 57:628–40.

Cai, W.-J., L. Chen, B. Chen, Z. Gao, S. H. Lee, J. Chen, D. Pierrot, K. Sullivan, Y. Wang, X. Hu, W.-J. Huang, Y. Zhang, S. Xu, A. Murata, J. M. Grebmeier, E. P. Jones, and H. Zhang. 2010. "Decrease in the CO2 Uptake Capacity in an Ice-Free Arctic Ocean Basin." *Science* 329:556–59.

Caldeira, K. 2007. "What Corals are Dying to Tell Us About CO2 and Ocean Acidification." *Oceanography* 20:188–95.

Calkins, L. B. 2011. "BP Spill Fine May Undercount Dead Turtles, Birds, Group Says." *Bloomberg*, 12 April. http://www.bloomberg.com/news/2011-04-12/bp-spill-fine -may-undercount-dead-turtles-dolphins-group-says.html.

Calmet, D. P. 1989. "Ocean Disposal of Radioactive Waste: Status Report." *IAEA Bulletin*, April, 47–50.

CalorieKing. 2011. "Calories in Fresh Fish: Jellyfish, Dried & Salted." CalorieKing .com. http://www.calorieking.com/foods/calories-in-fresh-fish-jellyfish-dried -salted_f-ZmlkPTEyODQ1Ng.html.

Cameron, F. 2002. "Turkish Feast due to Jellyfish Wars." *IntraFish*, 7 January. http:// www.intrafish.com/global/news/article1156266.ece.

Cargo, D. G., and D. R. King. 1990. "Forecasting the Abundance of the Sea Nettle, *Chrysaora quinquecirrha*, in the Chesapeake Bay." *Estuaries* 13:486–91.

Carl, C., J. Günther, and L. M. Sunde. 2011. "Larval Release and Attachment Modes of the Hydroid *Ectopleura larynx* on Aquaculture Nets in Norway." *Aquaculture Research* 42:1056–60.

Carlton, J. T. 1995. "Marine Invasions and the Preservation of Coastal Diversity." *Endangered Species Update* 12 (4/5): 1–3.

———. 2001. *Introduced Species in U.S. Coastal Waters: Environmental Impacts and Management Priorities*. Arlington, VA: Pew Oceans Commission.

———. 2009. "Deep Invasion Ecology and the Assembly of Communities in Historical Time." In *Biological Invasions in Marine Ecosystems: Ecological, Management and Geographic Perspectives*, edited by G. Rilov and J. Crooks, 13–56. Heidelberg, Germany: Springer.

Carlton, J. T., and J. B. Geller. 1993. "Ecological Roulette: The Global Transport of Nonindigenous Marine Organisms." *Science* 261:78–82.

Carlton, J. T., J. B. Geller, M. L. Reaka-Kudla, and E. A. Norse. 1999. "Historical Extinctions in the Sea." *Annual Review of Ecology and Systematics* 30:515–38.

Carvajal, P. 2002. "Jellyfish Attacks in Chiloé Cause More Salmon Deaths." *IntraFish*, 8 April. http://www.intrafish.com/incoming/article1156597.ece.

Casey, J. M., and R. A. Myers. 1998. "Near Extinction of a Large, Widely Distributed Fish." *Science* 281:690–92.

"Caspian Seal Origin, Life History, Threats and Conservation." 2011. Caspian Seal Project. Accessed 26 February. http://www.caspianseal.org/info.

Chapman, P. 2012. "Entire Nation of Kiribati to be Relocated over Rising Sea Level Threat." *Telegraph*, 7 March. http://www.telegraph.co.uk/news/worldnews /australiaandthepacific/kiribati/9127576/Entire-nation-of-Kiribati-to-be-relocated -over-rising-sea-level-threat.html.

Chosun Ilbo. 2001. "Jellyfish Knock Out Nuclear Plant Again." *Chosun Ilbo*, 26 August. http://english.chosun.com/site/data/html_dir/2001/08/26/2001082661140 .html.

Chunichi Shimbun. 2012. "Jellyfish Swarms in Danger of Clogging Ise Thermal Power Plants." *Japan Times*, 4 August. http://www.japantimes.co.jp/text/nn20120804cc .html.

CIESM. 2007. "Impact of Mariculture on Coastal Ecosystems." Paper read at CIESM [Mediterranean Science Commission] Workshop Monographs No. 32, Lisbon.

Cloern, J. E. 2001. "Our Evolving Conceptual Model of the Coastal Eutrophication Problem." *Marine Ecology Progress Series* 210:223–53.

CNN. 2011. "Utility: Radioactive water leak from reactor stopped." *CNN*, 5 April. http://articles.cnn.com/2011-04-05/world/japan.nuclear.reactors_1_radioactive -water-fukushima-daiichi-nuclear-plant-tokyo-electric?_s=PM:WORLD.

Coghlan, A. 2002. "Extreme Mercury Levels Revealed in Whalemeat." *New Scientist*, 6 June. http://www.newscientist.com/article/dn2362-extreme-mercury-levels -revealed-in-whalemeat.html.

Cohen, A. N., and J. T. Carlton. 1995. *Nonindigenous Aquatic Species in a United States Estuary: A Case Study of the Biological Invasions of the San Francisco Bay and Delta*. Washington DC: United States Fish and Wildlife Service and The National Sea Grant College Program Connecticut Sea Grant.

———. 1998. "Accelerating Invasion Rate in a Highly Invaded Estuary." *Science* 279:555–58.

Cole, S. 2012. "Satellites See Unprecedented Greenland Ice Sheet Surface Melt." NASA, 24 July. http://www.nasa.gov/home/hqnews/2012/jul/HQ_12-249 _Greenland_Ice_Sheet_Melt.html.

Coleman, F. C., W. F. Figueira, J. S. Ueland, and L. B. Crowder. 2004. "The Impact of United States Recreational Fisheries on Marine Fish Populations." *Science* 305:1958–60.

Colin, S. P., J. H. Costello, L. J. Hansson, J. Titelman, and J. O. Dabiri. 2010. "Stealth Predation and the Predatory Success of the Invasive Ctenophore *Mnemiopsis leidyi*." *Proceedings of the National Academy of Sciences U S A* 107:17223–27.

Colwell, R. R. 1996. "Global Climate and Infectious Disease: The Cholera Paradigm." *Science* 274:2025–31.

Condon, R. H., M. B. Decker, and J. E. Purcell. 2001. "Effects of Low Dissolved Oxygen on Survival and Asexual Reproduction of Scyphozoan Polyps (*Chrysaora quinquecirrha*)." *Hydrobiologia* 451:89–95.

Condon, R. H., W. M. Graham, C. M. Duarte, K. A. Pitt, C. H. Lucas, S. H. D. Haddock, K. R. Sutherland, K. L. Robinson, M. N. Dawson, M. B. Decker, C. E. Mills, J. E. Purcell, A. Malej, H. Mianzan, S.-I. Uye, S. Gelcich, and L. P. Madin. 2012. "Questioning the Rise of Gelatinous Zooplankton in the World's Oceans." *BioScience* 62:160–69.

Condon, R. H., D. K. Steinberg, P. A. Giorgio, T. C. Bouvier, D. A. Bronk, W. M. Graham, and H. W. Ducklow. 2011. "Jellyfish Blooms Result in a Major Microbial Respiratory Sink of Carbon in Marine Systems." *Proceedings of the National Academy of Sciences U S A* 108:10225–30.

Cone, M. 2000. "Aleutian Islands: A Wilderness Ecosystem in Collapse." *Los Angeles Times*, 28 October. http://www.commondreams.org/headlines.shtml?/headlines01/0128-01.htm.

Conley, D. J., J. Carstensen, R. Vaquer-Sunyer, and C. M. Duarte. 2009. "Ecosystem Thresholds with Hypoxia." *Hydrobiologia* 629:21–29.

Conley, D. J., C. L. Schelske, and E. F. Stoermer. 1993. "Modification of the Bio-geochemical Cycle of Silica with Eutrophication." *Marine Ecology Progress Series* 101:179–92.

Coonfield, B. R. 1936. "Regeneration in *Mnemiopsis leidyi*, Agassiz." *Biological Bulletin* (Woods Hole) 71:421–28.

Courchamp, F., E. Angulo, P. Rivalan, R. J. Hall, L. Signoret, L. Bull, and Y. Meinard. 2006. "Rarity Value and Species Extinction: The Anthropogenic Allee Effect." *PLoS Biology* 4:e415.

Cox, K. W. 1962. "California Abalones, Family Haliotidae." *Fish Bulletin* 118:1–131.

Cudmore, W. W. 2009. *Declining Expectations—The Phenomenon of Shifting Baselines*. Salem, OR: Northwest Center for Sustainable Resources.

Damanaki, M. 2011. "How Plastic Bags Pollute the Future of Our Seas." Round table discussion on pollution of the Mediterranean from marine litter, Athens, 8 April. http://ec.europa.eu/commission_2010-2014/damanaki/headlines/speeches /2011/04/20110408-speech-marinelitter_en.pdf.

Darby, A. 2006. "Trawled Fish on Endangered List." *Sydney Morning Herald*, 10 November. http://www.smh.com.au/news/environment/trawled-fish-on -endangered-list/2006/11/09/1162661830462.html.

———. 2008. "The Krilling Fields: Study Fears Catastrophe in Antarctic Food Chain." *The Age*, 14 October. http://www.theage.com.au/environment/the-krilling-fields -study-fears-catastrophe-in-antarctic-food-chain-20081013-4zxo.html.

Daryanabard, R., and M. N. Dawson. 2008. "Jellyfish Blooms: *Crambionella orsini* (Scyphozoa: Rhizostomeae) in the Gulf of Oman, Iran, 2002–2003." *Journal of the Marine Biological Association of the United Kingdom* 88:477–83.

Daskalov, G. M., and E. V. Mamedov. 2007. "Integrated Fisheries Assessment and Pos-sible Causes for the Collapse of Anchovy Kilka in the Caspian Sea." *ICES Journal of Marine Science* 64:503–11.

Davenport, J. 1982. "Oil and Planktonic Ecosystems." *Philosophical Transactions of the Royal Society of London, Series B, Biological Sciences* 297:369–84.

Davis, G. E., P. L. Haaker, and D. V. Richards. 1996. "Status and Trends of White Abalone at the California Channel Islands." *Transactions of the American Fisheries Society* 125:42–48.

Dawson, M. N., L. E. Martin, and L. K. Penland. 2001. "Jellyfish Swarms, Tourists, and the Christ-Child." *Hydrobiologia* 451:131–44.

Dayton, P. K., M. J. Tegner, P. B. Edwards, and K. L. Riser. 1998. "Sliding Baselines, Ghosts, and Reduced Expectations in Kelp Forest Communities." *Ecological Applications* 8:309–22.

Dayton, P. K., S. F. Thrush, M. T. Agardy, and R. J. Hofman. 1995. "Environmental Effects of Marine Fishing." *Aquatic Conservation: Marine and Freshwater Ecosystems* 5:205–32.

Dayton, P. K., S. Thrush, and F. C. Coleman. 2002. *Ecological Effects of Fishing in Marine Ecosystems of the United States.* Arlington, VA: Pew Oceans Commission.

De'ath, G., J. M. Lough, and K. E. Fabricius. 2009. "Declining Coral Calcification on the Great Barrier Reef." *Science* 323:116–19.

Delannoy, C. M. J., J. D. R. Houghton, N. E. C. Fleming, and H. W. Ferguson. 2011. "Mauve Stingers (*Pelagia noctiluca*) as Carriers of the Bacterial Fish Pathogen *Tenacibaculum maritimum*." *Aquaculture* 311:255–57.

Delano, F. 2006. "A Nettle-Some Problem on Potomac." *Fredericksburg.com*, 15 July. http://fredericksburg.com/News/FLS/2006/072006/07152006/206385.

Derraik, J. G. B. 2002. "The Pollution of the Marine Environment by Plastic Debris: A Review." *Marine Pollution Bulletin* 44:842–52.

DFO. 2012. "Fjords." Department of Fisheries and Oceans Canada, 30 March. http://www.glf.dfo-mpo.gc.ca/Gulf/By-The-Sea-Guide/Fjords.

Diaz, R. J. 2001. "Overview of Hypoxia around the World." *Journal of Environmental Quality* 30:275–81.

Diaz, R. J., and R. Rosenberg. 1995. "Marine Benthic Hypoxia—Review of Ecological Effects and Behavioral Responses on Macrofauna." *Oceanography and Marine Biology, Annual Review* 33:245–303.

———. 2008 "Spreading Dead Zones and Consequences for Marine Ecosystems." *Science* 321:926–29.

Dieckmann, U., M. Heino, and A. D. Rijnsdorp. 2009. "The Dawn of Darwinian Fishery Management." *ICES Insight* 46:34–43.

Digges, C. 2012. "Russia Announces Enormous Finds of Radioactive Waste and Nuclear Reactors in Arctic Seas." Bellona Foundation, 28 August. http://www.bellona.org/articles/articles_2012/Russia_reveals_dumps.

DiSavino, S. 2008. "PG&E Calif. Diablo Canyon Reactors Start Back." *Reuters*, 24 October. http://www.reuters.com/article/2008/10/24/utilities-operations-pge-diablo-idUSN2450595320081024.

DISL. 2007. "Invasive Australian Jellyfish Sighted in Gulf of Mexico." Dauphin Island Sea Lab via *ScienceDaily*, 18 August. http://www.sciencedaily.com/releases/2007/08/070817130118.htm.

Done, T., P. Whetton, R. Jones, R. Berkelmans, J. Lough, W. Skirving, and S. Wooldridge. 2003. *Global Climate Change and Coral Bleaching on the Great Barrier Reef.* Final report to the State of Queensland Greenhouse Taskforce through the Department of Natural Resources and Mines. Townsville, Queensland, and Aspendale, Victoria: CSIRO, Australian Institute of Marine Science, and CRC Reef.

Doney, S. C. 2006. "The Dangers of Ocean Acidification." Scientific American, February, 58–65.

———. 2010. "The Growing Human Footprint on Coastal and Open-Ocean Biogeochemistry." *Science* 328:1512–16.

Doney, S. C., V. J. Fabry, R. A. Feely, and J. Kleypas. 2009. "Ocean Acidification: The Other CO2 problem." *Annual Review of Marine Science* 1:169–92.

Dong, J., L.-x. Jiang, K.-f. Tan, H.-y. Liu, J. E. Purcell, P.-j. Li, and C.-c. Ye. 2009. "Stock Enhancement of the Edible Jellyfish (*Rhopilema esculentum* Kishinouye) in Liaodong Bay, China: A Review." *Hydrobiologia* 616:113–18.

Dong, Z., D. Liu, and J. K. Keesing. 2010. "Jellyfish Blooms in China: Dominant Species, Causes and Consequences." *Marine Pollution Bulletin* 60:954–63.

Doyle, T. K., H. D. Haas, D. Cotton, B. Dorschel, V. Cummins, J. D. R. Houghton, J. Davenport, and G. C. Hays. 2008. "Widespread Occurrence of the Jellyfish *Pelagia noctiluca* in Irish Coastal and Shelf Waters." *Journal of Plankton Research* 30:963–68.

Dr. Karl [Karl Kruszelnicki]. 2006. "Cockroaches and Radiation." Dr. Karl's Great Moments in Science, *ABC Radio*, 23 February. http://www.abc.net.au/science /articles/2006/02/23/1567313.htm?site=science/greatmomentsinscience.

Dugan, K. 2011. "Jellyfish Clog Waters for Fishermen in Gulf of Mexico." *Reuters*, 20 September. http://www.reuters.com/article/2011/09/20/us-jellyfish-gulf -idUSTRE78J5PX20110920.

Dumont, H. J. 2001. "Possible Consequences of the *Mnemiopsis* Invasion for the Bio-diversity of the Caspian Sea." Attachment 17. Caspian Environment Programme: *Mnemiopsis leidyi* in the Caspian Sea, First International Meeting, 24–26 April. http://www.caspianenvironment.org/newsite/mnemiopsis/mnem_attach17.htm.

Dunstan, W. M., P. P. Atkinson, and J. Natoli. 1975. "Stimulation and Inhibition of Phytoplankton Growth by Low Molecular Weight Hydrocarbons." *Marine Biology* 31:305–10.

Dybas, C. L. 2002. "Jellyfish 'Blooms' Could Be Sign of Ailing Seas." *Washington Post*, 6 May.

Dybern, B. I. 1967. "The Influence of the Medusa, *Tima bairdii*, on the Deep Sea Prawn Fishery in the Skagerak in 1966–1967." *Meddelande fran Havsfiskelaboratoriet-Lysekil* 34:1–6.

Edwards, M., and A. J. Richardson. 2004. "Impact of Climate Change on Marine Pelagic Phenology and Trophic Mismatch." *Nature* 430:881–84.

Egge, J. K., and D. L. Aksnes. 1992. "Silicate as Regulating Nutrient in Phytoplankton Competition." *Marine Ecology Progress Series* 83:281–89.

Eggleston, D. B., E. G. Johnson, G. T. Kellison, and D. A. Nadeau. 2003. "Intense Removal and Non-Saturating Functional Responses by Recreational Divers on Spiny Lobster *Panulirus argus*." *Marine Ecology Progress Series* 257:197–207.

Eiane, K., D. L. Aksnes, E. Bagøien, and S. Kaartvedt. 1999. "Fish or Jellies—A Question of Visibility?" *Limnology and Oceanography* 44:1352–57.

EPA. 2011. *RadNet Overview*. 1 April. http://www.epa.gov/enviro/facts/radnet/index .html.

Eremeev, V. N., and G. V. Zuyev. 2007. "Commercial Fishery Impact on the Modern Black Sea Ecosystem: A Review." *Turkish Journal of Fisheries and Aquatic Sciences* 7:75–82.

Eskenazi, B., J. Chevrier, L. G. Rosas, H. A. Anderson, M. S. Bornman, H. Bouwman, A. Chen, B. A. Cohn, C. de Jager, D. S. Henshel, F. Leipzig, J. S. Leipzig, E. C. Lorenz, S. M. Snedeker, and D. Stapleton. 2009. "The Pine River Statement: Human Health Consequences of DDT Use." *Environmental Health Perspectives* 117:1359–67.

Estes, J. A., J. Terborgh, J. S. Brashares, M. E. Power, J. Berger, W. J. Bond, S. R. Car-penter, T. E. Essington, R. D. Holt, J. B. C. Jackson, R. J. Marquis, L. Oksanen,

T. Oksanen, R. T. Paine, E. K. Pikitch, W. J. Ripple, S. A. S. M. Scheffer, T. W. Schoener, J. B. Shurin, A. R. E. Sinclair, M. E. Soulé, R. Virtanen, and D. A. Wardle. 2011. "Trophic Downgrading of Planet Earth." *Science* 333:301–6.

Estes, J. A., M. T. Tinker, T. M. Williams, and D. F. Doak. 1998. "Killer Whale Predation on Sea Otters Linking Oceanic and Nearshore Ecosystems." *Science* 282:473–76.

Faasse, M. A., and K. M. Bayha. "The Ctenophore *Mnemiopsis leidyi* A. Agassiz 1865 in Coastal Waters of The Netherlands: An Unrecognized Invasion?" *Aquatic Invasions* 1:270–77.

FAO. 2005. "Review of the State of World Marine Fishery Resources." FAO Marine Resources Service—Fishery Resources Division. FAO Fisheries Technical Paper no. 457. Rome: FAO.

Faris, S. 2009. "Jellyfish: A Gelatinous Invasion." *Time*, 2 November. http://www .time.com/time/magazine/article/0,9171,1931659,00.html.

Feely, R. A., C. L. Sabine, J. M. Hernandez-Ayon, D. Ianson, and B. Hales. 2008. "Evidence for Upwelling of Corrosive 'Acidified' Water onto the Continental Shelf." *Science* 320:1490–92.

Feely, R. A., C. L. Sabine, K. Lee, W. Berelson, J. Kleypas, V. J. Fabry, and F. J. Millero. 2004. "Impact of Anthropogenic CO2 on the CaCO3 System in the Oceans." *Science* 305:362–66.

Fenner, P., J. Lippmann, and L. Gershwin. 2010. "Fatal and Non-Fatal Severe Jellyfish Stings in Thai Waters." *Journal of Travel Medicine* 17 (2): 133–38.

Ferguson, H. W., C. M. J. Delannoy, S. Hay, J. Nicolson, D. Sutherland, and M. Crumlish. 2010. "Jellyfish as Vectors of Bacterial Disease for Farmed Salmon." *Journal of Veterinary Diagnostic Investigation* 22:376–82.

Ferretti, F., R. A. Myers, F. Serena, and H. K. Lotze. 2008. "Loss of Large Predatory Sharks from the Mediterranean Sea." *Conservation Biology* 22:952–64.

Finenko, G. A., A. E. Kideys, B. E. Anninsky, T. A. Shiganova, A. Roohi, M. R. Tabari, H. Rostami, and S. Bagheri. 2006. "Invasive Ctenophore *Mnemiopsis leidyi* in the Caspian Sea: Feeding, Respiration, Reproduction and Predatory Impact on the Zooplankton Community." *Marine Ecology Progress Series* 314:171–85.

Flannery, T. 2005. The Weather Makers. Melbourne: Text Publishing.

Forsyth, M. A., S. Kennedy, S. Wilson, T. Eybatov, and T. Barrett. 1998. "Canine Distemper Virus in a Caspian Seal." *Veterinary Record* 143:662–64.

Fosså, J. H. 1992. "Mass Occurrence of *Periphylla periphylla* (Scyphozoa, Coronatae) in a Norwegian Fjord." *Sarsia* 77:237–51.

Fuentes, V. L., D. L. Angel, K. M. Bayha, D. Atienza, D. Edelist, C. Bordehore, J.-M. Gili, and J. E. Purcell. 2010. "Blooms of the Invasive Ctenophore, *Mnemiopsis leidyi*, Span the Mediterranean Sea in 2009." *Hydrobiologia* 645:23–37.

Fuentes, V. L., D. Atienza, J.-M. Gili, and J. E. Purcell. 2009. "First Records of *Mnemiopsis leidyi* A. Agassiz 1865 off the NW Mediterranean Coast of Spain." *Aquatic Invasions* 4:671–74.

Fulton, E. A., A. D. M. Smith, D. C. Smith, and I. E. van Putten. 2011. "Human Behaviour: The Key Source of Uncertainty in Fisheries Management." *Fish and Fisheries* 12:2–17.

Galil, B., L. Gershwin, J. Douek, and B. Rinkevich. 2010. "*Marivagia stellata* gen. et sp. nov. (Scyphozoa: Rhizostomeae: Cepheidae), Another Alien Jellyfish from the Mediterranean Coast of Israel." *Aquatic Invasions* 5:331–40.

Galil, B., and A. Zenetos. 2002. "A Sea Change—Exotics in the Eastern Mediterranean Sea." In *Invasive Aquatic Species of Europe: Distribution, Impacts and Management*, edited by E. Leppakoski, S. Gollasch and S. Olenin, 1–19. Dordrecht: Kluwer Academic Publishers.

Galil, B. S. 2007. "Loss or Gain? Invasive Aliens and Biodiversity in the Mediterranean Sea." *Marine Pollution Bulletin* 55:314–22.

Galil, B. S., N. Kress, and T. A. Shiganova. 2009. "First Record of *Mnemiopsis leidyi* A. Agassiz, 1865 (Ctenophora; Lobata; Mnemiidae) off the Mediterranean Coast of Israel." *Aquatic Invasions* 4:357–60.

Galil, B. S., L. Shoval, and M. Goren. 2009. "*Phyllorhiza punctata* von Lendenfeld, 1884 (Scyphozoa: Rhizostomeae: Mastigiidae) Reappeared off the Mediterranean Coast of Israel." *Aquatic Invasions* 4:481–83.

Galil, B. S., E. Spanier, and W. W. Ferguson. 1990. "The Scyphomedusae of the Mediterranean Coast of Israel, Including Two Lessepsian Migrants New to the Mediterranean." *Zoologische Mededelingen* 64:95–105.

Gershwin, L. 1999. "Clonal and Population Variation in Jellyfish Symmetry." *Journal of the Marine Biological Association of the United Kingdom* 79:993–1000.

———. 2001. "Systematics and Biogeography of the Jellyfish *Aurelia labiata* (Cnidaria: Scyphozoa)." *Biological Bulletin* (Woods Hole) 201:104–19.

Gershwin, L., M. De Nardi, P. J. Fenner, and K. D. Winkel. 2009. "Marine Stingers: Review of an Under-Recognized Global Coastal Management Issue." *Journal of Coastal Management* 38:22–41.

GESAMP. 2010. Proceedings of the GESAMP International Workshop on Plastic Particles as a Vector in Transporting Persistent, Bio-Accumulating and Toxic Substances in the Oceans, edited by T. Bowmer and P. Gershaw. *GESAMP Reports and Studies* no. 82. Paris: UNESCO-IOC.

Gesner, J., M. Chebanov, and J. Freyhof. 2010. *Huso huso*. IUCN Red List of Threatened Species. Version 2012.1. Accessed 25 August 2012. http://www.iucnredlist.org/details/10269/0.

Ghabooli, S., T. A. Shiganova, A. Zhan, M. E. Cristescu, P. Eghtesadi-Araghi, and H. J. MacIsaac. 2011. "Multiple Introductions and Invasion Pathways for the Invasive Ctenophore *Mnemiopsis leidyi* in Eurasia." *Biological Invasions* 13:679–90.

Gibbs, S. J., P. R. Bown, J. A. Sessa, T. J. Bralower, and P. A. Wilson. 2006. "Nannoplankton Extinction and Origination across the Paleocene-Eocene Thermal Maximum." *Science* 314:1770–73.

Gille, D. 2003. "Seawater Intakes for Desalination Plants." *Desalination* 156:249–56.

GISP. 2010. *Invasive Species, Climate Change and Ecosystem-Based Adaptation: Addressing Multiple Drivers of Global Change*. Washington DC, and Nairobi, Kenya: Global Invasive Species Programme.

GloBallast. 2011. "Ten of the Most Unwanted." GloBallast and International Maritime Organization. http://globallast.imo.org/poster4_english.pdf.

———. 2012. "The Issue." GloBallast and International Maritime Organization. http://globallast.imo.org/index.asp?page=problem.htm&menu=true.

Goldburg, R. J., M. S. Elliott, and R. L. Naylor. 2001. *Marine Aquaculture in the United States: Environmental Impacts and Policy Options*. Philadelphia, PA: Pew Oceans Commission.

Gosselin, L. A., and P. Y. Qian. 1997. "Juvenile Mortality in Benthic Marine Invertebrates." *Marine Ecology Progress Series* 146:265–82.

Govan, F. 2010. "Jellyfish Invasion Closes Beaches across Spain." *Telegraph*, 2 August. http://www.telegraph.co.uk/earth/wildlife/7922422/Jellyfish-invasion-closes-beaches-across-Spain.html.

Goy, J., P. Morand, and M. Etienne. 1989. "Long-Term Fluctuations of *Pelagia noctiluca* (Cnidaria, Scyphomedusa) in the Western Mediterranean Sea: Prediction by Climatic Variables." *Deep-Sea Research Part A Oceanographic Research Papers* 36:269–79.

Grady, M., and J. Brook. 2000. "Senate Inquiry into Gulf St. Vincent." Conservation Council of South Australia, Inc., 3 February. http://www.ccsa.asn.au/submissions/gulfstvincent.htm.

Graham, E. R., and J. T. Thompson. 2009. "Deposit- and Suspension-Feeding Sea Cucumbers (Echinodermata) Ingest Plastic Fragments." *Journal of Experimental Marine Biology and Ecology* 368:22–29.

Graham, W. M., D. L. Martin, D. L. Felder, V. L. Asper, and H. M. Perry. 2003. "Ecological and Economic Implications of a Tropical Jellyfish Invader in the Gulf of Mexico." *Biological Invasions* 5:53–69.

Greenpeace. 2012a. "Coal." Greenpeace International. Accessed 28 April. http://www.greenpeace.org/international/en/campaigns/climate-change/coal/.

———. 2012b. "Greenpeace International Seafood Red List." Greenpeace International. Accessed 27 August 2012. http://www.greenpeace.org/international/en/campaigns/oceans/seafood/red-list-of-species/.

Gregory, M. 2009. "Environmental Implications of Plastic Debris in Marine Settings—Entanglement, Ingestion, Smothering, Hangers-On, Hitch-Hiking and Alien Invasions." *Philosophical Transactions of the Royal Society of London, Series B, Biological Sciences* 364:2013–25.

Greve, W., and T. R. Parsons. 1977. "Photosynthesis and Fish Production: Hypothetical Effects of Climatic Change and Pollution." *Helgoländer wissenschaftliche Meeresuntersuchungen* 30:666–72.

Griffin, D. B., and T. M. Murphy. 2012. "Cannonball Jellyfish: *Stomolophus meleagris*." South Carolina Department of Natural Resources. Accessed 1 September 2012. www.dnr.sc.gov/cwcs/pdf/Cannonballjellyfish.pdf.

Guenther, J., E. Misimi, and L. M. Sunde. 2010. "The Development of Biofouling, Particularly the Hydroid *Ectopleura larynx*, on Commercial Cage Nets in Mid-Norway." *Aquaculture* 300:120–27.

Guest, W. C. 1959. "The Occurrence of the Jellyfish *Chiropsalmus quadrumanus* in Matagorda Bay, TX." *Bulletin of Marine Science of the Gulf and Caribbean* 9:79–83.

Gunter, L., P. Gunter, S. Cullen, and N. Burton. 2001. *Licensed to Kill—How the Nuclear Power Industry Destroys Endangered Marine Wildlife and Ocean Habitat to Save Money.* Washington DC: Nuclear Information and Resource Service.

Haaker, P. 1998. "White Abalone—Off the Deep End . . . Forever?" *Outdoor California,* January–February, 17–20.

Hackett, J. D., D. M. Anderson, D. L. Erdner, and D. Bhattacharya. 2004. "Dino-flagellates: A Remarkable Evolutionary Experiment." *American Journal of Botany* 91:1523–34.

Haddad, M. A., and M. Nogueira Jr. 2006. "Reappearance and Seasonality of *Phyllorhiza punctata* von Lendenfeld (Cnidaria, Scyphozoa, Rhizostomeae) Medusae in Southern Brazil." *Revista Brasileira de Zoologia* 23:824–31.

Hall-Spencer, J. M., R. Rodolfo-Metalpa, S. Martin, E. Ransome, M. Fine, S. M. Turner, S. J. Rowley, D. Tedesco, and M.-C. Buia. 2008. "Volcanic Carbon Dioxide Vents Show Ecosystem Effects of Ocean Acidification." *Nature* 454:96–99.

Hallegraeff, G. 1993. "A Review of Harmful Algal Blooms and Their Apparent Global Increase." *Phycologia* 32:79–99.

Hallowell, C. 1998. "Save The Swordfish." *Time*, 26 January. http://www.time.com /time/magazine/article/0,9171,987715,00.html.

Halpern, B. S., S. Walbridge, K. A. Selkoe, C. V. Kappel, F. Micheli, C. D'Agrosa, J. F. Bruno, K. S. Casey, C. Ebert, H. E. Fox, R. Fujita, D. Heinemann, H. S. Lenihan, E. M. P. Madin, M. T. Perry, E. R. Selig, M. Spalding, R. Steneck, and R. Watson. 2008. "A Global Map of Human Impact on Marine Ecosystems." *Science* 319:948–52.

Hamer, H. H., A. M. Malzahn, and M. Boersma. 2011. "The Invasive Ctenophore *Mnemiopsis leidyi*: A Threat to Fish Recruitment in the North Sea?" *Journal of Plankton Research* 33:137–44.

Hamner, W. H., and M. N. Dawson. 2009. "A Review and Synthesis on the Systematics and Evolution of Jellyfish Blooms: Advantageous Aggregations and Adaptive Assemblages." *Hydrobiologia* 616:161–91.

Hamner, W. M., and R. M. Jenssen. 1974. "Growth, Degrowth, and Irreversible Cell Differentiation in *Aurelia aurita*." *American Zoologist* 14:833–49.

Hansson, H. G. 2006. "Ctenophores of the Baltic and Adjacent Seas—The Invader *Mnemiopsis leidyi* Is Here!" *Aquatic Invasions* 1:295–98.

Harashima, A., T. Kimoto, T. Wakabayashi, and T. Toshiyasu. 2006. "Verification of the Silica Deficiency Hypothesis Based on Biogeochemical Trends in the Aquatic Continuum of Lake Biwa—Yodo River—Seto Inland Sea, Japan." *Ambio* 35:36–42.

Hardt, M. J., and C. Safina. 2010. "Threatening Ocean Life from the Inside." *Scientific American*, August, 66–73.

Harrabin, R. 2007. "China Building More Power Plants." *BBC*, 19 June. http://news .bbc.co.uk/2/hi/6769743.stm.

Harvey, F. 2011. "Fishermen to Catch Plastic in EU Plan to Protect Fish." *Guardian Weekly*, 10 June, 8.

Haslob, H., C. Clemmesen, M. Schaber, H. Hinrichsen, J. O. Schmidt, R. Voss, G. Kraus, and F. W. Köster. 2007. "Invading *Mnemiopsis leidyi* as a Potential Threat to Baltic Fish." *Marine Ecology Progress Series* 349:303–6.

Havenhand, J. N., F. R. Buttler, M. C. Thorndyke, and J. E. Williamson. 2008. "Near-Future Levels of Ocean Acidification Reduce Fertilization Success in a Sea Urchin." *Current Biology* 18:R651–52.

Head, J. 2005. "Japan Pushes Whale Meat Revival." *BBC News*, 19 June. http://news .bbc.co.uk/2/hi/asia-pacific/4106688.stm.

Hedgpeth, J. W. 1979. "The Oceans: World Sump." In *Environmental Problems: Principles, Readings, and Comments*, 2nd edition, edited by W. H. Mason and G. W. Folkerts, 146–53. Dubuque, IA: Wm. C. Brown Company.

HELCOM. 2007. Towards a Baltic Sea Unaffected by Hazardous Substances. HELCOM Overview for Ministerial Meeting. Krakow: Helsinki Commission.

Helly, J., and L. A. Levin. 2004. "Global Distribution of Naturally Occurring Marine Hypoxia." *Deep-Sea Research* 51:1159–68.

Henderson, C. 2006. "Ocean Acidification: The Other CO2 problem." *New Scientist*, 5 August, 28–33.

Heron, A. C. 1972. "Population Ecology of a Colonizing Species: The Pelagic Tunicate *Thalia democratica* I. Individual Growth Rate and Generation Time." *Oecologia* 10:269–93.

Hewitt, C. L., M. L. Campbell, R. E. Thresher, R. B. Martin, S. Boyd, B. F. Cohen, D. R. Currie, M. F. Gomon, M. J. Keough, J. A. Lewis, M. M. Lockett, N. Mays, M. A. McArthur, T. D. O'Hara, G. C. B. Poore, D. J. Ross, M. J. Storey, J. E. Watson, and R. S. Wilson. 2004. "Introduced and Cryptogenic Species in Port Phillip Bay, Victoria, Australia." *Marine Biology* 144:183–202.

Higgins, A. 2011. "Japan Nuclear Radiation Rainwater Update—Idaho Iodine Levels 14,066% above EPA Limit." Alexander Higgins Blog, 23 April. http://blog .alexanderhiggins.com/2011/04/23/japan-nuclear-radiation-rainwater-update -idaho-iodine-levels-14066-epa-limit-19907/.

Hiromi, J., T. Kasuya, and H. Ishii. 2005. "Impacts of Massive Occurrence of Jellyfish on Pelagic Ecosystem" [in Japanese]. *Bulletin of Plankton Society of Japan* 52:82–90.

Hoegh-Guldberg, O. 1999. "Climate Change: Coral Bleaching and the Future of the World's Coral Reefs." *Marine and Freshwater Research* 50:839–66.

———. 2005. "Climate Change and Marine Ecosystems." In *Climate Change and Biodiversity*, edited by T. E. Lovejoy and L. J. Hannah, 256–73. New Haven, CT: Yale University Press.

———. 2010. "Dangerous Shifts in Ocean Ecosystem Function?" *ISME Journal* 4:1090–92.

Hoegh-Guldberg, O., P. J. Mumby, A. J. Hooten, R. S. Steneck, P. Greenfield, E. Gomez, C. D. Harvell, P. F. Sale, A. J. Edwards, K. Caldeira, N. Knowlton, C. M. Eakin, R. Iglesias-Prieto, N. Muthiga, R. H. Bradbury, A. Dubi, and M. E. Hatziolos. 2007. "Coral Reefs under Rapid Climate Change and Ocean Acidification." *Science* 318:1737–42.

Hoegh-Guldberg, O., and J. S. Pearse. 1995. "Temperature, Food Availability, and the Development of Marine Invertebrate Larvae." *American Zoologist* 35:415–25.

Hoff, G. R. 2006. "Biodiversity as an Index of Regime Shift in the Eastern Bering Sea." *Fishery Bulletin* 104:226–37.

Holst, S., and G. Jarms. 2007. "Substrate Choice and Settlement Preferences of Planula Larvae of Five Scyphozoa (Cnidaria) from German Bight, North Sea." *Marine Biology* 151:863–71.

———. 2010. "Effects of Low Salinity on Settlement and Strobilation of Scyphozoa (Cnidaria): Is the Lion's Mane *Cyanea capillata* (L.) Able to Reproduce in the Brackish Baltic Sea?" *Hydrobiologia* 645:53–68.

Hood, M. 2004. "A Carbon Sink that Can no Longer Cope?" World of Science 2:2–5.

Hooper, C., P. Hardy-Smith, and J. Handlinger. 2007. "Ganglioneuritis Causing High Mortalities in Farmed Australian Abalone (*Haliotis laevigata* and *Haliotis rubra*)." *Australian Veterinary Journal* 85:188–93.

Hoover, R. A., and J. E. Purcell. 2009. "Substrate Preferences of Scyphozoan *Aurelia labiata* Polyps among Common Dock-Building Materials." *Hydrobiologia* 616:259–67.

Hough, A. 2011. "British Explorers Row 450 Miles to North Pole in World First Voyage." *Telegraph*, 26 August. http://www.telegraph.co.uk/earth/earthnews /8724098/British-explorers-row-450-miles-to-North-Pole-in-world-first-voyage .html.

Houghton, J. D. R., T. K. Doyle, J. Davenport, and G. C. Hays. 2006. "Developing a Simple, Rapid Method for Identifying and Monitoring Jellyfish Aggregations from the Air." *Marine Ecology Progress Series* 314:159–70.

Hoyt, S. F. 1912. "The Name of the Red Sea." *Journal of the American Oriental Society* 32:115–19.

HSI. 2010. "The Threat of Shark Control Programs on Marine Species." Humane Society International, 9 December. http://www.hsi.org.au/?catID=116.

Huang, Y. 1986. "The Processing of Cannonball Jellyfish and Its Utilization." SST 11th Annual Conference, Seafood Science and Technology Society of the Americas, Tampa, Florida, 13–16 January.

Hughes, T. P. 1994. "Catastrophes, Phase Shifts, and Large-Scale Degradation of a Caribbean Coral Reef." *Science* 265:1547–51.

Hughes, T. P., A. H. Baird, D. R. Bellwood, M. Card, S. R. Connolly, C. Folke, R. Grosberg, O. Hoegh-Guldberg, J. B. C. Jackson, J. Kleypas, J. M. Lough, P. Marshall, M. Nyström, S. R. Palumbi, J. M. Pandolfi, B. Rosen, and J. Roughgarden. 2003. "Climate Change, Human Impacts, and the Resilience of Coral Reefs." *Science* 301:929–33.

Huq, A., E. B. Small, P. A. West, M. I. Huq, R. Rahman, and R. R. Colwell. 1983. "Ecological Relationships between *Vibrio cholerae* and Planktonic Crustacean Copepods." *Applied and Environmental Microbiology* 45:275–83.

Hutchings, J. A. 2000. "Collapse and Recovery of Marine Fishes." Nature 406:882–85.

Hutchings, J. A., and J. D. Reynolds. 2004. "Marine Fish Population Collapses: Consequences for Recovery and Extinction Risk." *BioScience* 54:297–309.

Huwer, B., M. Storr-Paulsen, H. U. Riisgård, and H. Haslob. 2008. "Abundance, Horizontal and Vertical Distribution of the Invasive Ctenophore *Mnemiopsis leidyi* in the Central Baltic Sea, November 2007." *Aquatic Invasions* 3:113–24.

Iglesias-Rodriguez, M. D., P. R. Halloran, R. E. M. Rickaby, I. R. Hall, E. Colmenero-Hidalgo, J. R. Gittins, D. R. H. Green, T. Tyrrell, S. J. Gibbs, P. V. Dassow, E. Rehm, E. V. Armbrust, and K. P. Boessenkool. 2008. "Phytoplankton Calcification in a High-CO2 World." *Science* 320:336–40.

IMO. 2004. "International Convention for the Control and Management of Ships' Ballast Water and Sediments (BWM)." International Maritime Organization, 13 February. http://www.imo.org/about/conventions/listofconventions/pages /international-convention-for-the-control-and-management-of-ships'-ballast-water -and-sediments-(bwm).aspx.

IPCC. 2007. *Climate Change 2007: Synthesis Report.* Geneva: Intergovernmental Panel on Climate Change.

ISSG. 2008. "100 of the World's Worst Invasive Alien Species." Invasive Species Specialist Group. Accessed 5 June. http://www.issg.org/worst100_species.html.

Ivanov, V. P., A. M. Kamakin, V. B. Ushivtzev, T. Shiganova, O. Zhukova, N. Aladin, S. I. Wilson, G. R. Harbison, and H. J. Dumont. 2000. "Invasion of the Caspian Sea by the Comb Jellyfish *Mnemiopsis leidyi* (Ctenophora)." *Biological Invasions* 2:255–58.

Jackson, J. B. C. 2001. "What Was Natural in the Coastal Oceans?" *Proceedings of the National Academy of Sciences U S A* 98:5411–18.

———. 2008. "Ecological Extinction and Evolution in the Brave New Ocean." *Proceedings of the National Academy of Sciences U S A* 105:11458–65.

———. 2010. "The Future of the Oceans Past." *Philosophical Transactions of the Royal Society of London, Series B, Biological Sciences* 365:3765–78.

Jackson, J. B. C., M. X. Kirby, W. H. Berger, K. A. Bjorndal, L. W. Botsford, B. J. Bourque, R. H. Bradbury, R. Cooke, J. Erlandson, J. A. Estes, T. P. Hughes, S. Kidwell, C. B. Lange, H. S. Lenihan, J. M. Pandolfi, C. H. Peterson, R. S. Steneck, M. J. Tegner, and R. R. Warner. 2001. "Historical Overfishing and the Recent Collapse of Coastal Ecosystems." *Science* 293:629–37.

Jacquet, J., D. Pauly, D. Ainley, P. Dayton, and J. Jackson. 2010. "Seafood Stewardship in Crisis." *Nature* 467:28–29.

Jacups, S. P. 2010. "Warmer Waters in the Northern Territory—Herald an Earlier Onset to the Annual *Chironex fleckeri* Stinger Season." *Ecohealth* 7:14–17.

Jaspers, C., T. Kiørboe, K. Tönnesson, and M. Haraldsson. 2011. "The Physical Characteristics of the Baltic Sea Might Act as a Bottleneck for the *Mnemiopsis leidyi* Population Expansion in This Newly Invaded Area." In 5th International Zooplankton Production Symposium, Pucón, Chile. http://www.pices.int/publications /presentations/zoop-2011/s3/S3-1450-Jaspers.pdf.

Javidpour, J., J. C. Molinero, A. Lehmann, T. Hansen, and U. Sommer. 2009a. "Annual Assessment of the Predation of *Mnemiopsis leidyi* in a New Invaded Environment, the Kiel Fjord (Western Baltic Sea): A Matter of Concern?" *Journal of Plankton Research* 31:729–38.

Javidpour, J., J. C. Molinero, J. Peschutter, and U. Sommer. 2009b. "Seasonal Changes and Population Dynamics of the Ctenophore *Mnemiopsis leidyi* after Its First Year of Invasion in the Kiel Fjord, Western Baltic Sea." *Biological Invasions* 11:873–82.

Javidpour, J., U. Sommer, and T. A. Shiganova. 2006. "First Record of *Mnemiopsis leidyi* A. Agassiz 1865 in the Baltic Sea." *Aquatic Invasions* 1:299–302.

"Jellyfish Take On U.S. Warship (USS Ronald Reagan)." 2006. *News Limited* via *Democratic Underground* (message board), 27 January. http://www.democratic underground.com/discuss/duboard.php?az=view_all&address=105x4655746.

Johansson, S., U. Larsson, and P. Boehm. 1980. "The Tsesis Oil Spill: Impact on the Pelagic Ecosystem." *Marine Pollution Bulletin* 11:284–93.

Johnson, C. R., S. D. Ling, D. J. Ross, S. Shepherd, and K. J. Miller. 2005. *Establishment of the Long-Spined Sea Urchin* (Centrostephanus rodgersii) *in Tasmania: First Assessment of Potential Threats to Fisheries*. Project report. Hobart, Tasmania: School of Zoology and Tasmanian Aquaculture and Fisheries Institute.

Johnston, N. 2002. "Sea Urchins Invade Tasmanian Waters." *World Today*, 23 August. http://www.abc.net.au/worldtoday/stories/s657194.htm.

Jones, K. 2010. "Giant Coral Die-Off Found—Gulf Spill "'Smoking Gun?'" *National Geographic News*, 5 November. http://news.nationalgeographic.com/news/2010 /11/101105-deepwater-coral-dieoff-gulf-oil-spill-science-environment/.

Jowit, J. 2011. "Paul Ehrlich, A Prophet of Global Population Doom Who Is Gloomier than Ever." *Guardian*, 23 October. http://www.guardian.co.uk/environment/2011 /oct/23/paul-ehrlich-global-collapse-warning.

Joyce, E. 2010. "Black Abalone Disappearing Off California Coast: Endangered Shellfish Threatened by Climate Change?" *KPBS*, 6 May. http://www.kpbs.org/news /2010/may/06/black-abalone-disappearing-california-coast/.

Kahn, J., and J. Yardley. 2007. "Choking on Growth: As China Roars, Pollution Reaches Deadly Extremes." *New York Times*, 26 August. http://www.nytimes .com/2007/08/26/world/asia/26china.html.

Karpov, K. A., P. L. Haaker, I. K. Taniguchi, and L. Rogers-Bennett. 2000. "Serial Depletion and the Collapse of the California Abalone (*Haliotis* spp.) Fishery." *Canadian Special Publication Fisheries and Aquatic Sciences* 130:11–24.

Kawabe, A., and F. W. Traplin. 1986. "Control of Macrofouling in Japan." *Existing and Experimental Methods, Research Report Section* 23:1–44.

Kawahara, M., S.-i. Uye, K. Ohtsu, and H. Iizumi. 2006. "Unusual Population Explosion of the Giant Jellyfish *Nemopilema nomurai* (Scyphozoa: Rhizostomeae) in East Asian Waters." *Marine Ecology Progress Series* 307:161–73.

Kearney, A. 2010. "Environmental Impacts of Shark Nets on the KwaZulu-Natal Coast." Undergraduate honors thesis, Indiana University.

Kempton, H. 2011. "Alarm at Derwent Invaders." Mercury, 21 April. http://www .themercury.com.au/article/2011/04/21/224301_tasmania-news.html.

Kennedy, A. D. 1997. "Case Study 2: Recent Changes to the Tuggerah Lakes System as a Result of Increasing Human Pressures." In *Nutrients in Marine and Estuarine Environments: State of the Environment Technical Paper Series* (Estuaries and the Sea), edited by P. R. Cosser. Canberra: Department of the Environment.

Kennedy, C. 2009. "An Upwelling Crisis: Ocean Acidification." *NOAA ClimateWatch Magazine*, 30 October. http://www.climatewatch.noaa.gov/2009/articles/an -upwelling-crisis/1.

Kennedy, S., T. Kuiken, P. D. Jepson, R. Deaville, M. Forsyth, T. Barrett, M. W. G. van de Bildt, A. D. M. E. Osterhaus, T. Eybatov, C. Duck, A. Kydyrmanov, I. Mitrofanov, and S. Wilson. 2000. "Mass Die-Off of Caspian Seals Caused by Canine Distemper Virus." *Emerging Infectious Diseases* 6:637–39.

Kerr, R. A. 2010. "Ocean Acidification Unprecedented, Unsettling." *Science* 328:1500–1.

Kideys, A. E., and A. C. Gücü. 1995. "*Rhopilema nomadica*: A Lessepsian Scyphomedusan New to the Mediterranean Coast of Turkey." *Israel Journal of Zoology* 41:615–17.

Kideys, A. E., and M. Moghim. 2003. "Distribution of the Alien Ctenophore *Mnemiopsis leidyi* in the Caspian Sea in August 2001." *Marine Biology* 142:163–71.

Kideys, A. E., and U. Niermann. 1994. "Occurrence of *Mnemiopsis* along the Turkish Coast." *ICES Journal of Marine Science* 51:423–27.

Kideys, A. E., A. Roohi, S. Bagheri, G. Finenko, and L. Kamburska. 2005. "Impacts of Invasive Ctenophores on the Fisheries of the Black Sea and Caspian Sea." *Oceanography* 18:76–85.

Kingsford, M. J., K. A. Pitt, and B. M. Gillanders. 2000. "Management of Jellyfish Fisheries, with Special Reference to the Order Rhizostomeae." *Oceanography and Marine Biology: An Annual Review* 38:85–156.

Kirby, A. 2003. "Nets 'Kill 800 Cetaceans a Day.'" *BBC News*, 13 June. http://news.bbc.co.uk/2/hi/science/nature/2985630.stm.

Kirby, R. R., and G. Beaugrand. 2009. "Trophic Amplification of Climate Warming." *Proceedings of the Royal Society, Series B, Biological Sciences* 276:4095–103.

Kirby, R. R., G. Beaugrand, and J. A. Lindley. 2009. "Synergistic Effects of Climate and Fishing in a Marine Ecosystem." *Ecosystems* 12:548–61.

Kishinouye, K. 1922. "On a New Rhizostome Medusa. *Nemopilema nomurai* ng. nsp" [in Japanese]. *Dobutsugaku Zassi* 34:343–46.

Kloser, R., C. Sutton and K. Krusic-Golub. 2012. *Australian Spawning Population of Orange Roughy: Eastern Zone Acoustic and Biological Index Fished from 1987 to 2010.* Hobart, Tasmania: CSIRO Marine and Atmospheric Research.

Koslow, J. A., G. W. Boehlert, J. D. M. Gordon, R. L. Haedrich, P. Lorance, and N. Parin. 2000. "Continental Slope and Deep-Sea Fisheries: Implications for a Fragile Ecosystem." *ICES Journal of Marine Science* 57:548–57.

Koslow, J. A., K. Gowlett-Holmes, J. K. Lowry, T. O'Hara, G. C. B. Poore, and A. Williams. 2001. "Seamount Benthic Macrofauna off Southern Tasmania: Community Structure and Impacts of Trawling." *Marine Ecology Progress Series* 213:111–25.

Kremer, P. 1979. "Predation by the Ctenophore *Mnemiopsis leidyi* in Narragansett Bay, Rhode Island." *Estuaries* 2:97–105.

KrillCount. 2011. "The Issue: Increasing Demand for Krill." The Antarctic Krill Conservation Project. Accessed 16 April 2011. http://www.krillcount.org/issues.html.

Laidre, K., M. P. Heide-Jørgensen, H. Stern, and P. Richard. 2012. "Unusual Narwhal Sea Ice Entrapments and Delayed Autumn Freeze-Up Trends." *Polar Biology* 35:149–54.

Lamarck, J. B. P. A. 1809. *Philosophie Zoologique, ou exposition des considérations relatives à l'histoire naturelle des animaux*. Paris: Dentu.

Land, T. 1999. "Pollution and Politics in the Black Sea." *Contemporary Review*, May, 1–7.

Lang, T. 2008. *Biological Effects of Contaminants and Fish Diseases in the North Sea and Baltic Sea*. Cuxhaven, Germany: Institut für Fischereiökologie.

Larson, R. J. 1987. "A Note on the Feeding, Growth, and Reproduction of the Epipelagic Scyphomedusa *Pelagia noctiluca* (Forskal)." *Biological Oceanography* 4:447–54.

———. 1988. "Feeding and Functional Morphology of the Lobate Ctenophore *Mnemiopsis mccradyi*." *Estuarine Coastal and Shelf Science* 27:495–502.

Lavers, J. 2011. "Marine Debris (or Plastic Pollution) Is a Global Problem." Jennifer Lavers. Accessed 2 September 2011. http://www.jenniferlavers.org/plastic-pollution/.

Lavers, J. L., and I. L. Jones. 2007. "Factors Affecting Rates of Intraspecific Kleptoparasitism and Breeding Success of the Razorbill at the Gannet Islands, Labrador." *Marine Ornithology* 35:1–7.

LeGrand, M. 2010. "The Sting." *Chesapeake Bay Magazine*, March. http://www .chesapeakeboating.net/Publications/Chesapeake-Bay-Magazine/2010/March -2010/The-Sting.aspx.

Lehtiniemi, M., J.-P. Pääkkönen, J. Flinkman, T. Katajisto, E. Gorokhova, M. Karjalainen, S. Viitasalo, and H. Björk. 2007. "Distribution and Abundance of the American Comb Jelly (*Mnemiopsis leidyi*)—a Rapid Invasion to the Northern Baltic Sea during 2007." *Aquatic Invasions* 2:445–49.

Leppäkoski, E., S. Gollasch, P. Gruszka, H. Ojaveer, S. Olenin, and V. Panov. 2002. "The Baltic: A Sea of Invaders." *Canadian Journal of Fisheries and Aquatic Sciences* 59:1175–88.

Leschin-Hoar, C. 2011. "Fleeced Again: How Microplastic Causes Macro Problems for the Ocean." *Grist*, 7 December. http://grist.org/living/2011-12-07-how -microplastics-cause-macro-problems-for-the-ocean/.

Lewison, R. L., L. B. Crowder, A. J. Read, and S. A. Freeman. 2004. "Understanding Impacts of Fisheries Bycatch on Marine Megafauna." *Trends in Ecology and Evolution* 19:598–604.

Licandro, P., D. V. P. Conway, M. N. D. Yahia, M. L. F. de Puelles, S. Gasparini, J. H. Hecq, P. Tranter, and R. R. Kirby. 2010. "A Blooming Jellyfish in the Northeast Atlantic and Mediterranean." *Biology Letters* 6:688–91.

Link, J. S., and M. D. Ford. 2006. "Widespread and Persistent Increase of Ctenophora in the Northeast US Shelf Ecosystem." *Marine Ecology Progress Series* 320:153–59.

Lippmann, J. M., P. J. Fenner, K. Winkel, and L. Gershwin. 2011. "Fatal and Severe Box Jellyfish Stings, Including Irukandji Stings, in Malaysia, 2000–2010." *Journal of Travel Medicine* 18 (4): 275–81.

Litchman, E., A. Klausmeier, and K. Yoshiyama. 2009. "Contrasting Size Evolution in Marine and Freshwater Diatoms." *Proceedings of the National Academy of Sciences U S A* 106:2665–70.

Liu, P., Y. Yu, and C. Liu. 1991. "Studies on the Situation of Pollution and Countermeasures of Control of the Oceanic Environment in Zhoushan Fishing Ground— The Largest Fishing Ground in China." *Marine Pollution Bulletin* 23:281–88.

Llanos, M. 2005. "44 Oil Spills Found in Southeast Louisiana." *MSNBC*, 19 September. http://www.msnbc.msn.com/id/9365607/.

Lo, W.-T., J. E. Purcell, J.-J. Hung, H.-M. Su, and P.-K. Hsu. 2008. "Enhancement of Jellyfish (*Aurelia aurita*) Populations by Extensive Aquaculture Rafts in a Coastal Lagoon in Taiwan." *ICES Journal of Marine Science* 65:453–61.

Lochbaum, D. 2011. "Jellyfish Put Nuclear Plant in a Jam." *All Things Nuclear*. 4 January. http://allthingsnuclear.org/post/2595096798/fission-stories-27-jellyfish-put-nuclear-plant-in-a#.

Loeb, V., V. Siegel, O. Holm-Hansen, R. Hewitt, W. Fraser, W. Trivelpiece, and S. Trivelpiece. 1997. "Effect of Sea-Ice Extent and Krill or Salp Dominance on the Antarctic Food Web." *Nature* 387:897–900.

Lotan, A., R. Ben-Hillel, and Y. Loya. 1992. "Life cycle of *Rhopilema nomadica*: A New Immigrant Scyphomedusan in the Mediterranean." *Marine Biology* 112:237–42.

Lotze, H. K., and B. Worm. 2009. "Historical Baselines for Large Marine Animals." *Trends in Ecology and Evolution* 24:254–62.

Lu, N., C. Liu, and P. Guo. 1989. "Effect of Salinity of Larva of Edible Medusae (*Rhopilema esculenta Kishinouye*) at Different Development Phases and a Review on the Cause of Jellyfish Resources Falling Greatly in Liaodong Bay" [in Chinese]. *Acta Ecologica Sinica* 9:304–9.

Lubchenco, J., S. R. Palumbi, S. D. Gaines, and S. Andelman. 2003. "Plugging a Hole in the Ocean: The Emerging Science of Marine Reserves." *Ecological Applications* 13:S3–S7.

Ludlam, S. 2012. *Let the Facts Speak*. Fremantle: Australian Greens Senator for Western Australia.

Lutz, P. L. 1990. "Studies on the Ingestion of Plastic and Latex by Sea Turtles." Paper read at Proceedings of the Second International Conference on Marine Debris, 2–7 April 1989, Honolulu, Hawaii.

Lynam, C. P., M. J. Gibbons, B. E. Axelsen, C. A. J. Sparks, J. Coetzee, B. G. Heywood, and A. S. Brierley. 2006. "Jellyfish Overtake Fish in a Heavily Fished Ecosystem." *Current Biology* 16:R492–R493.

Lynam, C. P., S. J. Hay, and A. S. Brierley. 2004. "Interannual Variability in Abundance of North Sea Jellyfish and Links to the North Atlantic Oscillation." *Limnology and Oceanography* 49:637–43.

———. 2005. "Jellyfish Abundance and Climatic Variation: Contrasting Responses in Oceanographically Distinct Regions of the North Sea, and Possible Implications for Fisheries." *Journal of the Marine Biological Association of the United Kingdom* 85:435–50.

Ma, X., and J. E. Purcell. 2005. "Effects of Temperature, Salinity, and Predators on Mortality of and Colonization by the Invasive Hydrozoan *Moerisia lyonsi*." *Marine Biology* 147:215–24.

Macfadyen, G., T. Huntington, and R. Cappell. 2009. "Abandoned, Lost or Otherwise Discarded Fishing Gear." In *UNEP Regional Seas Reports and Studies*, no. 185: FAO Fisheries and Aquaculture Technical Paper, no. 523. Rome: UNEP/FAO.

MacKenzie, B. R., H. Mosegaard, and A. A. Rosenberg. 2009. "Impending Collapse of Bluefin Tuna in the Northeast Atlantic and Mediterranean." *Conservation Letters* 2:25–34.

Malej, A. 1989. "Behaviour and Trophic Ecology of the Jellyfish *Pelagia noctiluca* (Forsskal, 1775)." *Journal of Experimental Marine Biology and Ecology* 126:259–70.

Malins, D. C., B. B. McCain, D. W. Brown, M. S. Myers, M. M. Krahn, and S. L. Chan. 1987. "Toxic Chemicals, Including Aromatic and Chlorinated Hydrocarbons and Their Derivatives, and Liver Lesions in White Croaker (*Genyonemus lineatus*) from the Vicinity of Los Angeles." *Environmental Science & Technology* 21:765–70.

MaltaMedia. 2006. "30,000 Jellyfish Stings in Mediterranean." *MaltaMedia News*, 9 August. http://www.maltamedia.com/news/2005/ln/article_11047.shtml.

Mancuso, R. 2006. "Mighty Warship Feels the Sting (Australian Jellyfish Attack USS Ronald Reagan!)." *Free Republic*, 27 January. http://www.freerepublic.com/focus /f-news/1566004/posts.

"Marine Turtles in Fuerteventura." 2011. Informational pamphlet [in Spanish]. *CanariasMedioambiente*. Accessed 2 May. http://www.canariasmedioambiente.com.

Marshall, M. 2010. "Super Goby Helps Salvage Ocean Dead Zone." *New Scientist*, 15 July. http://www.newscientist.com/article/dn19182-super-goby-helps-salvage -ocean-dead-zone.html.

Masilamoni, J. G., K. S. Jesudoss, K. Nandakumar, K. K. Satpathy, K. V. K. Nair, and J. Azariah. 2000. "Jellyfish Ingress: A Threat to the Smooth Operation of Coastal Power Plants." *Current Science* 79:567–69.

Mason, W. H., and G. W. Folkerts. 1979. *Environmental Problems: Principles, Readings, and Comments*. 2nd ed. Dubuque, IA: Wm. C. Brown Company.

"Mass Fish Farm Mortalities and Escapes Threaten the Survival of Wild Salmon." 2003. *The Salmon Farm Monitor*. Accessed 13 February 2011. http://www.salmon farmmonitor.org/pr010803.shtml.

Matear, R. 2006. "Carbon Dioxide Uptake in the Oceans." *The Science Show*, ABC Radio National, 27 May. http://www.abc.net.au/radionational/programs/scienceshow /carbon-dioxide-uptake-in-the-oceans/3324758.

Mato, Y., T. Isobe, H. Takada, H. Kanehiro, C. Ohtake, and T. Kaminuma. 2001. "Plastic Resin Pellets as a Transport Medium for Toxic Chemicals in the Marine Environment." *Environmental Science & Technology* 35:318–24.

Matsueda, N. 1969. "Presentation of *Aurelia aurita* at Thermal Power Station." *Bulletin of the Marine Biological Station of Asamushi* 13:187–91.

Matsuura, F., N. Fujisawa, and S. Ishikawa. 2007. "Detection and Removal of Jellyfish Using Underwater Image Analysis." *Journal of Visualization* 10:259–60.

McClenachan, L. 2009. "Documenting Loss of Large Trophy Fish from the Florida Keys with Historical Photographs." *Conservation Biology* 23:636–43.

McKay, B., and K. Mulvaney. 2001. "A Review of Marine Major Ecological Disturbances." *Endangered Species UPDATE* 18:14–24.

McLeod, D. J., G. M. Hallegraeff, G. W. Hosie, and A. J. Richardson. 2012. "Climate-Driven Range Expansion of the Red-Tide Dinoflagellate *Noctiluca scintillans* into the Southern Ocean." *Journal of Plankton Research* 34(4): 332–37.

McNeil, B. I., and R. J. Matear. 2008. "Southern Ocean Acidification: A Tipping Point at 450-ppm Atmospheric CO2." *Proceedings of the National Academy of Sciences USA* 105:18860–64.

McPhee, D. P., D. Leadbitter, and G. A. Skilleter. 2002. "Swallowing the Bait: Is Recreational Fishing in Australia Ecologically Sustainable?" *Pacific Conservation Biology* 8:40–45.

McTaggart, L. 2000. "Twentieth Century Plague." *Ecologist*, July. http://findarticles .com/p/articles/mi_m2465/is_5_30/ai_63859395/.

Mee, L. D. 1992. "The Black Sea in Crisis: The Need for Concerted International Action." *Ambio* 21:278–86.

Menzel, D. W., J. Anderson, and A. Randke. 1970. "Marine Phytoplankton Vary in Their Response to Chlorinated Hydrocarbons." *Science* 167:1724–26.

Miller, D. 2011. "Jellyfish Force Torness Nuclear Reactor Shutdown." *BBC News*, 30 June. http://www.bbc.co.uk/news/uk-scotland-edinburgh-east-fife-13971005.

Miller, D. J., J. E. Hardwick, and W. A. Dahlstrom. 1975. "Pismo Clams and Sea Otters." Marine Resources Technical Report no. 31. Long Beach: California Department of Fish and Game.

Milligan, P. 2012. "Costa del Sting: Swarms of Jellyfish Invade Spain's Beaches as More than a Thousand Holidaymakers are Treated by First Aiders." *Daily Mail*, 4 August. http://www.dailymail.co.uk/news/article-2183572/Malaga-Swarms -jellyfish-invade-Spains-beaches-1000-holidaymakers-treated-aiders.html.

Mills, C. E. 2001. "Jellyfish Blooms: Are Populations Increasing Globally in Response to Changing Ocean Conditions?" *Hydrobiologia* 451:55–68.

———. 2005. "Marine Conservation." 28 August. http://faculty.washington.edu /cemills/Conservation.html.

Mingliang, Z., Q. Shide, and L. Ming. 1993. "Investigation of Coelenterare Stings in North China Sea." Abstract. *Acta Academiae Qingdao Universitatis* 4. doi: CNKI:ISSN:1001-1047.0.1993-04-002.

Möbius, K. 1880. "Medusen werden durch frost getödtet." *Zoologischer Anzeiger* 3:67–68.

Møller, A. P., and T. A. Mousseau. 2009. "Biological Consequences of Chernobyl: 20 Years On." *Trends in Ecology and Evolution* 21:200–7.

Møller, A. P., T. A. Mousseau, F. d. Lope, and N. Saino. 2007. "Elevated Frequency of Abnormalities in Barn Swallows from Chernobyl." *Biology Letters* 3:414–17.

Möller, H. 1984. "Effects on Jellyfish Predation by Fishes." In Proceedings of the Workshop on Jellyfish Blooms in the Mediterranean, Athens 1983, 45–59. Athens: UNEP.

Møller, L. F., and H. U. Riisgård. 2007. "Impact of Jellyfish and Mussels on Algal Blooms caused by Seasonal Oxygen Depletion and Nutrient Release from the Sediment in a Danish Fjord." *Journal of Experimental Marine Biology and Ecology* 351:92–105.

Moloney, C. L. 2010. "The Humble Bearded Goby Is a Keystone Species in Namibia's Marine Ecosystem." *South African Journal of Science* 106. doi: 10.4102/sajs .v106i9/10.407.

Monash University. 2011. "Researcher Discovers New Dolphin Species in Victoria." Press release, Monash University, 15 September. http://www.monash.edu.au/news /releases/show/researcher-discovers-new-dolphin-species-in-victoria.

"Moon Jellyfish Kill Tons of Goliath Groupers at St. Lucie Nuclear Power Plant." 2011. *Huffington Post*, 9 December. http://www.huffingtonpost.com/2011/12/08 /jellyfish-grouper-power-plant_n_1136811.html.

Moore, C. J., S. L. Moore, M. K. Leecaster, and S. B. Weisberg. 2001. "A Comparison of Plastic and Plankton in the North Pacific Central Gyre." *Marine Pollution Bulletin* 42:1297–300.

Moore, C. J., S. L. Moore, S. B. Weisberg, G. L. Lattin, and A. F. Zellers. 2002. "A Comparison of Neustonic Plastic and Zooplankton Abundance in Southern California's Coastal Waters." *Marine Pollution Bulletin* 44:1035–38.

Morell, V. 2009. "Can Science Keep Alaska's Bering Sea Pollock Fishery Healthy?" *Science* 326:1340–41.

Morita, M., R. Suwa, A. Iguchi, M. Nakamura, K. Shimada, K. Sakai, and A. Suzuki. 2010. "Ocean Acidification Reduces Sperm Flagellar Motility in Broadcast Spawning Reef Invertebrates." *Zygote* 18:103–7.

Moy, A. D., W. R. Howard, S. G. Bray, and T. W. Trull. 2009. "Reduced Calcification in Modern Southern Ocean Planktonic Foraminifera." *Nature Geoscience* 2:276–80.

Mrosovsky, N., G. D. Ryan, and M. C. James. 2009. "Leatherback Turtles: The Menace of Plastic." *Marine Pollution Bulletin* 58:287–89.

Mulvaney, K. 2011. "Polar Bears and Cannibalism Revisited." *Discovery News*, 8 December. http://news.discovery.com/earth/when-polar-bears-attack-other-polar-bears.html.

Munday, P. L., D. L. Dixson, J. M. Donelson, G. P. Jones, M. S. Pratchett, G. V. Devitsina, and K. B. Døving. 2009. "Ocean Acidification Impairs Olfactory Discrimination and Homing Ability of a Marine Fish." *Proceedings of the National Academy of Sciences U S A* 106:1848–52.

Munday, P. L., D. L. Dixson, M. I. McCormick, M. Meekan, M. C. O. Ferrari, and D. P. Chivers. 2010. "Replenishment of Fish Populations Is Threatened by Ocean Acidification." *Proceedings of the National Academy of Sciences U S A* 107:12930–34.

Muscatine, L., and R. E. Marian. 1982. "Dissolved Inorganic Nitrogen Flux in Symbiotic and Nonsymbiotic Medusae." Limnology and Oceanography 27:910–17.

Myers, R. A., J. K. Baum, T. D. Shepherd, S. P. Powers, and C. H. Peterson 2007. "Cascading Effects of the Loss of Apex Predatory Sharks from a Coastal Ocean." *Science* 315:1846–50.

Myers, R. A., and B. Worm. 2003. "Rapid Worldwide Depletion of Predatory Fish Communities." *Nature* 423:280–83.

Nagata, R. M., M. A. Haddad, and M. Nogueira Júnior. 2009. "The Nuisance of Medusae (Cnidaria, Medusozoa) to Shrimp Trawls in Central Part of Southern Brazilian Bight, from the Perspective of Artisanal Fishermen." *Pan-American Journal of Aquatic Sciences* 4:312–25.

Naidoo, M. 2009. "Jelly Babies Booming." *Mercury*, 14 January. http://www.themer
cury.com.au/article/2009/01/14/49581_tasmania-news.html.

Nasrollahzadeh, A. 2010. "Caspian Sea and its Ecological Challenges." *Caspian
Journal of Environmental Sciences* 8:97–104.

NewsCore. 2011. "Australian Spotted Jellyfish, *Phyllorhiza punctata*, Invade Spanish
Beaches." *Courier Mail*, 22 July. http://www.couriermail.com.au/news/world
/australian-spotted-jellyfish-phyllorhiza-punctata-invade-spanish-beaches/story
-e6freoox-1226099789817.

Nomura, H., and M. Murano. 1992. "Seasonal Variation of Meso- and Macrozooplankton in Tokyo Bay, Central Japan" [in Japanese]. *La Mer* (Tokyo) 30:49–56.

Norse, E. A., S. Brooke, W. W. L. Cheung, M. R. Clark, I. Ekeland, R. Froese, K. M.
Gjerde, R. L. Haedrich, S. S. Heppell, T. Morato, L. E. Morgan, D. Pauly, R. Su-
maila, and R. Watson. 2012. "Sustainability of Deep-Sea Fisheries." *Marine Policy*
36:307–20.

NPS. 2009. *20 Years Later . . . Exxon Valdez Oil Spill*. Seward, AK: National Park
Service.

NRC. 1995. *Understanding Marine Biodiversity*. Edited by the National Research Council Committee on Biological Diversity in Marine Systems. Washington DC: National
Research Council.

———. 2003. *Understanding Climate Change Feedbacks*. Washington DC: Panel on
Climate Change Feedbacks, Climate Research Committee, National Research
Council.

NSF. 2011. "Jellyfish Gone Wild!" National Science Foundation. Last updated 3 March
2011. http://www.nsf.gov/news/special_reports/jellyfish/index.jsp.

Obama, B. 2010. "Remarks by the President to the Nation on the BP Oil Spill." White
House, Office of the Press Secretary. 15 June. http://www.whitehouse.gov/the
-press-office/remarks-president-nation-bp-oil-spill.

Ocean Conservancy. 2010. *Trash Travels. From Our Hands to the Sea, Around the Globe,
and Through Time*. Washington DC: Ocean Conservancy.

Officer, C. B., R. B. Biggs, J. L. Taft, L. E. Cronin, M. A. Tyler, and W. R. Boynton.
1984. "Chesapeake Bay Anoxia: Origin, Development, and Significance." *Science*
223:22–27.

Officer, C. B., and J. H. Ryther. 1980. "The Possible Importance of Silicon in Marine
Eutrophication." *Marine Ecology Progress Series* 3:83–91.

Oguz, T., B. Fach, and B. Salihoglu. 2008. "Invasion Dynamics of the Alien Ctenophore *Mnemiopsis leidyi* and Its Impact on Anchovy Collapse in the Black Sea."
Journal of Plankton Research 30:1385–97.

Okey, T. A., S. A. Shepherd, and P. Martinez. 2003. "A New Record of Anemone Barrens in the Galápagos." *Noticias de Galápagos* 62:17–20.

Oliveira, O. M. P. 2007. "The Presence of the Ctenophore *Mnemiopsis leidyi* in the
Oslofjorden and Considerations on the Initial Invasion Pathways to the North and
Baltic Seas." *Aquatic Invasions* 2:185–89.

Omori, M., and E. Nakano. 2001. "Jellyfish Fisheries in Southeast Asia." *Hydrobiologia*
451:19–26.

Orr, J. C., V. J. Fabry, O. Aumont, L. Bopp, S. C. Doney, R. A. Feely, A. Gnanadesikan, N. Gruber, A. Ishida, F. Joos, R. M. Key, K. Lindsay, E. Maier-Reimer, R. Matear, P. Monfray, A. Mouchet, R. G. Najjar, G.-K. Plattner, K. B. Rodgers, C. L. Sabine, J. L. Sarmiento, R. Schlitzer, R. D. Slater, I. J. Totterdell, M.-F. Weirig, Y. Yamanaka, and A. Yool. 2005. "Anthropogenic Ocean Acidification over the Twenty-First Century and Its Impact on Calcifying Organisms." *Nature* 437:681–86.

Österblom, H., S. Hansson, U. Larsson, O. Hjerne, F. Wulff, R. Elmgren, and C. Folke. 2007. "Human-Induced Trophic Cascades and Ecological Regime Shifts in the Baltic Sea." *Ecosystems* 10:877–89.

OTA. 1993. *Harmful Non-Indigenous Species in the United States*. Report by the US Congress, Office of Technology Assessment, OTA-F-565. Washington DC: US Government Printing Office.

Overington, C. 2007. "Green Answer Is Blowing in the Wind." *Australian*, 17 February. http://www.theaustralian.com.au/news/nation/green-answer-is-blowing-in-the -wind/story-e6frg6nf-1111113010110.

Özdemir, S. 2007. *Decreasing Methods of Jellyfish Bycatch on the Trawl Fishery*. Sinop, Turkey: Sinop University Fisheries Faculty.

Paavola, M., S. Olenin, and E. Leppäkoski. 2005. "Are Invasive Species Most Successful in Habitats of Low Native Species Richness Across European Brackish Water Seas?" *Estuarine, Coastal and Shelf Science* 64:738–50.

Painter, A. 2011. "On this day: 23 February 1968 Torrens Island power station." Professional Historians Association (South Australia) Accessed 1 May 2011. http://www .sahistorians.org.au/175/chronology/february/23-february-1968-torrens-island -power-station.shtml.

Palumbi, S. R. 2004. "Why Mothers Matter." *Nature* 430:621–22.

Pandolfi, J. M. 2009. "Evolutionary Impacts of Fishing: Overfishing's 'Darwinian Debt.'" *F1000 Biology Reports* 1:43.

Pandolfi, J. M., S. R. Connolly, D. J. Marshall, and A. L. Cohen. 2011. "Projecting Coral Reef Futures Under Global Warming and Ocean Acidification." *Science* 333:418–22.

Park, J.-s. 2009. "Predator Released to Kill Jellyfish at Beaches." *Korea Times*, 22 July. http://www.koreatimes.co.kr/www/news/nation/2009/07/113_48864.html.

Parliament of Australia. 1997. "The Management of Water and Biological Nutrients in Australia." Senate Environment, Recreation, Communications and the Arts References Committee—Meeting at Glenelg, 14 February. http://parlinfo.aph.gov .au/parlInfo/search/display/display.w3p;orderBy=_fragment_number;query= ((Dataset%3Acommsen)%20SearchCategory_Phrase%3A%22committees%22)%20 CommitteeName_Phrase%3A%22senate%20environment,%20recreation,%20 communications%20and%20the%20arts%20references%20committee%22%20 Questioner_Phrase%3A%22senator%20chapman%22;rec=3.

Parsons, S. 2010. "Japanese Schools Serve Kids Mercury-Filled Whale Meat." *Change .org*, 23 September. http://news.change.org/stories/japanese-schools-serve-kids -mercury-filled-whale-meat.

Parsons, T., W. K. W. Li, and R. Waters. 1976. "Some Preliminary Observations on the

Enhancement of Phytoplankton Growth by Low Levels of Mineral Hydrocarbons."
Hydrobiologia 51:85–89.

Parsons, T. R., and C. M. Lalli. 2002. "Jellyfish Population Explosions: Revisiting a Hypothesis of Possible Causes." *La Mer* 40:111–21.

Patton, L. 2010. "Gulf of Mexico 'Dead Zone' Grows as Spill Impact Is Studied." *Bloomberg*, 12 August. http://www.bloomberg.com/news/2010-08-12/crude -marred-gulf-of-mexico-s-dead-zone-grows-as-spill-impact-is-studied.html.

Pauly, D. 1995. "Anecdotes and the Shifting Baseline Syndrome of Fisheries." *Trends in Ecology & Evolution* 10:430.

Pauly, D., V. Christensen, J. Dalsgaard, R. Froese, and J. F. Torres. 1998. "Fishing Down Marine Food Webs." *Science* 279:860–63.

Pauly, D., and R. Watson. 2003. "Counting the Last Fish." *Scientific American*, July, 43–47.

Pauly, D., R. Watson, and J. Alder. 2005. "Global trends in world fisheries: impacts on marine ecosystems and food security." *Philosophical Transactions of the Royal of London, Series B, Biological Sciences* 360:5–12.

Pearse, J. S. 2006. "Ecological Role of Purple Sea Urchins." Science 314:940–41.

Pennisi, E. 2010. "How a Little Fish Keeps Overfished Ecosystem Productive." *Science* 329:268.

Perolina, F. 2005. "Jellyfish Attack Downs 127 Policemen." *Philippine Star*, 13 May. http://www.philstar.com/Article.aspx?articleId=277319&publicationSubCategory Id=67.

Peterson, C. H., S. D. Rice, J. W. Short, D. Esler, J. L. Bodkin, B. E. Ballachey, and D. B. Irons. 2003. "Long-Term Ecosystem Response to the Exxon Valdez Oil Spill." *Science* 302:2082–86.

PEW. 2003. *America's Living Oceans: Charting a Course for Sea Change*. A Report to the Nation. Arlington, VA: PEW Oceans Commission.

Pickrell, J. 2004. "Trawlers Destroying Deep-Sea Reefs, Scientists Say." *National Geographic News*, 19 February. http://news.nationalgeographic.com/news/2004/ 02/0219_040219_seacorals.html.

Piraino, S., F. Boero, B. Aeschbach, and V. Schmid. 1996. "Reversing the Life Cycle: Medusae Transforming into Polyps and Cell Transdifferentiation in *Turritopsis nutricula* (Cnidaria, Hydrozoa)." *Biological Bulletin* (Woods Hole) 190:302–12.

Pitt, K. A., D. T. Welsh, and R. H. Condon. 2009. "Influence of Jellyfish Blooms on Carbon, Nitrogen and Phosphorus Cycling and Plankton Production." *Hydrobiologia* 616:133–49.

Portier, M. M., and C. Richet. 1902. "De l'action anaphylactique de certains venins." *Comptes rendus de la Société de biologie* (Paris) 54:170–72.

Prasad, R. R. 1958. "Swarming of *Noctiluca* in the Palk Bay and Its Effect on the 'Choo-dai' Fishery, with a Note on the Possible Use of *Noctiluca* as an Indicator Species." *Proceedings of the Indian Academy of Sciences* 38:40–47.

Prevagen. 2011. "Prevagen Professional, Jellyfish Fight Aging, Stroke, Alzheimer's." *Rejuvenation Science*. Accessed 21 August 2011. http://www.rejuvenation-science .com/prevagen.html.

Purcell, J. E. 1992. "Effects of Predation by the Scyphomedusan *Chrysaora quinquecirrha* on Zooplankton Populations in Chesapeake Bay, USA." *Marine Ecology Progress Series* 87:65–76.

———. 2005. "Climate Effects on Formation of Jellyfish and Ctenophore Blooms: A Review." *Journal of the Marine Biological Association of the United Kingdom* 85:461–76.

———. 2007. "Environmental Effects on Asexual Reproduction Rates of the Scyphozoan, *Aurelia labiata*." *Marine Ecology Progress Series* 348:183–96.

———. 2012. "Jellyfish and Ctenophore Blooms Coincide with Human Proliferations and Environmental Perturbations." *Annual Review of Marine Science* 4:209–35.

Purcell, J. E., U. Bamstedt, and A. Bamstedt. 1999. "Prey, Feeding Rates, and Asexual Reproduction Rates of the Introduced Oligohaline Hydrozoan *Moerisia lyonsi*." *Marine Biology* 134:317–25.

Purcell, J. E., D. L. Breitburg, M. B. Decker, W. M. Graham, M. J. Youngbluth, and K. A. Raskoff. 2001a. "Pelagic Cnidarians and Ctenophores in Low Dissolved Oxygen Environments: A Review." In *Coastal Hypoxia: Consequences for Living Resources and Ecosystems*, edited by N. N. Rabalais and R. E. Turner, *Coastal and Estuarine Studies Series* 58:77–100. Washington DC: American Geophysical Union.

Purcell, J. E., and M. B. Decker. 2005. "Effects of Climate on Relative Predation by Scyphomedusae and Ctenophores on Copepods in Chesapeake Bay During 1987–2000." *Limnology and Oceanography* 50:376–87.

Purcell, J. E., and D. A. Nemazie. 1992. "Quantitative Feeding Ecology of the Hydromedusan *Nemopsis bachei* in Chesapeake Bay." *Marine Biology* 113:305–11.

Purcell, J. E., D. A. Nemazie, S. E. Dorsey, E. D. Houde, and J. C. Gamble. 1994. "Predation Mortality of Bay Anchovy *Anchoa mitchilli* Eggs and Larvae Due to Scyphomedusae and Ctenophores in Chesapeake Bay." *Marine Ecology Progress Series* 114:47–58.

Purcell, J. E., T. A. Shiganova, M. B. Decker, and E. D. Houde. 2001b. "The Ctenophore *Mnemiopsis* in Native and Exotic Habitats: U.S. Estuaries versus the Black Sea Basin." *Hydrobiologia* 451:145–76.

Purcell, J. E., J. R. White, D. A. Nemazie, and D. A. Wright. 1999. "Temperature, Salinity and Food Effects on Asexual Reproduction and Abundance of the Scyphozoan *Chrysaora quinquecirrha*." *Marine Ecology Progress Series* 180:187–96.

Quammen, D. 1998. "The Weeds Shall Inherit the Earth." *Independent*, 22 November. http://www.independent.co.uk/arts-entertainment/the-weeds-shall-inherit-the-earth-1186702.html.

Rabalais, N. N., R. E. Turner, R. J. Díaz, and D. Justić. 2009. "Global Change and Eutrophication of Coastal Waters." *ICES Journal of Marine Science* 66:1528–37.

Rabalais, N. N., R. E. Turner, and J. Wiseman. 2002. "Gulf of Mexico Hypoxia, a.k.a. 'The Dead Zone.'" *Annual Review of Ecology and Systematics* 33:235–63.

Raj, R. 2011. "Jellyfish Phenomena Seasonal." *Malay Mail*, 15 March. http://envdev malaysia.wordpress.com/2011/03/15/jellyfish-phenomena-seasonal/.

Rajagopal, S., K. V. K. Nair, and J. Azariah. 1989. "Some Observations on the Problem of Jelly Fish Ingress in a Power Station Cooling System at Kalpakkam, East Coast of India." *Mahasagar* 22:151–58.

Rakha, K. A., K. Al-Banaa, A. Al-Ragum, and F. Al-Hulail. 2008. "Hydrodynamic Study on the Currents at an Intake Basin in Kuwait Bay." Paper #227, presented at COPEDEC VII, Dubai, UAE, February.

Raloff, J. 2001. "Plastic Debris Picks Up Ocean Toxics." Science News 159:79.

Raskoff, K. A. 2001. "The Impact of El Niño Events on Populations of Mesopelagic Hydromedusae." *Hydrobiologia* 451:121–29.

Ray, G. C. 1988. "Ecological Diversity in Coastal Zones and Oceans." In *Biodiversity*, ed. E. O. Wilson, 36–49. Washington, DC: National Academy Press.

Read, A. J., P. Drinker, and S. Northridge. 2003. *By-Catches of Marine Mammals in U.S. Fisheries and a First Attempt to Estimate the Magnitude of Global Marine Mammal By-Catch*. Report for the International Whaling Commission. Beaufort, NC: Duke University Marine Laboratory.

Reeve, M. R., M. A. Walter, and T. Ikeda. 1978. "Laboratory Studies of Ingestion and Food Utilization in Lobate and Tentaculate Ctenophores." *Limnology and Oceanography* 23:740–51.

Reusch, T. B. H., S. Bolte, M. Sparwel, A. G. Moss, and J. Javidpour. 2010. "Microsatellites Reveal Origin and Genetic Diversity of Eurasian Invasions by One of the World's Most Notorious Marine Invader, *Mnemiopsis leidyi* (Ctenophora)." *Molecular Ecology* 19:2690–99.

Reuters. 1984. "Jellyfish Invasion Closes Nuclear Plant in Florida." *New York Times*, 2 September. http://www.nytimes.com/1984/09/02/us/around-the-nation-jelly fish-invasion-closes-nuclear-plant-in-florida.html.

———. 2007. "Whalemeat in Japanese School Lunches Found Toxic." *Reuters*, 1 August. http://www.reuters.com/article/2007/08/01/us-japan-whalemeat -idUST6359120070801.

Richardson, A. J., A. Bakun, G. C. Hays, and M. J. Gibbons. 2009. "The Jellyfish Joyride: Causes, Consequences and Management Responses to a More Gelatinous Future." *Trends in Ecology and Evolution* 24:312–22.

Richardson, A. J., and M. J. Gibbons. 2008. "Are Jellyfish Increasing in Response to Ocean Acidification?" *Limnology and Oceanography* 53:2040–45.

Richardson, A. J., D. McKinnon, and K. M. Swadling. 2009. "Zooplankton." In *A Marine Climate Change Impacts and Adaptation Report Card for Australia 2009*, edited by E. S. Poloczanska, A. J. Hobday and A. J. Richardson. Southport, Queensland: NCCARF Publication 05/09.

Riebesell, U. 2004. "Effects of CO2 Enrichment on Marine Phytoplankton." *Journal of Oceanography* 60:719–29.

———. 2008. "Climate Change—Acid Test for Marine Biodiversity." Nature 454:46–47.

Riisgård, H. U., P. Andersen, and E. Hoffmann. 2012. "From Fish to Jellyfish in the Eutrophicated Limfjorden (Denmark)." *Estuaries and Coasts* 1–13.

Riisgård, H. U., L. Bøttiger, C. V. Madsen, and J. E. Purcell. 2007. "Invasive Ctenophore *Mnemiopsis leidyi* in Limfjorden (Denmark) in Late Summer 2007— Assessment of Abundance and Predation Effects." *Aquatic Invasions* 2:395–401.

Roberts, C. M. 2007. *The Unnatural History of the Sea*. Washington, DC: Island Press.

Roberts, D., W. R. Howard, A. D. Moy, J. L. Roberts, T. W. Trull, S. G. Bray, and R. R. Hopcroft. 2011. "Interannual Pteropod Variability in Sediment Traps Deployed Above and Below the Aragonite Saturation Horizon in the Sub-Antarctic Southern Ocean." *Polar Biology* 34:1739–50.

Robinson, K. I. M. 1987. "Effects of Thermal Power Station Effluent on the Seagrass Benthic Communities of Lake Macquarie, a New South Wales Coastal Lagoon." *Wetlands* 7:1–12.

Robison, B., and J. Connor. 1999. *The Deep Sea*. Monterey, CA: Monterey Bay Aquarium Foundation.

Roemmich, D., and J. McGowan. 1995. "Climatic Warming and the Decline of Zooplankton of the California Current." *Science* 267:1324–26.

Rogan, W. J., and A. Chen. 2005. "Health Risks and Benefits of Bis(4-chlorophenyl)-1,1,1-trichloroethane (DDT)." *Lancet* 366:763–73.

Roger, L. M., A. J. Richardson, A. D. McKinnon, B. Knott, R. Matear, and C. Scadding. 2011. "Comparison of the Shell Structure of Two Tropical Thecosomata (*Creseis acicula* and *Diacavolinia longirostris*) from 1963 to 2009: Potential Implications of Declining Aragonite Saturation." *ICES Journal of Marine Science* 69:465–74.

Rogers-Bennett, L., P. L. Haaker, T. O. Huff, and P. K. Dayton. 2002. "Estimating Baseline Abundances of Abalone in California for Restoration." *CalCOFI Reports* 43:97–111.

Rogers, C. A., D. C. Biggs, and R. A. Cooper. 1978. "Aggregations of the Siphonophore *Nanomia cara* Agassiz 1865 in the Gulf of Maine: Observations from a Submersible." *Fishery Bulletin* 76:281–84.

Roohi, A., and A. Sajjadi. 2011. "*Mnemiopsis leidyi* Invasion and Biodiversity Changes in the Caspian Sea." In *Ecosystems Biodiversity*, ed. O. Grillo and G. Venora, 171–92. Rijeka, Croatia: InTech.

Roohi, A., Z. Yasin, A. E. Kideys, A. T. S. Hwai, A. G. Khanari, and E. Eker-Develi. 2008. "Impact of a New Invasive Ctenophore (*Mnemiopsis leidyi*) on the Zooplankton Community of the Southern Caspian Sea." *Marine Ecology* 29:421–34.

Roux, J.-P., C. van der Lingen, M. J. Gibbons, N. E. Moroff, L. J. Shannon, A. D. Smith, and P. M. Cury. In press. "Jellyfication of Marine Ecosystems as a Likely Consequence of Overfishing Small Pelagic Fish: Lessons from the Benguela." *Bulletin of Marine Science*.

Rosenthal, E. 2008. "Stinging Tentacles Offer Hint of Oceans' Decline." *New York Times*, 6 August. http://www.nytimes.com/2008/08/03/science/earth/03jellyfish.html.

Royal Society. 2005. "Ocean Acidification due to Increasing Atmospheric Carbon Dioxide." Policy Document 12/05. London: Royal Society.

Rudloe, J., and A. Rudloe. 2010. *Shrimp: The Endless Quest for Pink Gold*. New York: Pearson Education.

Russell, F. S. 1970. *Medusae of the British Isles. II. Pelagic Scyphozoa with a Supplement to the First Volume on Hydromedusae*, E.T. Browne monograph of the Marine Biological Association of the United Kingdom. Cambridge: Cambridge University Press.

Rutherford, L. D., and E. V. Thuesen. 2005. "Metabolic Performance and Survival of Medusae in Estuarine Hypoxia." *Marine Ecology Progress Series* 294:189–200.

Ryall, J. 2009. "Japanese Fishing Trawler Sunk by Giant Jellyfish." *Telegraph*, 2 November. http://www.telegraph.co.uk/earth/6483758/Japanese-fishing-trawler-sunk-by-giant-jellyfish.html.

Ryther, J. H. 1969. "Photosynthesis and Fish Production in the Sea." *Science* 166: 72–76.

Sàenz-Arroyo, A., C. M. Roberts, J. Torre, M. Cariño-Olvera, and R. R. Enríquez-Andrade. 2005. "Rapidly Shifting Environmental Baselines among Fishers of the Gulf of California." *Proceedings of the Royal Society, Series B, Biological Sciences* 272:1957–62.

Safina, C. 1994. "Where Have All the Fishes Gone?" *Science and Technology* (Spring): 37–43.

———. 1995. "The World's Imperiled Fish." *Scientific American*, November, 46–53.

Safina, C., and D. H. Klinger. 2008. "Collapse of Bluefin Tuna in the Western Atlantic." *Conservation Biology* 22:243–46.

Samuel, H. 2008. "Mauve Stingers to Invade the Riviera." *Telegraph*, 14 June. http://www.telegraph.co.uk/earth/3344475/Mauve-stingers-to-invade-the-Riviera.html.

———. 2012. "Jelly Fish 'Early Warning System' to be Launched in French Riviera." *Telegraph*, 1 July. http://www.telegraph.co.uk/earth/wildlife/9368181/Jelly-fish-early-warning-system-to-be-launched-in-French-Riviera.html.

Schaefer, J., P. Mickle, J. Spaeth, B. R. Kreiser, S. Adams, W. Matamoros, B. Zuber, and P. Vigueira. 2006. "Effects of Hurricane Katrina on the Fish Fauna of the Pascagoula River Drainage." *Proceedings of the 36th Annual Mississippi Water Resources Conference* (2006): 62–68.

Scheffer, M., and S. R. Carpenter. 2003. "Catastrophic Regime Shifts in Ecosystems: Linking Theory to Observation." *Trends in Ecology & Evolution* 18:648–56.

Scheffer, M., S. Carpenter, and B. de Young. 2005. "Cascading Effects of Overfishing Marine Systems." *Trends in Ecology and Evolution* 20:579–81.

Schelske, C. L., and E. F. Stoermer. 1971. "Eutrophication, Silica Depletion, and Predicted Changes in Algal Quality in Lake Michigan." *Science* 173:423–24.

Schiariti, A., M. Kawahara, S. Uye, and H. Mianzan. 2008. "Life Cycle of the Jellyfish *Lychnorhiza lucerna* (Scyphozoa: Rhizostomeae)." *Marine Biology* 156:1–12.

Scott, A. 1999. "Ecological History of the Tuggerah Lakes, Final Report." Canberra: CSIRO Land & Water.

Scripps News. 2008. "Oceans on the Precipice: Scripps Scientist Warns of Mass Extinctions and 'Rise of Slime.'" 13 August. Press release. http://scrippsnews.ucsd.edu/Releases/?releaseID=920.

Shaffer, L. H. 1967. "Solubility of Gypsum in Sea Water and Sea Water Concentrates at Temperatures from Ambient to 65°C." *Journal of Chemical and Engineering Data* 12:183–89.

Shiganova, T. A. 2001. "Development *Mnemiopsis leidyi* Population in the Black Sea and Other Seas of Mediterranean Basin." First meeting of the Caspian Environment Programme. Baku, Azerbaijan, 24–26 April.

Shiganova, T. A., Y. V. Bulgakova, S. P. Volovik, Z. A. Mirzoyan, and S. I. Dudkin. 2001a. "The New Invader *Beroe ovata* Mayer 1912 and Its Effect on the Ecosystem in the Northeastern Black Sea." *Hydrobiologia* 451:187–97.

Shiganova, T. A., and A. Malej. 2009. "Native and Non-Native Ctenophores in the Gulf of Trieste, Northern Adriatic Sea." *Journal of Plankton Research* 31:61–71.

Shiganova, T. A., Z. A. Mirzoyan, E. A. Studenikina, S. P. Volovik, I. Siokou-Frangou, S. Zervoudaki, E. D. Christou, A. Y. Skirta, and H. J. Dumont. 2001b. "Population Development of the Invader Ctenophore *Mnemiopsis leidyi*, in the Black Sea and in Other Seas of the Mediterranean Basin." *Marine Biology* 139:431–45.

Shiganova, T., A. F. Sokolsky, M. I. Kaptyuk, A. M. Kamakin, D. Tinenkova, and E. K. Kuraseva. 2001c. "Investigation of Invader Ctenophore *Mnemiopsis leidyi* and Its Effect on the Caspian Ecosystem in Russia in 2001." Report to the First Mnemiopsis Advisory Group Workshop, Baku, Azerbaijan, 3–4 December.

Shimomura, T. 1959. "On the Unprecedented Flourishing of 'Echizen-Kurage,' *Stomolophus nomurai* (Kishinouye), in the Tsushima Warm Current Regions in Autumn, 1958" [in Japanese]. *Bulletin of Japan Sea Regional Fisheries Research Laboratory* 7:85–107.

Shoji, J., R. Masuda, Y. Yamashita, and M. Tanaka. 2005. "Predation on Fish Larvae by Moon Jellyfish *Aurelia aurita* Under Low Dissolved Oxygen Concentrations." *Fisheries Science* 71:748–53.

Simenstad, C. A., J. A. Estes, and K. W. Kenyon. 1978. "Aleuts, Sea Otters, and Alternate Stable-State Communities." *Science* 200:403–11.

SIO. 1999. "Massive Pollution Documented Over Indian Ocean." Scripps Institution of Oceanography, 10 June. http://www.sciencedaily.com/releases/1999/06/990610074044.htm.

Sivan, A. A. 2011. "New Perspectives in Plastic Biodegradation." *Current Opinion in Biotechnology* 22:422–26.

SkyNews. 2011. "'Jellyfish Soup' Alert for Britain's Swimmers." *SkyNews*, 21 July. http://news.sky.com/home/uk-news/article/16034160.

Smayda, T. J. 2002. "Adaptive Ecology, Growth Strategies and the Global Bloom Expansion of Dinoflagellates." *Journal of Oceanography* 58:281–94.

Smith, A. D. M., C. J. Brown, C. M. Bulman, E. A. Fulton, P. Johnson, I. C. Kaplan, H. Lozano-Montes, S. Mackinson, M. Marzloff, L. J. Shannon, Y.-J. Shin, and J. Tam. 2011. "Impacts of Fishing Low-Trophic Level Species on Marine Ecosystems." *Science* 333:1147–50.

Solow, A. R. 2004. "Red Tides and Dead Zones: The Coastal Ocean Is Suffering from Overload of Nutrients" [adapted from *Oceanus Magazine*, 22 December]. Woods Hole Oceanographic Institution. http://vishnu.whoi.edu/services/communications/oceanusmag.050826/v43n1/solow-print.html.

Sommer, U. 1994. "Are Marine Diatoms Favoured by High Si:N Ratios?" *Marine Ecology Progress Series* 115:309–15.

Sommer, U., H. Stibor, A. Katechakis, F. Sommer, and T. Hansen. 2002. "Pelagic Food Web Configurations at Different Levels of Nutrient Richness and Their Implications for the Ratio Fish Production: Primary Production." *Hydrobiologia* 484:11–20.

Sørnes, T. A., D. L. Aksnes, U. Båmstedt, and M. J. Youngbluth. 2007. "Causes for Mass Occurrences of the Jellyfish *Periphylla periphylla*: A Hypothesis that Involves Optically Conditioned Retention." *Journal of Plankton Research* 29:157–67.

Sorte, C. J., S. L. Williams, and R. A. Zerebecki. 2010. "Ocean Warming Increases Threat of Invasive Species in a Marine Fouling Community." *Ecology* 91:2198–204.

Spangenberg, D. B. 1984. "Use of the *Aurelia* Metamorphosis Test System to Detect Subtle Effects of Selected Hydrocarbons and Petroleum Oil." *Marine Environmental Research* 14:281–303.

Spangenberg, D. B., T. Jernigan, C. Philput, and B. Lowe. 1994. "Graviceptor Development in Jellyfish Ephyrae in Space and on Earth." *Advances in Space Research* 14:317–25.

Speer, L., L. Lauck, E. Pikitch, S. Boa, L. Dropkin, and V. Spruill. 2000. *Roe to Ruin: The Decline of Sturgeon in the Caspian Sea and the Road to Recovery.* Silver Spring, MD: Sea Web.

Stachowicz, J. J., J. R. Terwin, R. B. Whitlatch, and R. W. Osman. 2002. "Linking Climate Change and Biological Invasions: Ocean Warming Facilitates Nonindigenous Species Invasions." *Proceedings of the National Academy of Sciences U S A* 99:15497–500.

Stebbing, A. R. D. 1991. "The Stimulation of Reproduction in Coelenterates by Low Levels of Toxic Stress." In *Jellyfish Blooms in the Mediterranean.* Proceedings of the II Workshop on Jellyfish in the Mediterranean Sea. MAP Technical Reports Series no. 47, 298–301. Athens: UNEP.

Stephens, P. A., W. J. Sutherland, and R. P. Freckleton. 1999. "What Is the Allee Effect?" *Oikos* 87:185–90.

Stibor, H., O. Vadstein, S. Diehl, A. Gelzleichter, T. Hansen, F. Hantzsche, A. Katechakis, B. Lippert, K. Løseth, C. Peters, W. Roederer, M. Sandow, L. Sundt-Hansen, and Y. Olsen. 2004. "Copepods Act as a Switch between Alternative Trophic Cascades in Marine Pelagic Food Webs." *Ecology Letters* 7:321–28.

Stockwell, C. A., A. P. Hendry, and M. T. Kinnison. 2003. "Contemporary Evolution Meets Conservation Biology." *Trends in Ecology & Evolution* 18:94–101.

Stokstad, E. 2010. "Down on the Shrimp Farm." *Science* 328:1504–5.

Stone, R. 2002. "Caspian Ecology Teeters on the Brink." *Science* 295:430–33.

——. 2010. "Chinese Initiative Aims to Comprehend and Combat a Slimy Foe." *Science* 330:1464–65.

Stoner, E. W., C. A. Layman, L. A. Yeager, and H. M. Hassett. 2011. "Effects of Anthropogenic Disturbance on the Abundance and Size of Epibenthic Jellyfish *Cassiopea* spp." *Marine Pollution Bulletin* 62:1109–14.

Stramma, L., S. Schmidtko, L. A. Levin, and G. C. Johnson. 2010. "Ocean Oxygen Minima Expansions and Their Biological Impacts." *Deep-Sea Research Part I Oceanographic Research Papers* 57:587–95.

Sullivan, L. J., and D. J. Gifford. 2007. "Growth and Feeding Rates of the Newly Hatched Larval Ctenophore *Mnemiopsis leidyi* A. Agassiz (Ctenophora, Lobata)." *Journal of Plankton Research* 29:949–65.

Sun, S., Y. Li and X. Sun. 2012. "Changes in the Small-Jellyfish Community in Recent

Decades in Jiaozhou Bay, China." *Chinese Journal of Oceanology and Limnology* 30(4): 507–18.

Takizawa, M. 2005. "Countermeasures for Jellyfish Attacks at Kashiwazaki Kariwa Nuclear Power Station" [in Japanese]. *Bulletin of Plankton Society of Japan* 52:36–38.

TEEB. 2010. *Mainstreaming the Economics of Nature: A Synthesis of the Approach, Conclusions and Recommendations of TEEB.* Geneva: The Economics of Ecosystems and Biodiversity.

Tegner, M. J., L. V. Basch, and P. K. Dayton. 1996. "Near Extinction of an Exploited Marine Invertebrate." *Trends in Ecology and Evolution* 11:278–80.

Tendal, O. S., K. R. Jensen, and H. U. Riisgård. 2007. "Invasive Ctenophore *Mnemiopsis leidyi* Widely Distributed in Danish Waters." *Aquatic Invasions* 2:272–77.

"Tens of Thousands of Jelly Fish Are Washing Ashore Florida East Coast." 2010. *Before It's News,* 1 July. http://beforeitsnews.com/story/91/422/Tens_of _thousands_of_Jelly_fish_are_washing_ashore_Florida_East_Coast.html.

Thomas, C. D., A. Cameron, R. E. Green, M. Bakkenes, L. J. Beaumont, Y. C. Collingham, B. F. N. Erasmus, M. F. de Siqueira, A. Grainger, L. Hannah, L. Hughes, B. Huntley, A. S. van Jaarsveld, G. F. Midgley, L. Miles, M. A. Ortega-Huerta, A. T. Peterson, O. L. Phillips, and S. E. Williams. 2004. "Extinction Risk from Climate Change." *Nature* 427:145–48.

Thomas, W. H., and D. L. R. Seibert. 1977. "Effects of Copper on the Dominance and the Diversity of Algae: Controlled Ecosystem Pollution Experiment." *Bulletin of Marine Science* 27:23–33.

Thompson, R., Y. Olsen, R. Mitchell, A. Davis, S. Rowland, A. John, D. McGonigle, and A. E. Russell. 2004. "Lost at Sea: Where Is All the Plastic?" *Science* 304:838.

Thrush, S. F., and P. K. Dayton. 2010. "What Can Ecology Contribute to Ecosystem-Based Management?" *Annual Review of Marine Science* 2:419–41.

Thuesen, E. V., L. D. Rutherford, P. L. Brommer, K. Garrison, M. A. Gutowska, and T. Towanda. 2005. "Intragel Oxygen Promotes Hypoxia Tolerance of Scyphomedusae." *Journal of Experimental Biology* 208:2475–82.

Tiemann, H., I. Sotje, A. Becker, G. Jarms, and M. Epple. 2006. "Calcium Sulfate Hemihydrate (bassanite) Statoliths in the Cubozoan *Carybdea* sp." *Zoologischer Anzeiger* 245:13–17.

Tirado, R. 2008. *Dead Zones: How Agricultural Fertilizers Kill our Rivers, Lakes and Oceans.* Amsterdam: Greenpeace.

Titelman, J., L. Riemann, T. A. Sørnes, T. Nilsen, P. Griekspoor, and U. Båmstedt. 2006. "Turnover of Dead Jellyfish: Stimulation and Retardation of Microbial Activity." *Marine Ecology Progress Series* 325:43–58.

Togias, A. G., J. W. Burnett, A. Kagey-Sobotka, and L. M. Lichtenstein. 1985. "Anaphylaxis after Contact with a Jellyfish." *Journal of Allergy and Clinical Immunology* 75:672–75.

Torné, R. 2009. "Hundreds of Loggerhead Turtles Released in Cabo de Gata." *Costa Almería News,* 21 August. http://www.costa-news.com/index.php?option=com_con tent&task=view&id=3574&Itemid=121.

Toxics Link. 2002. "Detergents under Scrutiny." *India Together*, October. http://www
.indiatogether.org/environment/articles/tlink-1002.htm.

Travis, J. 1993. "Invader Threatens Black, Azov Seas." *Science* 262:1366–67.

Tucker, A. 2010. "Jellyfish: The Next King of the Sea." *Smithsonian Magazine*, August.
http://www.smithsonianmag.com/specialsections/40th-anniversary/Jellyfish
-The-Next-Kings-of-the-Sea.html

Twaronite, L. 2011. "Whale Meat: A School Lunch Treat?" *Wall Street Journal*, 1 February. http://blogs.wsj.com/japanrealtime/2011/02/01/whale-meat-a-school
-lunch-treat/.

UN. 1992. *Convention on Biological Diversity*. New York: United Nations.

———. 2006. *The Impacts of Fishing on Vulnerable Marine Ecosystems*. New York: United
Nations.

UNEP. 2004. *Overfishing, A Major Threat to the Global Marine Ecology*. Geneva: United
Nations Environment Programme.

———. 2006. *The State of the Marine Environment: Trends and Processes*. The Hague:
United Nations Environment Programme.

———. 2010. *Environmental Consequences of Ocean Acidification: A Threat to Food Security*. Nairobi: United Nations Environmental Programme.

———. 2011. *UNEP Year Book 2011: Emerging Issues in Our Global Environment*. Nairobi: United Nations Environment Programme.

USDA. 2009. *Pesticide Data Program Annual Summary Calendar Year 2008, December
2009*. Washington DC: Agricultural Marketing Service.

USGS. 2009. "Decline of Shorebird Linked to Bait Use of Horseshoe Crabs." *United
States Geological Survey* via *ScienceDaily*, 2 March. http://www.sciencedaily.com
/releases/2009/02/090217151559.htm.

Utne-Palm, A. C., A. G. V. Salvanes, B. Currie, S. Kaartvedt, G. E. Nilsson, V. A.
Braithwaite, J. A. W. Stecyk, M. Hundt, M. van der Bank, B. Flynn, G. K. Sandvik,
T. A. Klevjer, A. K. Sweetman, V. Brüchert, K. Pittman, K. R. Peard, I. G. Lunde,
R. A. U. Strandabø, and M. J. Gibbons. 2010. "Trophic Structure and Community
Stability in an Overfished Ecosystem." *Science* 329:333–36.

Uye, S.-i. 1994, "Replacement of Large Copepods by Small Ones with Eutrophication
of Embayments: Cause and Consequence." *Hydrobiologia* 292/293:513–19.

Uye, S.-i., and T. Kasuya. 1999. "Functional Roles of Ctenophores in the Marine
Coastal Ecosystem" [in Japanese]. In *Update Progress in Aquatic Invertebrate Zoology*,
edited by T. Okutani, S. Ohta and R. Ueshima, 57–76. Tokyo: Tokai University
Press.

Uye, S.-i., and H. Shimauchi. 2005. "Population Biomass, Feeding, Respiration and
Growth Rates, and Carbon Budget of Scyphomedusa *Aurelia aurita* in the Inland
Sea of Japan." *Journal of Plankton Research* 27:237–48.

Uye, S.-i., and U. Ueta. 2004. "Recent Increase of Jellyfish Populations and Their Nuisance to Fisheries in the Inland Sea of Japan" [in Japanese]. *Bulletin of the Japanese
Society of Fisheries Oceanography* 68:9–19.

Vaidya, S. K. 2003. "Jellyfish Choke Oman Desalination Plants." *Gulf News*, 6 May.
http://gulfnews.com/news/gulf/uae/general/jellyfish-choke-oman-desalination-
plants-1.355525.

Vaquer-Sunyer, R., and C. M. Duarte. 2008. "Thresholds of Hypoxia for Marine Biodiversity." *Proceedings of the National Academy of Sciences U S A* 105:15452–57.

Verner, B. 1983. "Jellyfish Floatation by Means of Bubble-Barriers to Prevent Blockage of Cooling Water Supply and a Proposal for a Semi-Mechanical Barrier to Protect Bathing Beaches from Jellyfish." In *Report on the Workshop on Jellyfish Blooms in the Mediterranean*, 205–10. Athens: UNEP.

Veron, J. E. N., O. Hoegh-Guldberg, T. M. Lenton, J. M. Lough, D. O. Obura, P. Pearce-Kelly, C. R. C. Sheppard, M. Spalding, M. G. Stafford-Smith, and A. D. Rogers. 2009. "The Coral Reef Crisis: The Critical Importance of <350 ppm CO_2." *Marine Pollution Bulletin* 58:1428–36.

Viitasalo, S., M. Lehtiniemi, and T. Katajisto. 2008. "The Invasive Ctenophore *Mnemiopsis leidyi* Overwinters in High Abundances in the Subarctic Baltic Sea." *Journal of Plankton Research* 30:1431–36.

VIMS. 2011. "New Web-Based Map Tracks Marine 'Dead Zones' Worldwide." Virginia Institute of Marine Science, 20 January. http://www.vims.edu/newsandevents /topstories/archives/2011/wri_dead_zones_partnership.php.

Vinogradov, M. E., M. V. Flint, and E. A. Shushkina. 1985. "Vertical Distribution of Mesoplankton in the Open Area of the Black Sea." *Marine Biology* 89:95–107.

Waggett, R., and J. H. Costello. 1999. "Capture Mechanisms Used by the Lobate Ctenophore, *Mnemiopsis leidyi*, Preying on the Copepod *Acartia tonsa*." *Journal of Plankton Research* 21:2037–52.

Waldoks, E. Z. 2010. "Another Problematic Flotilla." *Jerusalem Post*, 28 June. http://www.jpost.com/HealthAndSci-Tech/ScienceAndEnvironment/Article .aspx?id=179711.

Walsh, B. 2009. "Still Digging Up Exxon Valdez Oil, 20 Years Later." *Time*, 4 June. http://www.time.com/time/health/article/0,8599,1902333,00.html.

Wan Maznah, W. O. 2011. "Curriculum Vitae, 2011." Plankton Research Laboratory blog. Accessed 7 September 2011. http://usmplanktonlab.blogspot.com/p/ curriculum-vitae.html.

Ward-Paige, C., C. Mora, H. K. Lotze, C. Pattengill-Semmens, L. McClenachan, E. Arias-Castro, and R. A. Myers. 2010. "Large-Scale Absence of Sharks on Reefs in the Greater-Caribbean: A Footprint of Human Pressures." *PLoS ONE* 5:e11968.

Wassmann, P. 2005. "Cultural Eutrophication: Perspectives and Prospects." Accessed July 2011. http://munin.uit.no/bitstream/handle/10037/2391/article .pdf?sequence=1.

Watanabe, I., T. Kunito, S. Tanabe, M. Amano, Y. Koyama, N. Miyazaki, E. A. Petrov, and R. Tatsukawa. 2002. "Accumulation of Heavy Metals in Caspian Seals (*Phoca caspica*)." *Archives of Environmental Contamination and Toxicology* 43:109–20.

Watanabe, M., S. Tanabe, R. Tatsukawa, M. Amano, N. Miyazaki, E. A. Petrov, and S. L. Khuraskin. 1999. "Contamination Levels and Specific Accumulation of Persistent Organochlorines in Caspian Seal (*Phoca caspica*) from the Caspian Sea, Russia." *Archives of Environmental Contamination and Toxicology* 37:396–407.

Watling, L., and E. Norse. 1998. "Disturbance of the Seabed by Mobile Fishing Gear: A Comparison to Forest Clear-Cutting." *Conservation Biology* 12:1180–97.

Watson, R., and D. Pauly. 2001. "Systematic Distortion in World Fisheries Catch Trends." *Nature* 424:534–36.

Weisman, A. 2008. *The World Without Us*. London: Virgin.

Weiss, K. R. 2006. "Altered Oceans: A Primeval Tide of Toxins." *Los Angeles Times*, 30 July. http://articles.latimes.com/2006/jul/30/local/me-ocean30.

Weiss, Y. 2010. "Jellyfish Endanger the Power Station" [in Hebrew]. *Ashdod News*, 5 July. http://www.ashdodnews.co.il/news_p2.asp?page_id=106&page_id_2= 12566.

Welch, C. 2009. "Oysters in Deep Trouble: Is Pacific Ocean's Chemistry Killing Sea Life?" *Seattle Times*, 14 June. http://seattletimes.nwsource.com/html/localnews /2009336458_oysters14m.html.

"Whale Meat Increasingly Back on Menu for School Lunches in Japan." 2011. *Japan Today*, 11 September. http://www.japantoday.com/category/national/view/whale -meat-menus-increasingly-back-for-school-lunches-in-japan.

White, M. A., N. S. Diffenbaugh, G. V. Jones, J. S. Pal, and F. Giorgi. 2006. "Extreme Heat Reduces and Shifts United States Premium Wine Production in the 21st Century." *Proceedings of the National Academy of Sciences U S A* 103:11217–22.

Whitehead, J., C. Coughanowr, J. Agius, J. Chrispijn, U. Taylor, and F. Wells. 2010. *State of the Derwent Estuary 2009: A Review of Pollution Sources, Loads and Environmental Quality Data from 2003–2009*. Hobart, Tasmania: Derwent Estuary Program, Department of Primary Industries, Parks, Water and Environment.

Whiteman, L. 2002. "The Blobs of Summer. Swarms of Jellyfish Are Invading Coasts around the World." *On Earth Magazine* (Natural Resources Defense Council), Summer. http://www.nrdc.org/onearth/02sum/jelly1.asp.

Wilkerson, E. P., and R. Dugdale. 1983. "Possible Connection between Sewage Effluent, Nitrogen Levels and Jellyfish Blooms." In Report on the Workshop Jellyfish Blooms in the Mediterranean, 195–201. Athens: UNEP.

Williams, C. 2009. "Jellyfish Sushi: Seafood's Slimy Future." *New Scientist*, 7 March, 40–43.

Winans, A. K., and J. E. Purcell. 2010. "Effects of pH on Asexual Reproduction and Statolith Formation of the Scyphozoan, *Aurelia labiata*." *Hydrobiologia* 645:39–52.

WLOX. 2010. "Mississippi Scientists Find Dead Jellyfish Covered with Oil." *WLOX-TV*, 12 May. http://www.wlox.com/global/story.asp?s=12463094.

Worm, B., E. B. Barbier, N. Beaumont, J. E. Duffy, C. Folke, B. S. Halpern, J. B. C. Jackson, H. K. Lotze, F. Micheli, S. R. Palumbi, E. Sala, K. A. Selkoe, J. J. Stachowicz, and R. Watson. 2006. "Impacts of Biodiversity Loss on Ocean Ecosystem Services." *Science* 314:787–90.

Xian, W., B. Kang, and R. Liu. 2005. "Jellyfish Blooms in the Yangtze Estuary." *Science* 307:41.

Youngbluth, M. J., and U. Båmstedt. 2001. "Distribution, Abundance, Behavior and Metabolism of *Periphylla periphylla*, a Mesopelagic Coronate Medusa in a Norwegian Fjord." *Hydrobiologia* 451:321–33.

Zachos, J. C., U. Röhl, S. A. Schellenberg, A. Sluijs, D. A. Hodell, D. C. Kelly,

E. Thomas, M. Nicolo, I. Raffi, L. J. Lourens, H. McCarren, and D. Kroon. 2005. "Rapid Acidification of the Ocean during the Paleocene-Eocene Thermal Maximum." *Science* 308:1611–15.

Zaitsev, Y., and V. Mamaev. 1997. *Biological Diversity in the Black Sea: A Study of Change and Decline*. New York: United Nations Publications.

Zaitsev, Y. P. 1992. "Recent Changes in the Trophic Structure of the Black Sea." *Fisheries Oceanography* 1:180–89.

Zavodnik, D. 1987. "Spatial Aggregations of the Swarming Jellyfish *Pelagia noctiluca* (Scyphozoa)." *Marine Biology* 94:265–70.

Zeitlin, J. A. 2003. "Deep Trouble: The Other Gulf." *Naples Daily News Saturday*, 4 October. http://web.naplesnews.com/03/10/naples/d974822a.htm.

Zimmer, M. 2011a. "Green Flourescent Protein." Connecticut College website. Accessed 14 August 2011. http://www.conncoll.edu/ccacad/zimmer/GFP-ww/GFP-1.htm.

———. 2011b. "Brainbow." Connecticut College website. Accessed 25 August 2011. http://www.conncoll.edu/ccacad/zimmer/GFP-ww/cooluses0.html.

———. 2011c. "Glowing Salamanders: The Road to Limb Regeneration?" Connecticut College website. Accessed 25 August 2011. http://www.conncoll.edu/ccacad/zimmer/GFP-ww/cooluses24.html.

The letter t *following a page number denotes a table.*